Methanol As An Alternative Fuel Choice: An Assessment

Wilfrid L. Kohl
Editor

International Energy Program
Foreign Policy Institute
The Paul H. Nitze
School of Advanced International Studies
The Johns Hopkins University
Washington, D.C.

Library of Congress Cataloging-in-Publication Data

Methanol as an alternative fuel choice : an assessment / Wilfrid L.
  Kohl, editor.
        p.   cm.
     Papers presented at a conference organized at the Paul H. Nitze
School of Advanced International Studies on Dec. 4–5, 1989, and
co-sponsored by the American Petroleum Institute and the World
Resources Institute.
     Includes bibliographical references.
     ISBN 0-941700-64-X (pbk.)
     1. Methanol as fuel—Congresses.   I. American Petroleum
Institute.   II. World Resources Institute.
     TP 358.M478   1990   90–4950
     333.79'68—dc20   CIP

July 31, 1990

Dear Colleague:

The revised Clean Air Act that is expected to be passed by Congress in 1990 will call for reduced motor vehicle emissions to diminish air pollution over the next ten years, especially in the cities with the worst urban smog. While reformulated gasoline in some form is likely to meet most of the first phase criteria, the standards for fleet vehicles and later more stringent standards for general vehicles are likely to require cleaner alternative fuels such as methanol, ethanol, compressed natural gas, or electric vehicles. There is some evidence to suggest that heretofore methanol has been the Bush administration's preferred candidate. The state of California has already made a major commitment to methanol fuels.

As a contribution to the public policy debate on alternative fuels, we are pleased to present you with this conference volume on methanol. The book contains revised versions of papers

Resources Institute, the conference was convened to evaluate the
pros and cons of methanol as an alternative fuel.

Written by recognized government, industry, and academic experts, the papers address such questions as: What are the air quality benefits that might result from the use of methanol fuels? What are the economics of methanol fuels? Is methanol a cost-effective alternative fuel option? What are the implications of methanol for energy security and global warming? What are the health and safety benefits or risks? Finally, what are some of the problems that might be anticipated regarding public acceptance and implementation of methanol as an alternative fuel?

We hope that you will find this publication of interest. Any comments would be welcome. A limited number of further copies is also available upon request.

Yours sincerely,

Wilfrid L. Kohl
Director
International Energy Program

encl.

Table of Contents

Part III
Sources of Methanol Supply, Energy Security, and Global Warming

Part IV
Health and Safety

Part V
Costs of Methanol Systems

Part VI
Public Acceptance and Implementation

Part VII
Summary and Conclusions

Preface

The Bush administration's Clean Air Act proposals, announced in June 1989, contained an ambitious program to introduce alternative fuel vehicles into a number of the nation's heavily polluted metropolitan areas beginning in 1995. The proposals have sparked considerable political debate in Congress and elsewhere. While the approach of mandating specified numbers of alternative fuel vehicles was subsequently modified in early 1990, the question of reducing future vehicle emissions via the use of alternative fuels remains one of the key issues in the effort to revise the Clean Air Act. A number of observers have suggested that methanol is the administration's preferred candidate. The state of California has also made a major commitment to methanol fuels.

In order to illuminate the costs and benefits of methanol as an alternative fuel, and to explore related issues, a conference was organized at the Paul H. Nitze School of Advanced International Studies on December 4 and 5, 1989 on "Methanol As An Alternative Fuel Choice: An Assessment." The conference was convened by the International Energy Program of the Johns Hopkins Foreign Policy Institute. Co-sponsors were the American Petroleum Institute and the World Resources Institute.

The objective of the conference was to address such questions as: What are the air quality benefits that might result from the use of methanol fuels? What are the economics of methanol fuels, and is methanol a cost-effective alternative fuel option? What are the implications of methanol for energy security and global warming? What are the health and safety benefits or risks? What are some of the problems that might be anticipated regarding public acceptance of methanol as an alternative fuel and implementation of the President's program?

In order to analyze these questions, a number of experts from government, industry, consulting firms, and academia were invited to give papers and provide commentary. This book contains the revised conference papers and formal comments. It is published as a contribution to the public policy debate on the question of alternative fuels.

We express our appreciation to the American Petroleum Institute for financial support of the conference and this publication. The World Resources Institute, and especially Dr. James MacKenzie, provided important intellectual input into the conference agenda and assisted in its organization. We are grateful to Carol W. Rendall, an associate of the SAIS International Energy Program, for serving as chief editor and coordinator of this publication. Nancy McCoy assisted in editing the manuscript for style and consistency. Thanks also go to A. Camille Richardson, our program assistant, and to a number of SAIS students who, as part-time assistants, shared the extensive word processing tasks which were essential to the production of this book. The cover design and speedy production of the book by The Magazine Group are also gratefully acknowledged.

Introduction and Overview

WILFRID L. KOHL

Interest in alternative fuels developed as a possible way to enhance U.S. energy security following the oil supply and price shocks in the 1970s and early 1980s. Later as U.S. oil production plummeted after the 1985–86 oil price collapse, and as oil imports rose sharply, attention focused once again on alternative fuels. The U.S. transportation sector (which includes jet airplanes as well as light- and heavy-duty motor vehicles) is heavily dependent on oil and accounts for two-thirds of the U.S. oil requirement, more oil than the United States produces. To help stabilize future oil supplies and prices, it would be desirable to reduce U.S. oil consumption and imports. The development of alternative fuels for highway vehicles is a logical focus of such an effort.

A second reason for interest in alternative fuels—and the principal concern driving the current debate—is urban air pollution. According to a recent report from the Office of Technology Assessment, about 100 cities comprising about two-thirds of the U.S.

population still do not meet the federal ozone standard (set at a peak one hour concentration of 0.12 parts per million).[1] The pollution in some cities, such as Los Angeles, New York, and Houston, is so severe that these cities have no hope of attaining compliance in this century.

Exposure to high levels of ozone interferes with lung functioning and causes shortness of breath, coughing, and chest pains. Moreover, ozone is suspected of playing a role in the long-term development of chronic lung diseases and permanent lung structure damage. Ozone also has adverse effects on crops such as soybeans, wheat, and corn, is damaging forests in California, and is a suspected contributor to the deterioration of forests in the southeastern United States.

Ozone is produced when volatile organic compounds (VOCs) and nitrogen oxides (NO_x) react in the presence of sunlight. NO_x is principally the result of fossil fuel combustion. Principal sources include highway vehicles and electric utility boilers.[2] So far greater attention has been focused on reducing VOCs. The major sources of VOCs are motor vehicles (40–45 percent) and small stationary sources, e.g., bakeries, dry cleaners, and consumer solvents (40 percent). Large point sources, e.g., petroleum refineries and industrial plants, and air, rail, and marine transport account for the rest.[3] Considerable progress was made under the Clean Air Act in reducing emissions from large point sources and from tailpipes of new vehicles. However, as the number of vehicles on the road continues to grow, so does air pollution. (Smaller area sources remain largely uncontrolled.)

An additional factor that needs to be considered is long-run concern about global climate change. One of the principal sources of CO_2 and other greenhouse gas emissions is carbon-based fossil fuels. While debate continues about the severity, regional effects, and timing of global warming due to the greenhouse effect, there is a widening consensus that we do have a global warming problem. International negotiations on the issue will intensify in 1990, and there is a distinct possibility that governments may be called on to limit the use of fossil fuels. If an alternative-fuels program is launched, care should be taken so that the fuels chosen lessen, or at a minimum not increase, greenhouse gas emissions.

The Bush administration's ambitious proposals to amend the Clean Air Act, announced in June 1989, set the framework for the current debate. The administration's original plan would mandate, beginning in 1995, increasing numbers of vehicles that would burn cleaner fuels, such as methanol, ethanol, or compressed natural gas, in the nine major urban areas that have the greatest concentration of ozone. The plan called for the sale of 500,000 alternative-fuel vehicles in 1995, 750,000 in 1996, and one million vehicles each year from 1997 through 2004. It was estimated that the alternative-fuels program would reduce VOC emissions in the selected urban areas by 2 to 5 percent, and also result in less toxic air emissions such as benzene, toluene, and xylene.[4]

Other provisions in the President's plan called for:
- tightening volatility requirements nationwide on conventional gasoline during the summer months to reduce evaporative emissions;
- reducing vehicle evaporative emissions caused by running losses;
- tightening hydrocarbon emission tailpipe standards for automobiles, and adding more stringent tailpipe standards for light-duty trucks;
- expanding vehicle inspection and maintenance programs;
- providing the Environmental Protection Agency (EPA) with authority to regulate VOC emissions from small sources and consumer products (e.g., solvents and paints) and authority to issue control technology guidelines to major stationary source emitters (e.g., factories and plants).[5]

In August 1989, a development of related interest was the Atlantic Richfield Company's announcement that it was introducing a reformulated gasoline into the California market. While the particular reformulated gasoline to be introduced was targeted for use in older vehicles that burn leaded gasoline, this action raised the question whether further research on the part of oil companies could produce other reformulated gasolines that might considerably lessen polluting emissions from conventional vehicles.[6]

While the administration's proposal did not mandate any particular fuel, the EPA would be required to set emissions standards for alternative-fuel vehicles. The standards that EPA said it would adopt for 1995 and 2000 could most easily be met by methanol, and possibly by compressed natural gas vehicles. Reformulated gasoline might meet the standard, depending on the progress of research and development over the next few years.

The Congressional debate on the administration's Clean Air Act proposals began in the fall of 1989 and continued into the spring of 1990. In the fall the action centered mainly in the House of Representatives Committee on Health and Environment, which reported H.R. 3030 including the Fields-Hall amendment on alternative fuels and the Waxman-Dingell compromise on tailpipe standards and other issues. H.R. 3030, which did not receive administration support, originally adopted an annual emission reduction target based on assumed emissions from increasing numbers of M85 vehicles (vehicles that can burn 85 percent methanol and 15 percent straight gasoline), but it did not mandate sales of vehicles using alternative fuels. Later, the new vehicle emissions standards in the bill were changed so they could be met by reformulated gasoline, which must be offered for sale in the nine cities with the worst ozone pollution beginning in 1995. The standards are to be tightened in the year 2000.

On March 1, 1990 a bipartisan agreement between members of the Senate and the administration was announced on Clean Air Act revisions.[7] This agreement signified the administration's willingness to give up its previously sought mandate for the sale of specified numbers of alternative-fuel vehicles. Instead, the agreement focuses on performance standards (an overall reduction to not more than .75 grams/mile of ozone-forming emissions) beginning in model year 1995 for all new vehicles sold in the nine cities with the most severe ozone nonattainment problems. The standards will tighten in phase 2 beginning in 1999. The EPA will issue performance standards which are to be met primarily by reformulated gasoline. Implementation of the program will be undertaken by the states. The agreement also requires the use of even cleaner fuels in fleet vehicles.

In early April 1990, the Senate passed a Clean Air bill based largely on the bipartisan compromise. An amendment which increased the oxygen content of gasoline in the most polluted cities appears to favor ethanol blends.

In May 1990 the House passed a bill setting more stringent standards on vehicle emissions in the most polluted cities for ten years—standards that can presumably still be met by reformulated gasoline. Unlike the Senate bill, there is a requirement for the sale of 300,000 cars per year in California by 1997 that can run on fuels 50 percent cleaner than conventional gasoline. Public buses and fleet vehicles would have to run on ultraclean fuels in the 27 smoggiest cities.[8]

As this book went to press, reconciliation of the Senate and House bills had not yet occurred. While methanol was no longer expected to play the central role in meeting the first stage general vehicle emissions standards, it will be a potential choice for fleet vehicles and in California programs. And methanol remains a serious alternative fuel candidate to meet even tighter phase 2 emissions standards after the year 2000.

Emissions from Gasoline and Methanol

Part I of this volume assesses prospective emissions from methanol versus conventionally fueled vehicles. Philip Lorang of EPA's Office of Mobile Sources explains that the performance standard proposed by the administration—an 80 percent reduction in per-vehicle emissions that contribute to ozone beginning in year 2000—could probably be achieved by methanol or compressed natural gas vehicles by that time. However, he believes that methanol is the alternative fuel with the best chance of being adopted within the administration's original time frame, which called for the production of about 10 million vehicles between 1995 and 2005.

Flexible-fuel vehicles that can burn M85 could provide a transition, since they do not require further substantial engine development. However, an M85 vehicle yields only a 30 percent net reduction in hydrocarbon equivalent emissions. The final performance standard that EPA may implement around the year 2000 would be based on M100 vehicles (which burn 100 percent methanol). Lorang acknowledges that so far only a few M100 test vehicles have been built, whereas many more experimental M85 vehicles exist today.

Mr. Lorang reviews the provisions of the administration's proposals and compares them with the earlier bills that have been introduced in the House of Representatives and the Senate, respectively. He provides a detailed analysis of tailpipe and other types of vehicle emissions (e.g., parking losses, running losses, refueling emissions). He concludes that optimized M100 vehicles could achieve an emissions equivalent of 0.19 grams/mile of non-methane hydrocarbons, or a reduction in hydrocarbon equivalent emissions per vehicle of between 67 and 80 percent.

An opposing view is offered by Thomas C. Austin of Sierra Research, Inc., who argues that EPA's forecasts of emissions from methanol-fueled vehicles are optimistic and based on relatively low mileage prototypes. He notes that the California Air Resources Board has found evaporative emissions from methanol to be higher than those of gasoline. Conversely, Austin finds that EPA assumptions concerning emissions from conventional vehicles are unduly pessimistic and do not take into account improvements likely to be made as a result of on-board diagnostics, controls on refueling emissions, and better inspection and maintenance programs. He concludes that it is premature to assume that a government mandate requiring the introduction of methanol-fueled vehicles will result in any significant reduction in vehicle emissions.

In his commentary, Jerry Horn of Chevron Research points out that the EPA presentation compares data on current in-use vehicles with data on M100 prototypes and does not appear to apply the same deterioration factors to M100 vehicles. Horn agrees with the Sierra Research analysis that significant further reductions in conventional gasoline vehicle emissions can probably be achieved if regulations such as those enacted recently in California (requirements for on-board diagnostic systems, more stringent emission standards, and more effective inspection and maintenance) are adopted elsewhere. Mr. Horn also questions why the EPA emissions presentation considered only M100, while M85 is most likely to be introduced first. Auto companies will not be in a position to manufacture M100 vehicles until further research and development is conducted to overcome cold-start problems and address other concerns such as flame luminosity and fuel tank vapor flammability. Mr. Horn also noted that M100 technology will have to be improved to reduce formaldehyde and NO_x emissions, especially at higher mileage.

The second commentator, Harry Schwochert of General Motors, highlights a point underlying both papers: similar amounts of emissions (total organic emissions, CO and NO_x) should be expected from automobile engines using either gasoline or methanol. It is the composition of the total organic emissions that will change and determine whether or not air quality will improve. The potential benefit of methanol vehicles will be highly dependent upon the level of formaldehyde emissions and on the NO_x levels of the ambient environment in which the vehicles will operate.

Methanol and Urban Ozone

The papers in Part II consider the impact of gasoline versus methanol emissions on air quality, specifically as causes of urban ozone. These studies rely on air quality models, which are very sensitive to starting assumptions, to simulate complex chemical reactions in the atmosphere. In a study undertaken specifically for our conference, Armistead G. Russell of Carnegie Mellon University uses an urban airshed model to simulate the likely air quality impacts of methanol fuel use in the city of Los Angeles in the year 2010, assuming that about half of the light-duty vehicles were powered by M85 by that time as called for in the administration's proposal. Methanol is assumed to be significantly less reactive than most of the organic compounds in gasoline vehicle exhaust. Peak ozone levels are predicted to decrease by up to 9 percent, and exposure to levels of ozone above the federal standard to decrease by about 19 percent. Formaldehyde exposure will not increase greatly.

Russell also observes that the effectiveness of methanol as an anti-ozone strategy is very dependent on the ratio of reactive organic gases to nitrogen oxides in the atmosphere.* Regions where switching to methanol has less effect in reducing ozone would benefit from decreasing emissions of nitrogen oxides. The same point is made by the authors of the next two papers.

A second study by T.Y. Chang and S.J. Rudy of the Ford Motor Company research staff utilizes a much different approach and obtains more limited results. Applying a simple trajectory model to the nine U.S. cities with the highest peak ozone levels, they find that full penetration of optimized M100 vehicles by the year 2015 would reduce peak ozone levels between 1 and 5 percent, while the reduction from M85 vehicles is estimated at less than 1 percent. Ozone reductions in the Los Angeles basin are one to three times larger than the nine-city average reductions, depending upon the ratio of non-methane organics to nitrogen oxides. Concern is expressed about increases in formaldehyde levels, especially in commuter tunnels and underground parking garages, while other air toxics are found to decrease with methanol use.

A third study by Sanford Sillman and Perry J. Samson of the University of Michigan applies an urban and regional scale model to the Detroit-Port Huron area and the downwind midwestern region. The use of methanol fuels is found to yield a modest 1 to 5 percent reduction in peak ozone concentration in Detroit, but to

---

*Editor's Note: A number of papers in this collection discuss emissions from gasoline fuels using the terms reactive organic gases (ROG), non-methane organic compounds (NMOC), and volatile organic carbon (VOC). These are equivalent terms and refer to organic substances which react in the presence of sunlight in the atmosphere and are major contributors to the formation of urban ozone.

have virtually no impact on the rural regional ozone level, since the latter is more sensitive to NO_x emissions than to VOCs.

Differences between the studies and approaches and their assumptions are highlighted in the commentary by Jana B. Milford of the University of Connecticut. She also emphasizes that the ratio of ROG (reactive organic gases) to NO_x is very important, and this could explain the relatively large benefit predicted by Russell for Los Angeles in 2010 where that ratio is projected to be low. Some areas may require reductions of NO_x emissions in order to meet the ozone standard, as emphasized also in the paper by Sillman and Samson. The lower reactivity of methanol emissions is also sensitive to other factors, especially the level to which formaldehyde emissions can be controlled. Ms. Milford stresses that projections of emission rates from future-year vehicles are very uncertain.

In his comment Alan C. Lloyd, chief scientist of the California South Coast Air Quality Management District, observes that the Russell and Chang-Rudy studies, although quite different in their approaches and assumptions, both confirm that the use of methanol does improve air quality by reducing peak ozone concentrations. In cities such as Los Angeles with high smog concentration, he contends that even a few percentage points of peak ozone reduction can be significant in light of the limited means available to control the problem.

Methanol, Energy Security, and Global Warming

Part III examines the potential impacts of methanol use on U.S. energy security and global warming. Noting that oil use in the U.S. transportation sector now exceeds 10 million b/d, Carmen Difiglio of the Department of Energy argues that U.S. energy security could be enhanced if flexible-fuel methanol powered (presumably M 85) vehicles were introduced in sufficient quantities between now and the year 2010 to reduce the nation's gasoline consumption by one-seventh, or about one million b/d. This would require two million b/d of methanol, which would probably be imported from a diverse group of countries. In his view the world methanol market is likely to remain more competitive than the oil market. Large capital investments in methanol plants by supplier countries would increase their stake in uninterrupted revenues and therefore exports. If supply curtailments occurred, flexible-fuel vehicles could switch to gasoline. According to Difiglio, reduced oil demand resulting from increased use of methanol would very likely reduce world oil prices and weaken OPEC's hold on the oil market.

In commenting on Difiglio's paper, Michael Kelly of Jensen Associates is in general agreement with Difiglio's line of argument, but he questions some of its economic assumptions. Capital costs of methanol plants might well be higher than assumed by Difiglio, and it could take longer for a world methanol market to develop. In particular, Kelly questions who will take the necessary risks to finance and build large methanol plants in countries with large natural gas reserves but which are already short of investment capital.

Emissions of greenhouse gases will not be significantly reduced by the use of any fossil fuel feedstocks in the making of transportation fuels, contends Mark A. DeLuchi. Coal use as a feedstock would be the worst case, since it would result in considerably increased CO_2 emissions. The use of natural gas to produce methanol, compressed natural gas (CNG), liquefied natural gas (LNG), or electricity for electric vehicles will at best result in a small reduction of CO_2 equivalent emissions, but under other assumptions it could produce a small increase in greenhouse gases compared to the current gasoline-from-crude oil cycle. DeLuchi therefore concludes that methanol should be no more than a transitional fuel choice on the way to future nonfossil fuel sources (e.g., hydrogen- and electric-powered vehicles) if we are serious about limiting the prospect of future global climate change. Improving automobile fuel efficiency is also a strategy, he suggests, to reduce greenhouse gas emissions from the transportation sector.

Commenting on DeLuchi's paper, Michael Jackson expresses general agreement but points out that under some different assumptions methanol and natural gas as transportation fuels might yield a small reduction in greenhouse gas emissions compared to gasoline.

There could also be a greenhouse gain if engines of methanol vehicles could be designed to operate much more efficiently.

Health and Safety Effects of Methanol

Health and safety aspects of methanol fuels are considered in Part IV. Among the concerns that have been expressed about methanol are greater fuel tank flammability, increased fuel toxicity, and the fact that pure methanol burns with an invisible flame. Taking an optimistic view, Paul Machiele of the Environmental Protection Agency reminds us that gasoline use also involves risks. He goes on to discuss the lower volatility, lower flammable limit, and lower vapor density of M100, which if adopted as a fuel, could in his view result in a reduction in vehicle fires and related injuries and fatalities each year. According to Machiele, design modifications could lower the risk of fuel tank explosions and additives could be devised to provide flame luminosity. Methanol is more toxic but he asserts that the risks of ingestion or skin exposure will be manageable, since it is not anticipated that methanol will be used in household or garden applications. Exhaust emissions of air toxics will, he contends, be less with methanol than with gasoline.

The second paper by Lawrence Fishbein and the commentaries by Murphy and Spitzer point out some uncertainties and a lack of data in several areas, especially concerning methanol's physiological effects. For example, toxicity data obtained from rodent studies may not be transferable to humans. Certain sensitive populations, such as pregnant women and elderly people on poor diets, who have folic-acid deficiencies may be especially endangered by methanol exposure. There is still much to be learned about the health effects of methanol-gasoline mixtures, ocular toxicity, and dermal exposure.

Estimates of Methanol's Costs

In the discussion of the economics of methanol, in Part V, there is general agreement on the uncertainties of forecasting about a fuel industry that does not yet exist in a future energy market in which the prices of oil and other fuels are difficult to predict. The chapter by Michael F. Lawrence of Jack Faucett Associates, which reviews other studies as well as the work of his own firm, presents the most optimistic view. Lawrence concludes that there is an abundance of low-cost natural gas available in the world to serve as a feedstock for conversion into methanol, and that it could be brought to the U.S. market at prices competitive with the cost of gasoline. Assuming additional investment in vehicle conversion and in fuel-delivery infrastructure, he believes that the United States could displace up to 2.5 million barrels of gasoline consumption per day with methanol by the year 2000. He predicts that if a market for methanol develops in the United States, the Persian Gulf countries, Venezuela, and Ecuador would probably be the major suppliers.

Lawrence makes the important point, echoed by Albert McGartland, that private and public sector decision making are intrinsically different. Whereas private sector decisions are rooted in investment analysis internal to the firm and based primarily on profit maximization, public sector decisions must take account of internal and external factors that are sometimes difficult to quantify. In the case of methanol the potential benefits to be analyzed include its impact on environmental quality (both air quality and health effects), energy security, market efficiency, and effective utilization of resources. Most cost studies, including the ones in this volume, emphasize a narrow analysis of methanol as an investment. Lawrence stresses that more studies are needed that incorporate cost-benefit analysis, risk analysis, cost-effectiveness, and market failure analysis.

In his paper Albert McGartland, a former government analyst now with ABT Associates, attempts to compare the costs of a methanol strategy for controlling emissions with the cost of other control strategies. He finds that the methanol control cost is very sensitive to the future price spread between methanol and gasoline. The cost-effectiveness of methanol is therefore very uncertain and unpredictable. In view of this uncertainty, McGartland concludes that a government mandate for methanol is unwise. He favors the Bush administration's announced approach of setting pollution reduction targets for each industry and allowing trading so that the market can determine the most effective mix of fuels and technologies.

The very comprehensive chapter comparing production costs of methanol, ethanol, and compressed natural gas by James Sweeney of Stanford University draws on a larger study recently completed by the National Research Council of the National Academy of Sciences at the request of the U.S. Department of Energy. Agreeing with others that domestic U.S. production of methanol does not appear to be economic, he goes on to conclude that the costs of imported methanol from remote foreign locations will vary anywhere between $30/b to $45/b in oil equivalent costs, depending on the future price of natural gas in the foreign locations and the costs of investments in new methanol plants. Thus, Sweeney emphasizes the possibility that methanol may not be economically competitive in large-scale production, even when compared to compressed natural gas. He agrees with McGartland that a government mandate for methanol is not a good idea, urging instead that policy flexibility be maintained in order to adapt to future market conditions.

In commenting on the papers, Thomas J. Lareau of the American Petroleum Institute criticizes the analysis by Lawrence as being too limited and not taking sufficient account of important variables such as future fuel prices, the volume of production, and other market factors that could raise the actual selling price of methanol. He applauds Sweeney's sensitivity analysis of methanol's cost with respect to oil prices, investment costs, natural gas prices, and other factors, and agrees with Sweeney's conclusion that it is not appropriate to mandate methanol. While expressing disagreement with some of the assumptions of the McGartland paper, Lareau agrees with McGartland that the so-called "chicken-and-egg" argument, which says that markets cannot accept new complementary products, is unfounded.

In her comments Margaret A. Walls of Resources for the Future also underscores the uncertainty of future methanol prices and supports the administration's approach of setting tighter emissions standards and allowing emissions trading. She advocates a gasoline tax, which would have the side effect of making alternative fuels more competitive.

A third commentary by Glyn D. Short of ICI Americas, a domestic methanol company, notes that producing countries are likely to have incentives to price their gas at a point where methanol production will be profitable. With respect to costs of capital, Short observes that the three economic cost papers underestimate the savings that will accrue from advanced technology methanol plants. Disagreeing with McGartland, Short believes that a chicken-and-egg problem does exist with regard to the penetration of methanol fuel and methanol vehicles into the market. Government involvement of some kind, Short contends, will be required to facilitate a transition if and when the long-term production of high-volume methanol becomes feasible.

Implementing the Introduction of an Alternative Fuel

Leading off in Part VI, Carl B. Moyer of the Acurex Corporation in California discusses market-driven approaches to alternative-fuel implementation. Since it is impossible to foresee the economics and vehicle performance of alternative fuels and their marketability, he urges implementation plans that have built-in flexibility in order to respond to technological change. In his view dual- or flexible-fuel vehicles are likely to be introduced first. Moyer presents a matrix of implementation approaches that have been used in several countries, with varying degrees of government involvement. He then describes the flexible approach being developed by the California Air Resources Board combining emissions standards and "fuel pool averaging," which includes all motor fuels. Referring to the Bush administration's Clean Air Act proposals, he notes that while some sort of federal mandates for alternative-fuel vehicles appear likely, state implementation plans may determine how the mandates are carried out.

Daniel Sperling of the University of California at Davis and his co-authors analyze the various factors that might influence consumer demand for methanol, drawing upon experience in foreign countries and survey data from the United States. While fuel prices and availability are obviously important factors, he sees the role of government (along with industry) as critical in reducing uncertainty and creating confidence in the new fuel. In contrast to Moyer, Sperling and his co-authors believe that if the right marketing and investment strategies are chosen, a transition period of multi-fuel vehicles will not be necessary.

In his comments Charles E. Risch of Ford Motor Company notes a number of uncertainties that could complicate any transition to alternative fuels. Incentives will be needed to ease the transition for all parties involved—the auto industry, the fuel industry, and the consumer—but he stresses that any new fuel must eventually pay for itself. Mr. Risch favors the fuel pool averaging scheme described by Carl Moyer with its transfer of credits to allow the market to pick the best fuels. Citing Ford's experience in California and survey data, Risch disagrees with Sperling on flexible-fuel vehicles, which will be needed, he believes, before complete infrastructure for a new fuel is in place and dedicated vehicles can be introduced. To be accepted, Risch concludes, methanol vehicles need to provide value to the customer.

Expressing a skeptical view about the use of methanol, Jerrold L. Levine of Amoco Oil Company points out that while the administration may set quotas for production of alternative-fuel vehicles, it cannot force consumers to buy them. A small environmental benefit may not be sufficient to attract customers. A large infrastructure of methanol plants will be required, with each plant costing about a billion dollars. But can oil companies make that kind of investment and take the risk that methanol might soon be replaced by another alternative fuel? The implementation of fuel pool averaging described by Carl Moyer may be more complex than meets the eye, suggests Levine, who is concerned about cheating in the trading of credits. He urges that the government give the new joint oil-auto industry research program a chance to study the costs and benefits of reformulated gasoline in new automobiles, as well as methanol, CNG, and improved vehicles, so as to minimize the investment risks involved with rapid introduction of methanol or any other alternative fuel.

In Part VII two authors provide summaries as seen from different perspectives. The first of these, Jeffrey Alson from the EPA Emissions Control Laboratory in Ann Arbor, favors methanol—and specifically the introduction of M100—because it will improve urban air quality, health, and national energy security. Speaking for himself and not officially, he recognizes that in the long run a more radical shift to zero-polluting transportation fuels may be required if a few large U.S. cities with the worst smog problems are ever to achieve clean and healthy air, and if the problem of global warming is to be seriously confronted. However, he sees methanol as a good less-polluting transitional fuel on the way to more advanced technologies such as hydrogen fuel or electric fuel cells. Alson discusses the institutional factors that he believes will inhibit the introduction of methanol into the marketplace and urges that government take a strong leadership role.

James MacKenzie of the World Resources Institute, on the other hand, opposes the introduction of methanol on balance because it does not contribute sufficiently to the solution of long-term problems—specifically global warming, national energy security, and the balance of trade deficit. He is very skeptical that consumers will be willing to buy vehicles fueled by pure methanol and doubts that methanol will be competitive with gasoline in the near term. MacKenzie suggests a different strategy, including a concerted effort to remove older, more polluting vehicles from the road, improved vehicle efficiency, more stringent emission limits, improvements in gasoline, and stronger inspection and maintenance programs.

Notes

1. Office of Technology Assessment, U.S. Congress, *Catching Our Breath: Next Steps for Reducing Urban Ozone*, 1989.

2. *Ibid.*, pp. 17–19.

3. *Ibid.*, pp. 11–13.

4. "Fact Sheet: President Bush's Clean Air Plan," Office of the Press Secretary, the White House, June 12, 1989.

5. *Ibid.*

6. *New York Times*, August 16, 1989. See also "Arco EC-1 Cleaning Up in California," *Energy Daily* (October 18, 1989). In the fall of 1989, fourteen U.S. oil companies joined with three domestic automobile manufacturers to begin a joint research and testing program to generate more data on methanol, other alternative fuels, and reformulated gasoline.

7. Summary of Bi-Partisan Senate Clean Air Act Agreement on Nonattainment of Health Standards for Ozone, provided by Senate Committee on Environment and Public Works, March 1990.

8. *Washington Post*, May 24, 1990; *The Wall Street Journal*, May 25, 1990.

Part I

Emissions from Gasoline and Methanol

CHAPTER 2

Emissions Improvements From Methanol Fuels

PHILIP A. LORANG

Introduction

President Bush proposed a clean fuels program to Congress on June 12, 1989, as the opening step in an effort to reconcile the need for a cleaner environment and the ever growing number of automobiles in use in U.S. cities. The proposal raises issues of urban air quality, public health, highway traffic safety, fuel costs, national energy security, and global warming. Some observers have also been concerned about how a clean fuels program might influence public support for other possible solutions to current environmental, transportation, and energy problems. A central issue

---

The views expressed in this paper are those of the author and do not necessarily represent the estimates, views, or policies of EPA except where explicitly stated in the text.

is how the use of alternative fuels will affect urban ozone pollution, in which vehicle emissions play a major role.

Why the Special Interest in Methanol?

The President's proposal emphasizes a fuels-and-vehicle program, rather than simply a tightening of emissions standards for vehicles. The administration is convinced that ultra low-polluting vehicles can most cost-effectively be achieved if fuel composition is also a factor subject to optimization. A "chicken or egg" dilemma exists with respect to introducing a new fuel that would require separate handling and could be used only in specially designed vehicles. If a new fuel was deemed to be the socially superior approach to achieving a given level of vehicle emissions reductions, government action would be required to introduce it. Without government action to coordinate the fuel and vehicle sectors, vehicle manufacturers would have to assume that only conventional gasoline would be available for the vehicles they designed to meet stricter emissions standards.

The Bush administration proposed an 80 percent per-vehicle reduction in ozone-forming emissions, to begin with the 2000 model year. However, it is doubtful that this could be achieved using conventional gasoline; such a reduction on an individual vehicle basis probably exceeds the foreseeable capability of gasoline-fueled vehicle technology, even with reformulated gasoline.[1]

Yet, this level of reduction is feasible with some or all of the non-petroleum fuels: methanol, ethanol, compressed natural gas (CNG), propane, and electricity. Of these, methanol deserves special attention because it provides the "safety net" for the President's program. With the methanol option available, the proposal carries very little risk of public dissatisfaction. The other non-petroleum fuels would all face serious obstacles as the fuel source for about nine million mostly private vehicles produced between 1995 and 2005. Each of the others appears to have a significant disadvantage in terms of cost, potential supply, volume, or vehicle performance.

However, these fuels may be able to fill important specialized market segments. The administration would be pleased if, during the next 5 to 15 years, other fuels proved to be even more attractive than methanol in some or all applications.

What Should the Clean Fuels Debate Be About?

Two methanol fuels are being debated. The fuel that is the safety net for the administration's program is 100 percent methanol, or M100. The fuel for which a sizable number of experimental vehicles have been built is 85 percent methanol and 15 percent gasoline, or M85. This paper focuses on the emissions of M100 vehicles, particularly the emissions of organic species that are precursors to ozone.

The final performance standard that the administration's plan directs the U.S. Environmental Protection Agency (EPA) to set might not be achievable with M85. To set the final performance standard, EPA anticipates using the performance capability of M100 and CNG vehicles, which appear to be very close in capability, as the benchmark. The result would be about an 80 percent reduction in ozone precursor emissions per vehicle, rather than the 30 percent achievable with M85 flexible-fuel vehicles. EPA expects that such a standard would be implemented around 2000.

With this background, the relevant issues in the alternative-fuel debate emerge: How much will the emissions of future gasoline-fueled vehicles improve under each possible set of new statutory requirements that Congress may choose? Is it reasonable to expect that M100 vehicles will have the emission levels that EPA anticipates? What will be the cost, consumer utility, and other benefits (or negative impacts) of M100 vehicles? (Other authors in this collection will also consider some of these issues.)

The decision that must be made, based on answers to these questions, is whether the United States should undertake a limited-scale program that requires some vehicles in its most polluted cities to achieve M100 or CNG emissions levels (with M100, CNG, or any other fuel of the vehicle manufacturer's choosing). Experience with these vehicles may lead to private or public decisions for their wider use to increase environmental benefits. Alternatively, should the United States settle for some level of gasoline-fueled vehicle performance, as uncertain or costly as it may be, which gives less benefit and makes it more difficult for state and local governments to develop successful plans for ozone attainment?

Attainment of the ozone standard will be especially difficult in the Los Angeles-Riverside-San Bernardino-San Diego area, the Chicago-Milwaukee area, Houston, and portions of the Northeast. It appears likely that ozone attainment will in effect be unachievable in many of these areas unless vehicle emissions can be reduced approximately 80 percent beyond the administration's gasoline proposal. Otherwise, vehicle emissions alone will be so large a component of total allowable emissions that most economic activity will be constrained. Even with fuels that emit much less pollution, it will be difficult for states to control other sources sufficiently.[2]

A Caution on Prototype Data

To estimate what emissions reduction would result from changes in emissions standards or fuels, available test data from prototype vehicles must be consulted. None of the few prototypes specifically designed for M100, however, has seen daily in-use operation. Consequently, data from numerous M85 prototypes have been used as an implied indication of what can be expected of M100 vehicles. Unfortunately, some early M85 prototypes did not use the best gasoline technology of their day (which was not as good as that used in recent gasoline-fueled or M85 vehicles). Yet these older M85 prototypes are the only methanol vehicles that have accumulated significant real-world mileage.

Without taking this inferior technology into account, some have argued that the poor performance of some M85 prototypes at high mileages represents an inherent tendency of M85 and M100 vehicles to deteriorate, a tendency gasoline-fueled vehicles do not share.

In cases in which M85 or M100 prototypes have shown very low emissions levels, opponents of the administration's program minimize their significance on the grounds that they have been hand-built, that they have not accumulated a full 50,000 miles of service, or that they have been used in artificially benign environments. These criticisms have some merit, but they do not justify using data from poorly designed M85 prototypes.

Clean Air Act Proposals for Gasoline-Fueled Vehicles

As of March 1, 1990, three competing proposals (Table 1) for reducing gasoline-fueled vehicle emissions of ozone precursors in the next century were before the U.S. House of Representatives and Senate: the administration's proposal; House Resolution 3030, as amended by the Waxman-Dingell compromise; and Senate Bill 1630, as reported to the full Senate for floor action. The administration's proposal calls for reductions of about 30 percent in both the hydrocarbon (HC) and nitrogen oxides (NO_x) tailpipe standards, better control of evaporative HC emissions, control of running-loss HC emissions, and a federal requirement for Stage 2 control of refueling HC emissions. The plan also includes enhancements to inspection and maintenance (I & M) programs and a first-time federal requirement for emissions-control diagnostic systems on vehicles. Also, the Reid vapor pressure (RVP) level for gasoline is to be reduced to 9 psi during the ozone season.

The Phase 1 tailpipe standards in both the House and Senate versions are close to the single set of standards in the administration's proposal. However, they set the NO_x standard at 0.4 g/mi—60 percent lower than the current standard—instead of following the administration's proposal for a 30 percent reduction. Improvements in evaporative and running-loss emissions, as well as refueling HC control, I & M programs, emissions-control diagnostics, and gasoline RVP are essentially the same as in the administration's proposal. However, the Senate version would require on-board systems

Table 1
Current and Proposed Car Standards (g/mi)

	Useful Life	Current Standards
Non-Methane	5/50	(0.36)*
HC	7/75	—
	10/100	—
Total HC	5/50	0.41
	7/75	—
	1/100	—
CO	5/50	3.4
	7/75	—
	10/100	—
NO_x	5/50	1.0
	10/100	—
Evaporative Emissions		Some Control
Running Losses		No Control
Refueling Emissions		No Federal Control
I/M		Basic
RVP (psi)		10.5
Emission Warranties		5/50 for defects
Emissions System Diagnostics		No requirement intent to adopt something

*Rough equivalent to actual total HC standard, for comparison to proposals.

Source: Emissions Control Tech. Div., USEPA (1989)

Table 1
(Continued)

Recommended by Administration	Reported by House Subcommittee		Reported by Senate Subcommittee	
	Phase 1	Phase 2** (2004)	Phase 1	Phase 2 (2003)
0.25	0.25	—	—	—
—	0.31	—	—	—
—	—	0.125	0.25	—
0.41	—	—	—	—
—	—	—	—	—
—	—	—	0.31	0.125
3.4	3.4	—	—	—
—	4.2	—	—	—
—	—	1.7	3.4	1.7
0.7	0.4	—	—	—
—	—	0.2	0.4	0.2
Better Control	Better Control		Better Control	
Control	Control		Control	
Stage 2	Stage 2: On-board if safe		Stage 2 + On-board	
Enhanced	Enhanced		Enhanced	
9.0	9.0		9.0	
Same as current all parts	Same as current		—Extended to 8/80 for catalyst and computer —Reduced to 2/24 for all other parts	
Authority and something	EPA must adopt something		EPA must adopt	

**House Subcommittee Phase 2 standards are subject to EPA study and revision, not to exceed Phase 1 standards.

for refueling control, in addition to pump-based Stage 2 controls. According to the House version, EPA would have to make a finding concerning the safety of on-board refueling controls before imposing that requirement. S. 1630 also increases the useful life of automobiles from 50,000 to 100,000 miles. On the negative side, S. 1630 reduces emissions warranty coverage from 5 years/50,000 miles to 2 years/24,000 miles, except for emissions computers and catalytic converters, which would be extended to 8 years/80,000 miles.

The Senate Phase 2 proposal for 2003 and beyond would reduce the tailpipe HC and NO_x standards further, to only 30 percent and 20 percent, respectively, of currently allowed levels. A late proposal in the Senate would make Phase 2 standards contingent on a specified degree of progress in ozone attainment. The Waxman-Dingell compromise Phase 2 plan is very close to the Senate's, but it requires confirmation or relaxation of the tailpipe standards by an EPA rule making in 1999. Also, the tailpipe HC standard is somewhat less stringent than in the Senate version, since it is defined on the basis of non-methane HC instead of total HC.

How Clean Is Gasoline?

Estimates of in-use emissions of future gasoline-fueled vehicles differ on how often, and by how much, these vehicles would fail to meet future emissions standards. Understanding this issue requires that several sources of vehicle HC be examined: tailpipe emissions; evaporative fuel emissions of parked cars; evaporative fuel emissions of running cars; and refueling emissions of evaporated fuel. The three kinds of evaporative fuel emissions are similar in their formation and control. The need for new controls of evaporative emissions is widely agreed upon. The only difference in the proposals for these three areas is that the Senate version would apply on-board refueling controls in addition to Stage 2 controls.

Parking-Loss Emissions

A standard in effect since 1981 requires more than 95 percent control of the diurnal and hot-soak emissions of parked cars, as measured on a particular test procedure. The two grams of emissions permitted is equal to about 0.10 g/mi. If parking-loss emissions in the real world were this low, they would be of little concern. However, many cars that are certified to comply with official standards when new, do not continue to meet these standards for their full life. Since vehicles' parking-loss emissions can increase by as much as 2,000 percent, it does not take many malfunctioning or disabled vehicles to increase the fleet average to well above the permitted 0.10 g/mi.

Furthermore, true parking-loss emissions are not fully measured by the official test procedure, in which the temperature never exceeds 86 degrees F. In the real world, temperatures often reach or exceed 95 degrees F, and fuel evaporation increases. Also, in the test a vehicle is parked for only one "day" of diurnal temperature increase. The vapor that this increase in temperature loads into the charcoal canister has an opportunity to be purged into the engine before another vapor-generating episode occurs. In the real world, many cars sit unused for two or more days. Vapor generated during the first day fills up the charcoal canister, and the second or third day's vapors are released to the atmosphere.

EPA estimates,[3] based on tests of hundreds of cars borrowed from their owners, that the average for parking-loss emissions of today's vehicles is 0.28 g/mi—nearly three times the nominal standard of 0.10 g/mi. As part of the administration's plan, EPA is now developing an official test to measure fully and restrict evaporative emissions on hotter days and on cars parked for two or three days. Malfunctions and disablements, of course, cannot be directly addressed by a new test procedure. However, even improved control systems would not be able to limit vapors from cars parked for longer periods. Therefore, parking-loss emissions would be reduced only from 0.28 g/mi to 0.18 g/mi (Stout, 1989).

Running-Loss Emissions

EPA has discovered in only the last several years that many cars' fuel tanks release vapor during trips longer than 20 minutes or in slow-moving traffic. Although the evaporative parking-loss emissions-control system on any vehicle provides some control of these vapors, sufficiently adverse circumstances relative to vehicle design can cause fuel in the tank to warm up to the point that too much vapor develops to burn in the engine or store temporarily. EPA estimates that even after gasoline's RVP is reduced to 9.0 psi, HC running-loss emissions will be 0.55 g/mi.

It is clear that much lower running losses are possible, even under adverse conditions, and EPA will impose new standards. EPA estimates that its new standard would reduce fleet average running losses to 0.16 g/mi (Stout, 1989). Of the 0.16 g/mi, 0.10 g/mi represents the emissions of an average nondisabled vehicle operating over a typical mix of short and long, morning (cooler) and afternoon (hotter), and low-speed and high-speed trips on a 71 to 95 degree F day. The other 0.06 is due to excess emissions from vehicles with a disconnected vapor-control system hose or a missing gas tank cap. The 0.06 is beyond the control of EPA's new vehicle program.

The cars with the best emissions that have been tested by EPA for running losses have emissions well below 0.10 g/mi under the same circumstances of trip mix and temperature. Those who believe that the running losses of gasoline-fueled vehicles can be reduced to a very low level argue that EPA can achieve such a level by simply imposing a sufficiently stringent standard for all cars. However, it would be unreasonable to assume that fleet average running losses would be or could be near zero, even without excessive losses from disabled vehicles. It is unlikely that EPA will require a prototype of every vehicle model to demonstrate zero or near-zero running-loss emissions under the most adverse temperature, speed, and trip length. If the worst real-world conditions are not realistically reproduced in the test procedure, not all cars will be designed for complete control during them. Also, vehicles in actual use will not perform as well as vehicles in certification even on the same test procedure.

Parking-loss emissions from in-use vehicles are the closest parallel to what may be expected in the control of running-loss emissions. In both cases, vapors generated in the fuel tank should be released only through a hose leading to the engine and canister, where they should be stored and burned. EPA's testing of modern fuel-injected vehicles shows that the four or five vehicles with the lowest emissions had in-use parking-loss emissions that were only one-tenth of the standard for certification. A considerable number had emissions less than one-half of the standard. However, the average of all vehicles exceeded the standard by 40 percent. The average of all vehicles exceeded the best vehicles by a factor of 14. These data indicate that even if good technology is available, and a strong certification standard is enforced, in-use emissions can still be significant.

Refueling Emissions

EPA has performed extensive testing during the past several years to measure refueling emissions from gasoline-fueled vehicles. This effort was in connection with EPA's recent, but now suspended, proposal to require on-board systems to control these vapors. These vapors—spillage emissions and filling emissions—total 0.20 g/mi if there is no attempt at control.

Pump-based Stage 2 vapor control hardware and on-board control hardware could capture 95 percent or more of the filling emissions during refueling. The final version of amendments to the Clean Air Act, however, may allow Stage 2 exemptions for small gas stations and fleet pumps. Inspections would be largely the responsibility of state and local agencies, all of which may not be able to enforce a high level of compliance. Therefore, EPA believes that uncontrolled refueling emissions of 0.20 g/mi would be reduced to 0.07 g/mi in an average non-attainment area.

On-board refueling systems would likely be subject to some disablement and malfunction. However, unlike heavily used pump hardware, routine maintenance is unnecessary. EPA estimates that refueling emissions would be reduced to 0.02 g/mi with on-board control systems. A local agency should be able to achieve the same 0.02 level of control, via Stage 2, with sufficient effort.

Total Evaporative Emissions

Figure 1 shows the EPA estimates for each of the three categories just discussed and for total evaporative emissions. The bars represent the emissions of a fully turned-over fleet, under current standards or with the proposed new controls.

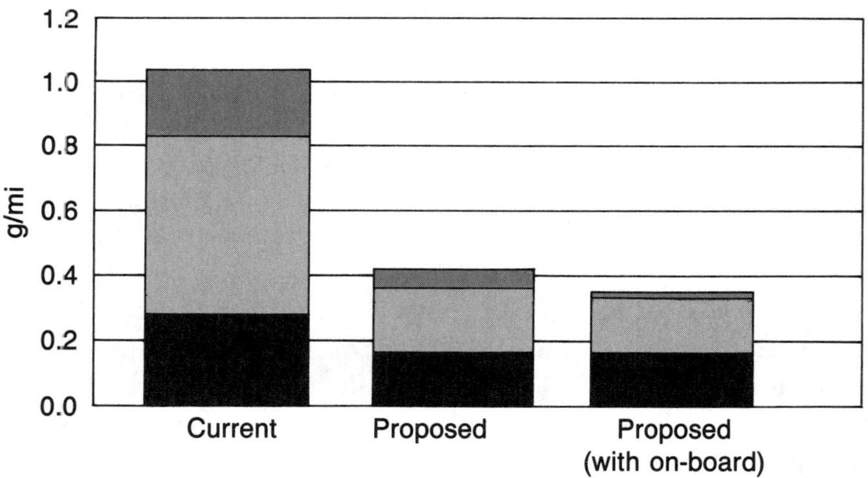

Figure 1
Emissions of Gasoline Vapors (HC) under Current and Proposed Controls

(Emissions of a fully turned over passenger car fleet with 9 psi RVP fuel on a day when the temperature increases from 71° to 95°)

Source: Emissions Control Tech. Div., USEPA (1989)

Tailpipe Emissions

Although well-designed vehicles, operating as they were intended to, can have very low evaporative emissions when tested for compliance, the above discussion showed that many in-use vehicles have sufficiently high losses to raise the fleet's average emissions considerably. The same may be said of tailpipe emissions. In addition, since exhaust emissions of even a properly operating vehicle increase with its length of service, the useful life over which compliance should be maintained becomes an additional issue.

Estimates of future gasoline-fueled vehicle emissions must begin with the in-use emissions of recently manufactured, fuel-injected vehicles. Figure 2 shows EPA's estimate of the average HC emissions of vehicles in 49 states as a function of age, the latter expressed

Figure 2
Tailpipe HC Emissions of Current Fuel Injected Vehicles: MOBILE4 versus an Estimate with Seriously Malfunctioning Vehicles Removed

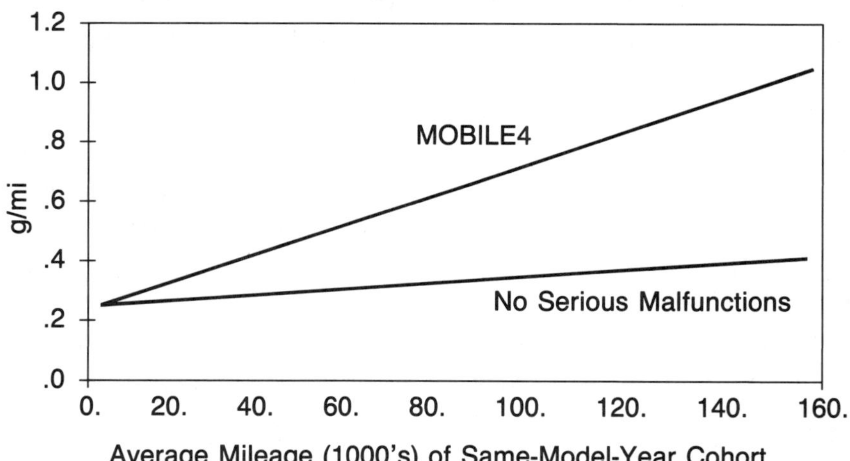

Fully turned-over passenger car fleet, enhanced I/M, standard EPA test conditions.

Source: Emissions Control Tech. Div., USEPA (1989)

as the average odometer value for a model year cohort. (The effect of an enhanced I & M program is included.) The second line—estimated from the same data but with seriously malfunctioning vehicles excluded—shows the important influence that malfunctioning vehicles have on fleet average emissions, even in the first 50,000 miles. It is important to note that these malfunctions were not owner-caused. Virtually all occurred in normal use, and were either not noticed or not fixed by the owner.

Future in-use emissions can be predicted for the administration's proposal on the tailpipe HC standard (which is almost identical to Phase 1 of the current House version, and to California standards). The proposed 30 percent reduction in the HC standard is sufficiently small that one can assume that vehicle hardware would not change and thereby affect the incidence of malfunctioning vehicles. In Figure 2 terms, the HC lines would start 30 percent lower but would stay parallel to their current position at all mileages.

A proportional drop in the starting HC level might also be anticipated as a result of the even stricter Phase 2 regulations. The 100,000-mile useful life in the Phase 2 proposals, however, raises the question of whether manufacturers must reduce the emissions of new vehicles more than proportionally, to allow for additional deterioration before the compliance test.

EPA's recall program, now and in the future, will have the most control over starting HC levels, since actual usage that applies in the recall program is a more stringent test of deterioration than the certification program's durability protocol, even though the proper maintenance of vehicles must be documented. Setting a useful life at 100,000 miles will force starting HC levels down an additional amount if manufacturers expect EPA to operate a recall testing program at higher mileages, and if that creates a threat sufficient to justify the expense involved.

High-mileage recall would be difficult to implement effectively. It would be hard to find vehicles that have well-documented usage and maintenance. Also, only a small percentage of owners would be likely to return high-mileage vehicles for a free emissions repair.

In addition, EPA may face a credibility issue if it presses hard for expensive recall campaigns of vehicles that are close to being scrapped. Thus, the longer useful life may not by itself force down starting HC levels very much, particularly for Phase 2 (when near-zero starting levels would be required anyway). It can be reasonably assumed that the longer useful life will have an additional effect on starting HC levels, somewhere between zero and the hypothetical effect of a further 17 percent reduction in the 50,000-mile standard. The 17 percent factor is suggested by the "No Serious Malfunctions" line in Figure 2, whose 50,000-mile point is 17 percent less than the 100,000-mile point.

A larger question is whether manufacturers could and would improve the durability of conventional components and thereby reduce the incidence of malfunctions. Automobile manufacturers already have an incentive to reduce component failures through 50,000 miles. There is no incentive after 50,000 miles except customer satisfaction. If recall testing was conducted after 50,000 miles, new incentives might be created.

Current regulations permit manufacturers to specify replacement of any component at 50,000 miles or later. Assuming that EPA modifies these regulations so that important components are not allowed to be replaced prior to a recall test even after 50,000 miles, each further mile of service before the recall test adds to manufacturers' incentive to build a better component, if they have the know-how. The outcome is not clear-cut. Even at 75,000 miles, EPA predicts that only one in six current vehicles would have a serious malfunction. Some of these are clustered into particular models, but most are scattered. On the one hand, manufacturers may find that they can live with these statistics, since they make a recall order unlikely and at 75,000 miles most owners might ignore a recall order. On the other hand, even if they greatly fear recall orders, manufacturers might find it difficult to predict which components would be vulnerable and need design improvements. They may learn which components have problems five years too late, or after the old design has already been changed for other reasons. The longer useful life might then produce more recall campaigns but no real improvement in future design.

If manufacturers had sufficient foresight and were inescapably responsible for the average emissions of *all* in-use vehicles, it would be reasonable to assume that they would produce designs with one-half the rate of increase in average emissions if made responsible for doubling useful life. Thus, in-use emissions at 100,000 miles would be equal to what they were at 50,000 miles. In-use malfunctions of all types at a given mileage would be cut in half. The slope of the MOBILE4 line in Figure 2 would also be cut in half.

A 100,000-mile useful-life requirement might (1) have no effect, since any effects depend on EPA's discretionary actions, which have no guarantee of success; (2) result in manufacturers designing new vehicles as if they were complying with a 50,000-mile standard, 17 percent less than the actual 100,000-mile standard (to allow room for more properly maintained vehicle deterioration before the recall test); or (3) reduce the incidence of all in-use malfunctions by 50 percent. Figure 3 shows what the in-use tailpipe HC of a fully turned-over fleet would be for each of the proposed amendments to the Clean Air Act, calculated for each of these possible effects. The Phase 2 proposals would provide a sizeable benefit, compared with the administration's proposal, only if one accepts the 50 percent reduction in all types of in-use malfunctions.

Total Gasoline Vehicle Emissions

Figure 4 combines the evaporative emissions estimates in Figure 1 with the tailpipe estimates in Figure 3.

Emissions-Control Diagnostics

The estimates of gasoline-fueled vehicle emissions in figures 1 through 4 contain no explicit adjustment for the new emissions-control diagnostics regulations that each Clean Air Act proposal

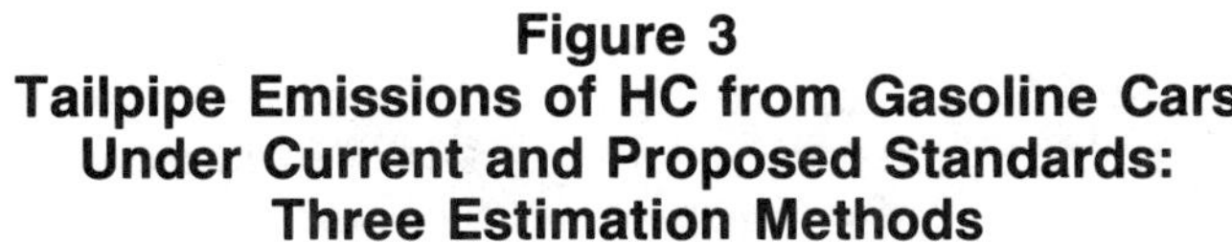

Figure 3
Tailpipe Emissions of HC from Gasoline Cars
Under Current and Proposed Standards:
Three Estimation Methods

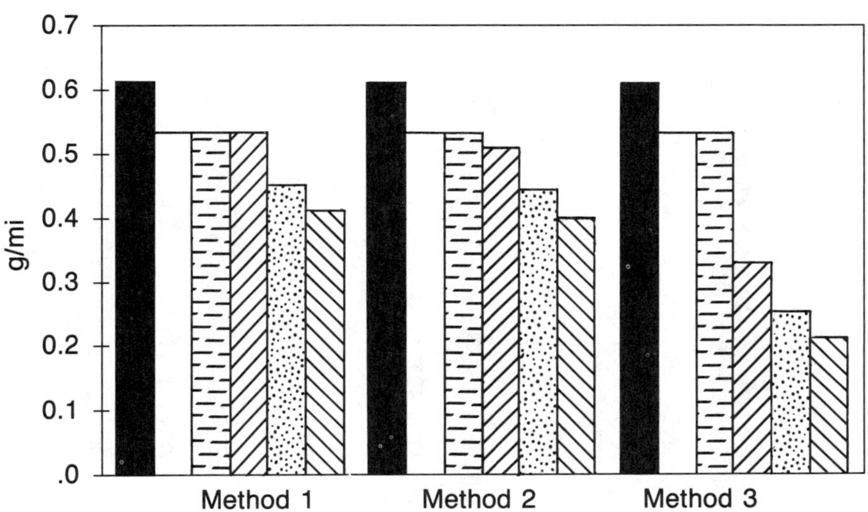

(Emissions of a fully turned over passenger car fleet with
Enhanced I & M on a day when the temperature increases from 71° to 95°)

■ Current Standards   ☐ Administration   ⊟ House Phase 1
▨ Senate Phase 1   ▦ House Phase 2   ◫ Senate Phase 2

Source: Emissions Control Tech. Div., USEPA (1989)

directs EPA to promulgate. Whether a better gasoline-fueled vehicle can be achieved with diagnostics must be considered.

Emissions-control diagnostics are a relatively new means of achieving better gasoline-fueled vehicle performance. Three benefits are possible: owners may seek emissions repairs earlier and more often in response to accurate warning lights; I & M programs may require a low-emission diagnostic status to pass, regardless of the idle test result; or mechanics may locate more defective components with the help of the diagnostic system, and the warning light may

Figure 4
Total Emissions of HC from Gasoline Cars
Under Current and Proposed Standards:
Three Estimation Methods

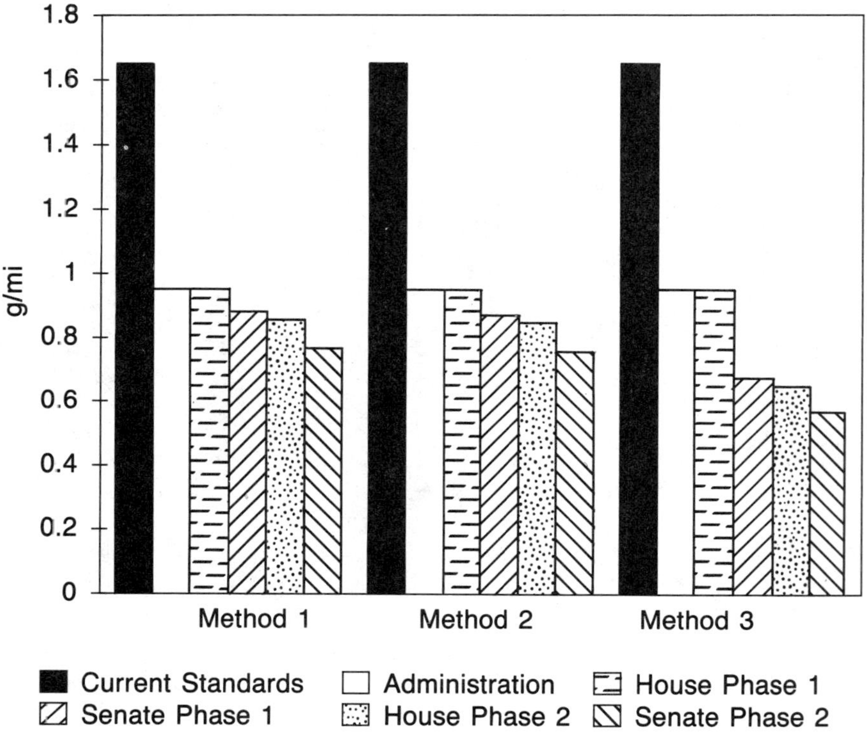

Source: Emissions Control Tech. Div., USEPA (1989)

convince owners of the need for the expense of replacement. Benefits could also be derived for tailpipe emissions and for parking-loss, running-loss, and refueling evaporative emissions.

No one can say with certainty whether diagnostic systems would make a difference or how much of a difference that could be. The state of California has adopted an ambitious set of enhanced diagnostic requirements. However, it has allowed a long lead time and left open the option of partial postponements if the required technology is slow to be perfected. The success of any regulations would depend strongly on manufacturers making warning lights accurate indicators of high emissions. Furthermore, owners of older, low-value vehicles might not seek repairs any more often than they do now. I & M programs cannot enforce a diagnostic status check unless there is a standardized, automated, non-human intervention method for doing so. In addition, most older vehicles (those that will have the most need for repair) are serviced at non-dealer facilities that do not stock a full line of parts. Non-dealer mechanics may have less of a business incentive for using the diagnostic system. If all the components involved were warranted for 100,000 miles—so that necessary replacements were, in effect, pre-paid with the vehicle purchase—diagnostic systems could be a powerful tool for lowering emissions. But that is not likely to be the case.

If one expects diagnostics to be successful, it would be prudent to consider them as further support for the more optimistic of the three tailpipe estimation methods, not for a still more optimistic estimate.

How Clean Is M100?

A great deal is now known about the emissions of M100 vehicles. Although only a few M100 prototypes have been fully tested, their results have been very consistent and in line with expectations based on the physical properties of methanol. If pure methanol is used as a fuel, only methanol vapor can be emitted from the fuel system, and about 90 percent of the organic emissions from the tailpipe is unburned methanol. The remainder consists of hydrocarbons and formaldehyde, with trace amounts of other aldehydes. The hydrocarbons are generally the same species as those emitted from gasoline-fueled vehicles, with a shift toward the paraffins.

Parking-Loss Emissions

The headspace of an M100 fuel tank contains only 20 percent as much mass of methanol vapor as there is hydrocarbon mass in the headspace of a gasoline tank of equal size. Diurnal warming of both tanks will expand the vapors and push some out, but (accounting also for different tank sizes) only 30 percent as much methanol will be released. If both vehicles are well designed and properly operating, all the vapor released will be captured in the carbon canister. Test results from gasoline-fueled vehicles in good condition and M100 vehicles can therefore show similarly low emission levels if the test conditions are the same as those for which the vehicle was designed.

Under other assumptions, the emissions of the two vehicles can actually exceed the underlying 70 percent difference in vapor generation. These conditions include any failure of the canister purge system, a hose disconnection or other type of tampering, a missing tank cap or leaky gasket, high temperatures, wide daily temperature swings, parking for several days, operating in low-speed traffic, or making frequent short trips. Most of the gasoline-fueled vehicle parking-loss emissions anticipated by EPA would arise from such unusual conditions. An assumption underlying EPA's support for M100 is that average mass of methanol hot-soak and diurnal losses will be 80 percent lower than the mass of hydrocarbon parking-loss emissions from a gasoline vehicle.

EPA recently tested four vehicles using M100 fuel and a temperature rise of about 71 to 95 degrees. The control system on each was completely disabled. Diurnal emissions were about three grams of methanol (Platte, 1989a). Under these conditions, hydrocarbon emissions from gasoline-fueled vehicles would have been about 15 grams.[4] Hot-soak emissions should follow a pattern similar to that of diurnal emissions. No vehicle manufacturer or charcoal supplier has indicated any concern about charcoal canisters being more adversely affected by long use with methanol than with gasoline.[5]

Running-Loss Emissions

Running-loss emissions from gasoline-fueled vehicles are generated at higher fuel tank temperatures than are parking-loss emissions. At these temperatures, gasoline boils and methanol does not. EPA tests of vehicles with gas caps removed have shown that an M100 car generates only about 3 percent as much vapor as a gasoline-fueled car during a one-hour trip (Platte, 1989). EPA assumes that overall running-loss emissions would be 85 percent less by mass from M100 vehicles and that these running losses will consist of methanol.

Refueling Emissions

A 70 percent automatic benefit in refueling emissions occurs because of methanol's lower vapor pressure and molecular weight. Small differences in estimated emissions can result from differences in fuel economy assumptions depending on whether on-board controls, or the use of Stage 2 controls, are assumed for M100 vehicles. Refueling emissions for methanol are estimated to be about 75 to 80 percent less by mass than EPA's estimate of gasoline refueling emissions with partially applied and enforced Stage 2 equipment.

Total Evaporative Emissions

Combining the above three estimates, total evaporative methanol mass from an M100 fleet would be 0.072 g/mi, or 82 percent less than the evaporative hydrocarbon mass from a gasoline-fueled fleet. If the lower ozone reactivity of methanol is considered, the benefit increases to 97 percent.[6] This benefit applies to the 44 percent of gasoline-fueled vehicle emissions that are evaporated fuel. In other words, M100 would give a 42 percent ozone benefit even if there were no benefit for tailpipe emissions.

Tailpipe Emissions

Table 2 shows the available data (Gabele, 1989a; Piotrowski, 1988; Smith and Urban, 1982; CARB, 1989; CARB, 1988; Gabele, 1989; Williams et al., 1989) on exhaust emissions of prototype

Table 2
M100 Prototype Data Versus EPA Estimates
(Gram/Mile)

	Non-Methane HC	Methanol	Formaldehyde
DEDICATED M100 VEHICLES			
1986 Toyota Carina[a]	0.004	0.240	0.014
1981 VW Rabbit[b]	0.010	0.490	0.006
(non-stock catalyst)			
Average	0.007	0.365	0.010
Carbon Percent	4%	93%	3%
M85 VEHICLES TESTED ON M100			
1981 Ford Escort[c]	0.003	0.151	0.003
1987 Ford Crown Victoria[d]	0.017	0.500	0.035
1988 Chev. Corsica[e]	0.030	0.490	0.053
1988 Chev. Corsica[f]	0.040	0.820	0.041
198? GM (model not given)[g]	0.008	0.761	0.040
Average	0.020	0.544	0.034
Carbon Percent	7%	87%	6%
EPA ESTIMATES (Special Report)	0.050	0.500	0.015
Carbon Percent	18%	79%	3%

Sources:

[a] Gabele (1989a)
[b] E. Piotrowski (1988)
[c] Smith and Urban (1982)
[d] CARB (1989)
[e] CARB (1988)
[f] Gabele (1989)
[g] Williams (1989)

vehicles operating on M100 versus EPA estimates of the average in-use emissions of an M100 fleet (EPA, 1989). The table shows two vehicles designed expressly for M100 and five vehicles designed for M85 but tested on M100. The vehicles all are gasoline-car derivatives, with no fundamental optimization for methanol. Except for the Toyota Carina, all have conventional stoichiometry-centered engines and emissions controls. The Carina operates lean most of the time. In some cases the M85 vehicles did not start normally on M100, even at the 75 degree test temperature of this data—a problem that could probably be corrected with a fuel-calibration change and that would definitely be eliminated by any solution to the M100 cold-starting problem. Poor starting probably contributed to the somewhat higher emissions from the M85 vehicles. In light of these circumstances, the data from the M85 vehicles are in line with those from the M100 vehicles. The experimental results from both prototype groups are close to EPA estimates of in-use emissions. Further design improvements (motivated by stringent standards) and the inevitable prototype-to-real-world slippage will work in opposite directions, thus making EPA's estimates quite plausible.

Plausibility, of course, is not proof. Somewhat more can be said, however. Of the seven vehicles shown in Table 2, four are flexible-fuel vehicles, which were also tested with gasoline. A fifth car was compared with an identical gasoline-fueled version. The M100 and gasoline results for organic carbon from these five cars are compared in Table 3. Considering that the vehicles were not fully optimized for M100, even within the limits of their physical design, Table 3 provides good support for a general assumption that stoichiometric M100 and gasoline engines, *with the same control systems*, can readily be designed to have equal emissions on a non-methane "organic material" basis. In other words, total exhaust emissions of non-methane carbon were about equal. Many air quality modeling studies of methanol fuels have assumed just this. (Some studies also incorrectly assume equal carbon in evaporative emissions.) This equality makes engineering sense also, in that both vehicles use about the same fuel carbon per mile and, at stoichiometry, the carbon density in a cylinder's unburnable quench layer is essentially the same for both fuels. This engineering explanation supports a further expectation that even under malfunction conditions, carbon emissions rates would be very similar. There are no data to test this expectation, however.

Table 3
M100 Versus Gasoline Organic Carbon Emissions
from Vehicles Designed for M85[*]

	Organic Carbon Emissions (g/mile)		M100/ Gasoline Ratio
	Gasoline	M100	
1981 Ford Escort[c]	0.281	0.258	0.92
1987 Ford Crown Victoria[d]	0.300	0.218	0.73
1988 Chev. Corsica[e]	0.175	0.231	1.32
1988 Chev. Corsica[f]	0.254	0.363	0.70
198? GM[g]	0.192	0.309	0.62
Average	0.240	0.276	1.15

[*]Gasoline result is from a test on a different but essentially identical vehicle except for fuel type.

Sources:
   [c] Smith and Urban (1982)
   [d] CARB (1989)
   [e] CARB (1988)
   [f] Gabele (1989)
   [g] Williams (1989)

EPA has been assuming that an optimized M100 fleet would have 50 percent as much tailpipe organic carbon as a gasoline fleet produced to meet the administration's proposals. An important question, then, is whether this estimate for an M100 fleet is a reasonable expectation for the year 2000 and beyond. It may well be true that a 50 percent reduction can be achieved only by applying technology not needed on a gasoline-fueled vehicle meeting the administration's proposed 0.25 g/mi non-methane HC standard. The administration intended for its proposal to be a challenge to industry, and some amount of technology forcing is appropriate. A fresh look at engine design with M100 would be appropriate, anyway, in order to select a long-term solution to M100's cold-start problem. The two development efforts, cold start and lower organic emissions, can be combined. It should be as easy to achieve the organics reduction from M100 vehicles as it would be to comply with the proposed Phase 2 HC standards with gasoline. It should be easier, since M100 offers the option of a lean-burn approach, which is not possible with gasoline. (It is uncertain whether a 0.4 g/mi or 0.2 g/mi NO_x standard, if adopted for gasoline, would apply to M100 and whether it would significantly constrain the lean-burn option.)

The composition of carbon-containing organic emissions also matters. For the two M100 vehicles in Table 2, 3 percent of the organic carbon was in the form of formaldehyde, 4 percent as hydrocarbon, and the other 93 percent as methanol. The proportion that EPA assumes will be formaldehyde is 3 percent, which is in complete agreement with the M100 data in Table 1. EPA appears to have been overestimating the proportion of hydrocarbon emissions, at about 18 percent instead of 4 percent. Correctly assigning the extra 14 percentage points to methanol would increase EPA's benefit estimate.

With the mix of non-methane HC, methane, and formaldehyde assumed by EPA, an average tailpipe carbon-atom in M100 exhaust has only 65 percent of the ozone reactivity as in gasoline exhaust. A 50 percent reduction in carbon would therefore represent a 67 percent reduction in tailpipe contribution to ozone. Since tailpipe emissions are 56 percent of the total gasoline-fueled vehicle base, the tailpipe differences with M100, under the administration's proposal, would contribute 38 percentage points to overall reduction.

Total M100 Vehicle Organic Emissions

Combining the above estimates, EPA expects the M100 fleet under the administration's plan to have steady-state emissions of

0.05 g/mi of non-methane HC, 0.572 g/mi of methanol, and 0.015 g/mi of formaldehyde. Applying reactivity factors, the ozone-equivalent non-methane HC level would be 0.19 g/mi.

Conclusions

Total gasoline-fueled vehicle non-methane HC emissions on a summer day, when ozone could be a problem, can reasonably be estimated at no higher than 0.95 g/mi under the administration's proposal, and no higher than 0.77 g/mi under the Senate's Phase 2 standards. It is difficult to conceive of gasoline-fueled vehicles using 9 psi fuel emitting any less pollution than 0.57 g/mi under the Senate's Phase 2 requirements, and even this value depends on an assumption that in-use malfunctions will be halved. A reasonable, but admittedly uncertain, expectation for achievable emissions from an M100 fleet would be the equivalent of 0.19 g/mi of non-methane HC. In percentage terms, the M100 benefit would be 67 or 80 percent, relative to the optimistic assessment of the Senate or to the administration's proposal, respectively.

A HC emissions rate of 0.57 g/mi to 0.95 g/mi for gasoline-fueled vehicles is low enough that total regional vehicle emissions would be very small, compared with the vehicle emissions that helped cause the high ozone levels of the last few years. However, compared with allowable VOC emissions for achieving attainment, vehicles would continue to be a significant source. The option of restricting vehicle use would always be suggesting itself. An eventual transition to M100 or to any other fuel that could achieve low emissions of approximately 0.19 g/mi would largely remove vehicles from the VOC control equation.

Notes

1. On March 1, 1990, a modified form of the administration's program was incorporated into a proposed compromise floor amendment to Senate 1630. In the modified plan, a required reduction may be obtained from all vehicle sales, rather than on an

individual basis from alternatively-fueled vehicles. Reformulated gasoline might allow compliance on these terms. A 75 percent reduction on an individual vehicle basis would still apply to some purchases by certain fleets.

2. Attainment requirements are outside the scope of this paper, but a simple example will illustrate the point. Assume that a certain urban area had daily volatile organic compounds (VOC) emissions of 400 tons in 1987, of which 200 tons came from motor vehicles. In 2010 motor vehicle emissions will be reduced about 75 percent under the President's gasoline vehicle proposal. If the VOC reduction requirement for attainment is 80 percent, motor vehicles will account for 50 of the 80 allowable tons of VOC, and other sources will have to be reduced 85 percent, net of growth. However, if all gasoline cars and light trucks were to be replaced with "clean" fuel vehicles meeting an 80 percent reduction performance requirement, other sources would only need to be reduced 68 percent, which translates into twice the amount of allowable emissions for them. In all of the amendments being considered by Congress the most severely polluted areas must demonstrate attainment with sophisticated ozone models, which likely will predict an 80 percent VOC requirement in at least several areas.

3. This is a MOBILE 4 prediction for CY 2012 and later, 9 psi fuel, 71 degrees minimum, 95 degrees maximum, with an I & M program but no anti-tampering program.

4. This is a MOBILE 4 value for the effect of disconnection on diurnal emissions of 1992 and later model cars.

5. SAE paper 801360 on a DOE-funded study by K. R. Stamper is sometimes cited. This study operated two vehicles for 50,000 miles on 10 percent methanol (M10), but no control vehicles on straight gasoline and no vehicles at all on M100.

6. Relative reactivities of organic species and mixtures in ozone formation are outside the scope of this paper. The reactivity-adjusted emissions numbers given here are based on EPA's estimates that a gram of methanol is equivalent to 0.19 grams of non-methane HC, and a gram of formaldehyde to 2.2 grams of HC.

References

CARB (1988). Alcohol Fueled Fleet Test Program, Eighth Interim Report. California Air Resources Board.

_____ (1989). Definition of a Low-Emission Motor Vehicle. California Air Resources Board.

EPA (1989). Analysis of the Economic and Environmental Effects of Methanol as an Alternative Fuel. U.S. Environmental Protection Agency.

Gabele, P. (1989). Characterization of Emissions from a Variable Gasoline/Methanol Fueled Car. U.S. Environmental Protection Agency, ORD.

_____ (1989a). Study of Lube Oil Emissions from Toyota Carina. EPA Memo to Karl Hellman (OMS), September 14, 1989.

Piotrowski, G. (1988). Methanol Vehicle Catalyst Evaluation: Phase III. U.S. Environmental Protection Agency, Technical Report No. EPA/AA/CTAB/88-10.

Platte, L. (1989). Compilation of Evaporative Data for M85 and M100 Fuels. EPA Memo to Charles L. Gray, September 21, 1989.

_____ (1989a). High Temperature Diurnals—M85, M100 Fuels. EPA Memo to Charles L. Gray, November 15, 1989.

Smith, L., and C. Urban (1982). Characterization of Exhaust Emissions from Methanol- and Gasoline-Fueled Automobiles. U.S. Environmental Protection Agency, SWRI, Report No. 460/3-82-004.

Stout, A. (1989). Reductions in Evaporative Emissions and Running Losses from Enhanced Vehicle-Based Control. EPA Memo to Charles L. Gray, December 19, 1989.

Williams, R., et al. (1989). Formaldehyde, Methanol, and Hydrocarbon Emissions from Methanol-Fueled Cars. General Motors, GMR-6782.

CHAPTER 3

Can Methanol Fuels Improve Vehicle Emissions?

THOMAS C. AUSTIN

Abstract

Based on recent reports (EPA, 1989) and articles (Gray and Alson, 1989) by officials of the Environmental Protection Agency, switching to methanol fuel is the only way to reduce motor vehicle emissions significantly while retaining personal mobility. Superficially, the air quality characteristics of methanol look attractive. Based on smog chamber studies, methanol produces about 80 percent less ozone than gasoline (per gram of unburned fuel emitted). In addition, the lower volatility of methanol would be expected to reduce evaporative emissions substantially. However, methanol-fueled vehicles emit more than just unburned methanol, and the total organic emissions tend to be higher than the emissions from gasoline vehicles using similar emission controls. In addition, available test data indicate that the evaporative emissions of methanol-fueled vehicles are higher than would be expected based only on fuel volatility. Analysis by Sierra Research, Inc., indicates that the case for methanol is built on the same overly optimistic forecasts for unproven alternative technology and the same pessimistic forecasts for conventional fuels and engines that led to the rise and fall of interest in steam engines and gas turbines almost twenty years ago.

Background

The emission characteristics of gasoline-fueled vehicles with no emission control systems provide a point of reference for a discussion of vehicle emissions control. Since the early 1960s, the control program has been focused on emissions of hydrocarbons, carbon monoxide, and nitrogen oxides. It is in the hydrocarbon category where the potential advantages of methanol are the most significant. Exhaust emissions of hydrocarbons consist of unburned and partially-reacted fuel and lubricating oil emitted in the exhaust and have received the greatest attention. However, crankcase, evaporative, and refueling emissions are also important.

Crankcase emissions include unburned fuel and other compounds in the engine blowby gases which, in older engines, were vented to the atmosphere through a road-draft tube. Evaporative emissions consist of fuel vapor emitted from the engine and fuel system. They are divided into diurnal emissions, which result from the "breathing" of the fuel tank as it is heated and cooled over the course of a day; hot-soak emissions produced by the heat from the engine after it is shut down; and running losses from the fuel system while the vehicle is in operation. Refueling emissions are produced when the addition of fuel displaces the gasoline vapor in a partially empty tank. Evaporation of leaks and spills (primarily associated with refueling) also contributes to emissions.

Figure 1 compares the total uncontrolled hydrocarbon emissions from gasoline-fueled passenger cars with EPA's 1989 estimate of emissions from gasoline-fueled cars meeting the emission standards proposed in the Clean Air Act amendments supported by

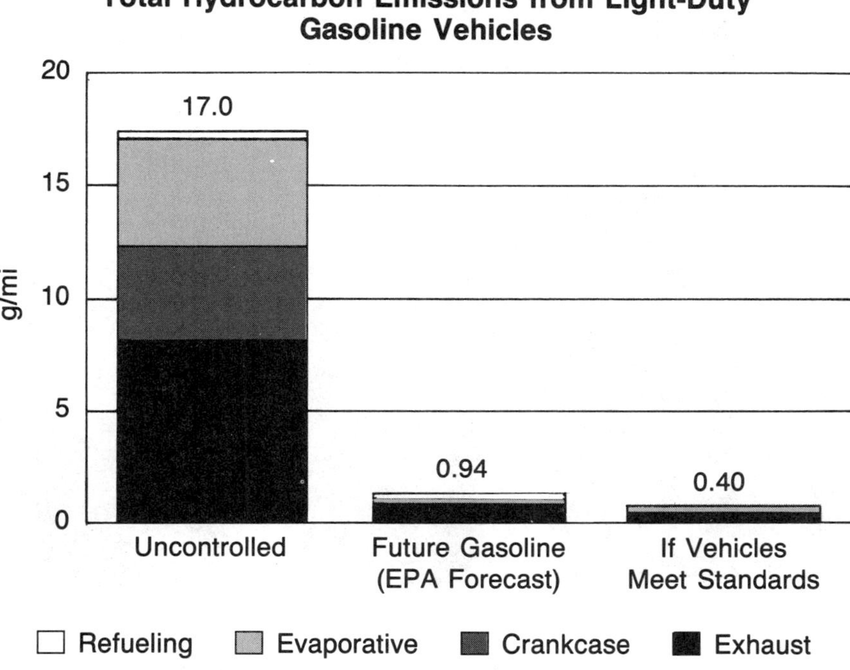

Figure 1
Total Hydrocarbon Emissions from Light-Duty Gasoline Vehicles

Source: EPA, 1989, and Sierra Research Inc.

the Bush administration (which includes a 0.25 g/mi non-methane hydrocarbon exhaust emissions standard). Also shown in the figure are estimates for emissions if all cars produced exactly the level allowed by the exhaust, crankcase, evaporative, and refueling emissions standards that apply in California. As the figure indicates, EPA does not project that cars in actual use will meet the standards that they are certified to meet.

Figure 2 shows how EPA's forecast of gasoline vehicle emissions compares with EPA's estimate of the emissions from vehicles fueled by a blend of 85 percent methanol and 15 percent gasoline (M85) and pure methanol fuel (M100). All of the forecasts have been presented using EPA's recommended technique for computing

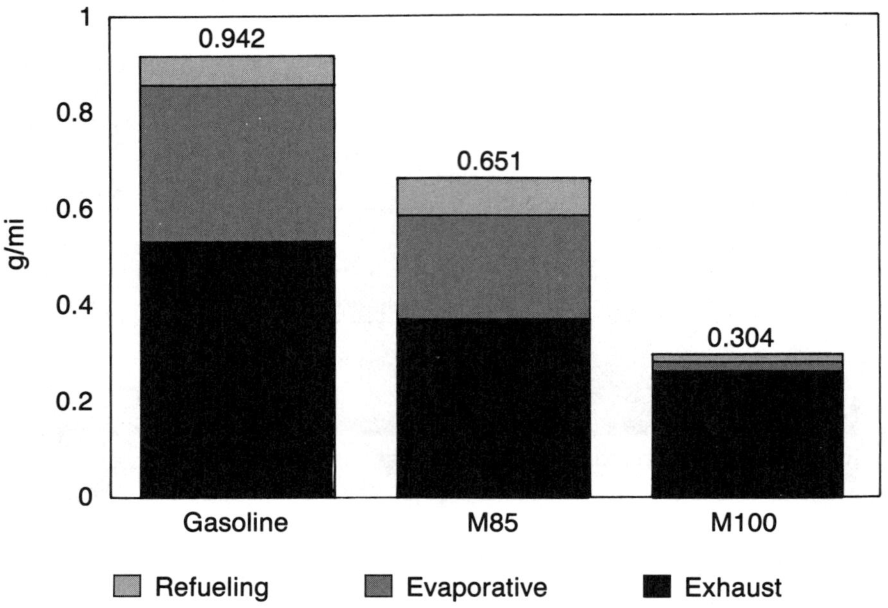

Source: EPA, 1989

"organic material, hydrocarbon equivalent" (OMHCE).* This technique subtracts the oxygen and some of the hydrogen from the oxygenated emissions (methanol and formaldehyde) from methanol
vehicles and is generally regarded as the most equitable way to compare emissions from the different fuels.

---

*As used in this paper, OMHCE = NMHC + $(0.46)CH_2O$ + $(0.43)CH_2OH$ where:
   NMHC = non-oxygenated, non-methane hydrocarbons;
   CH_2O = formaldehyde; and
   CH_3OH = methanol
Under EPA regulations, OMHCE *includes* methane. However, EPA frequently
reports only NMHC data and computes OMHCE on an NMHC basis.

Based on the comparison shown in Figure 2, methanol-fueled vehicles, especially those using M100, are projected to have far lower emissions. However, as discussed below, EPA's projections of methanol vehicle emissions are optimistic. More importantly, EPA's forecasts of emissions from gasoline-fueled vehicles are pessimistic.

Exhaust Emissions

The exhaust emissions component of EPA's projections is presented in Table 1. Based on Sierra Research's analysis of available test data, EPA's projections for methanol-fueled vehicle exhaust emissions are reasonable for what can be expected from vehicles produced during the mid-1990s, *assuming proper maintenance*. It should be noted, however, that the projected exhaust emissions for M100 represent a substantial improvement over the level of emission control that has been demonstrated to date. The performance assumed by EPA has not yet been achieved by vehicles that have completed the required 50,000-mile durability test. The level of NMHC, methanol, and formaldehyde exhaust emissions used in EPA's projections

Table 1
EPA Projections of Exhaust Emissions in Customer Service

| | | g/mi | |
	Gasoline	Optimized M85	Optimized M100
Non-oxygenated HC	0.530	0.150	0.050
Methanol	0.000	0.500	0.500
Formaldehyde	0.005	0.035	0.015
Total Organics	0.535	0.685	0.565
OMHCE	0.532	0.383	0.274
NMHC Standard	0.250		
OMHCE Emissions at the NMHC Standard	0.252		

Source: EPA, 1989

have thus far only been achieved by lower mileage prototype vehicles.

Table 1 also shows that EPA is projecting that gasoline-fueled vehicles will exceed the standard they are certified to meet by over 100 percent. It should be noted that the California Air Resources Board (CARB) is projecting compliance with a 0.25 g/mi NMHC standard in actual use starting with the 1995 model year (Russell et al., 1989). CARB's analysis indicates that the use of on-board diagnostic (OBD) systems in conjunction with vehicle inspection and maintenance (I & M) programs will virtually eliminate excess emissions. In contrast to EPA's estimates for methanol-fueled vehicles, every gasoline car model currently sold in the United States has already demonstrated substantially lower emission levels than projected by EPA during the 50,000-mile durability test required by EPA. In fact, CARB test data indicate that some gasoline-fueled models are *already* achieving exhaust emissions in actual use that are less than 0.25 g/mi (CARB, 1989). As noted above, CARB expects the average exhaust emissions of all new gasoline cars to meet 0.25 g/mi NMHC by the mid-1990s. If CARB is right, all of the OMHCE emissions benefit for methanol projected by EPA results from EPA's more pessimistic assumption regarding the actual performance of future gasoline vehicles.

Evaporative and Refueling Emissions

Table 2 summarizes the evaporative and refueling emissions forecasts that have been made by EPA for gasoline and methanol-fueled vehicles. As the table shows, the total evaporative and refueling emissions from methanol-fueled vehicles are projected to be lower. In the case of M100, EPA projects 92 percent lower OMHCE emissions than for gasoline. However, available test data indicate that EPA has underestimated the evaporative emissions from methanol-fueled vehicles. In addition, the benefits projected for methanol would be much less if gasoline-fueled vehicles were assumed (1) to comply with emission standards, (2) to be designed

Table 2
EPA Projections of Evaporative and Refueling Emissions
in Customer Service

	9.0 RVP Gasoline	————g/mi———— Optimized M85	Optimized M100
Hot-soak/Diurnal HC	0.180	0.058	0.000
Hot-soak/Diurnal Methanol	0.000	0.122	0.030
Running Loss HC	0.160	0.049	0.000
Running Loss Methanol	0.000	0.111	0.025
Refueling HC	0.070	0.053	0.000
Refueling Methanol	0.000	0.017	0.017
Total Organics	0.410	0.410	0.072
OMHCE	0.410	0.268	0.031
Sierra Estimate for Meeting Evap. Standard With Low Running Loss Designs and Refueling Controls...........	0.153		

Source: EPA, 1989

for low running-loss emissions, and (3) to use state-of-the-art controls for refueling emissions.

Diurnal and Hot-Soak Evaporative Emissions

Evaporative emissions are measured by capturing all vapors evaporating from vehicles placed in an enclosure called a Sealed Housing for Evaporative Determinations (SHED). Since the volume of the SHED is known, the mass of evaporative emissions can be calculated by measuring the hydrocarbon concentration in the SHED. Under the EPA and CARB emission test procedures, the approved instrument for measuring evaporative emissions is the flame ionization detector (FID). (The FID is the same instrument used to measure the hydrocarbons in vehicle exhaust emissions.) The FID is quite accurate for non-oxygenated hydrocarbons, but it doesn't respond as well to oxygenated materials like methanol. An 0.75 FID response factor for methanol measurement is typical (i.e., the FID reads about 25 percent low for methanol). No official test procedure for measuring evaporative emissions that contain methanol has yet been adopted by EPA or CARB. However, it is generally recognized that a gas chromatograph should be used to obtain accurate measurements of emissions from vehicles with oxygenated fuel. Because few laboratories are equipped to measure evaporative emissions with gas chromatographs, evaporative data on methanol-fueled vehicles are sparse. As a result, engineering estimates are often used to predict evaporative emissions from methanol-fueled vehicles.

The diurnal portion of the evaporative test measures emissions from the vehicle as the temperature in the gasoline tank is increased to simulate the warming of the tank that occurs as ambient temperature rises during the course of a day. The hot-soak portion of the evaporative test measures the emissions from the vehicle for one hour immediately following the exhaust emissions test. The vehicle is pushed from the dynamometer into the SHED as soon as the driving test is completed. The combination of diurnal and hot-soak emissions is termed the total grams per test. The individual diurnal and hot-soak test results can also be used to translate grams per test into grams per mile. The formula is:

$$\text{grams/mile} = \frac{\text{diurnal grams} + N \text{ (hot-soak grams)}}{\text{miles/day}}$$

where: N is the average number of trips per day; and miles/day is the average driving distance per day.

The evaporative emissions associated with the diurnal variation in ambient temperature and the evaporation from hot engines when the vehicle is turned off have been subject to control since the early 1970s. By venting the fuel metering system and gasoline tank to an activated charcoal canister, these emissions can be reduced by over 95 percent on gasoline-fueled vehicles. Vapors stored in the canister are then purged into the engine the next time the vehicle is started.

When vehicles meet the 2.0 gram per test evaporative emissions standard, evaporative loss per mile of travel is reduced to about 0.1 g/mi. However, EPA has estimated that advanced technology gasoline vehicles will not comply with the standard in customer service and will emit 0.18 g/mi of diurnal and hot-soak emissions (EPA, 1989). In contrast, CARB's official forecasts (from CARB's EMFAC model) project less tampering, and diurnal plus hot-soak emissions at 50,000 miles are projected to be under 0.1 g/mi.

CARB's test results also show that the same flexible-fuel vehicle operating on three different fuels had *higher* diurnal and hot-soak evaporative emissions on M85 than on straight gasoline, and *still higher* emissions on M100. Based on the volatility characteristics of M100, this result may seem to be surprising. However, as discussed in more detail below, it is more difficult to control evaporative emissions from vehicles with methanol in the fuel. CARB's test results for the same vehicle operating on three different fuels are presented in Table 3.

Although inherently lower evaporative emissions are frequently cited as a benefit of methanol fuels (particularly M100), the higher emissions reported by CARB are consistent with other available data on the difficulties of controlling methanol's evaporative emissions. Data reported by the Department of Energy (Stamper, 1980) indicate that the effectiveness of charcoal canisters is substantially degraded after 50,000 miles of operation on a blend of 10 percent methanol and 90 percent gasoline. The suspected mechanism for this deterioration in performance is related to adsorption of water vapor in canister purge air onto sites in the charcoal canister where methanol has been adsorbed. Over time, more of the available adsorption sites become tied up with methanol and water molecules that are more difficult to strip off during canister purge cycles. (This effect shows up during the accelerated durability

Table 3
Diurnal and Hot-Soak Emissions Reported by CARB

	Gasoline	M85	M100
	----------g/mi----------		
Non-oxygenated HC	0.045	0.036	0.011
Methanol	0.000	0.055	0.134
Total Organics	0.045	0.091	0.145
		(+102%)	(+222%)
Methanol as OMHCE	0.000	0.024	0.058
Total OMHCE	0.045	0.060	0.069
		(+ 33%)	(+ 53%)

Source: California Air Resources Board, 1989

mileage accumulation that occurs during a 50,000-mile certification test.)

Running-Loss Emissions

As shown earlier in Table 2, EPA has concluded that the average running-loss emissions from future gasoline vehicles will be no better than 0.16 g/mi. However, the average running-loss emissions measured on four specific 1981 and later model year vehicles already tested by EPA was much lower. At 95 degrees F with 9.0 Reid vapor pressure fuel, the running-loss emissions from the cleanest 1981 and later models were only 0.06 g/mi. Over a range of more representative ambient temperatures, available data (EPA, 1988) indicate that the average running-loss emissions from these vehicles should be at least 50 percent lower (i.e., 0.03 g/mi).

Running-loss emissions occur when vapor generation in the gasoline tank exceeds the capability of the evaporative emissions control system to purge the vapors from the charcoal canister. The high degree of vehicle-to-vehicle variability is in part due to the fact that differences in exhaust system and fuel tank design and location result in highly variable rates of heat transfer from the exhaust system and roadway to the fuel tank. Regulation of running-loss emissions will cause heat source/fuel tank proximity to become a design constraint for the first time. When this occurs, it is reasonable to expect all vehicles to achieve the same degree of running-loss control that currently is achieved by some vehicles in the fleet. Given the fact that 0.06 g/mi running-loss emission rates have already been demonstrated at 95 degrees F on vehicles that were not designed for minimum running-loss emissions, and given the fact that running-loss emissions are much lower over a range of more typical daily temperatures, 0.03 g/mi may be a much better estimate of running-loss emissions from late model vehicles under an I & M program designed for maximum effectiveness.

Refueling Emissions

The primary sources of hydrocarbon emissions associated with the vehicle refueling process are the vapors contained in vehicle fuel tanks which are displaced by fuel added during refueling operations at service stations. Additional emissions are associated with vehicle refueling operations as the result of "spitback" and spillage of liquid fuel from the fillpipe and dispensing nozzle and breathing losses from underground storage tanks at service stations. Finally, vapor is also displaced from the underground storage tank when it is refilled by a cargo truck. To account for all differences in emissions between various fuels, emissions from the filling of underground storage tanks were also categorized as refueling emissions in this analysis.

At some older gasoline stations, the emissions associated with the filling of underground storage tanks and tank breathing losses have been estimated by CARB to be 4.31 and 0.45 g/gal of gasoline delivered, respectively. When gasoline is pumped from the underground tank into vehicle fuel tanks, CARB's studies show that vapor emissions of 4.54 g/gal delivered occur at the nozzle-fillpipe interface. Field surveys of spillage emissions at service stations not equipped with vapor recovery systems indicate that the average emissions due to spillage are 0.318 g/gal dispensed (California Air Resources Board, 1983).

The combination of underground storage tank filling emissions, nozzle-fillpipe interface emissions, spillage, and breathing losses totals 9.62 g/gal dispensed. For vehicles which average 27.5 mi/gal (the corporate average-fuel efficiency [CAFE] standard), these emissions (with 9.0 psi gasoline) are 0.35 g/mi of vehicle operation. (EPA has estimated 0.20 g/mi refueling emissions without controls [EPA, 1989]; with the primary reason for the difference being the fact that EPA did not consider underground storage tank emissions.)

Throughout California's polluted metropolitan areas, refueling emissions have been controlled by Stage I and Stage II vapor recovery systems. Stage I vapor recovery controls the emissions associated with the filling of the underground storage tank by gasoline cargo trucks by the addition of a vapor return line connecting the underground storage tank with the gasoline cargo truck. Vapors which would normally be displaced from the underground tank are returned to the truck and subsequently reprocessed. The cargo truck tank is no longer vented during the delivery of gasoline because the underground tank vapors are ingested instead of ambient air. Based on CARB tests, Stage I vapor recovery is 95 percent efficient. Underground storage tank filling emissions are reduced from 4.31 to 0.22 g/gal dispensed.

With Stage II vapor recovery systems, gasoline vapors are collected at the vehicle fillpipe opening using a nozzle designed to capture the vapors and return them to the underground storage tank. Until recently, all Stage II vapor recovery nozzles were equipped with a flexible, tubular bellows positioned around the nozzle spout. The bellows is commonly referred to as a boot. The nozzle is also equipped with a vapor passage in the body of the nozzle which connects the inside of the boot to the vapor space in the underground storage tank. The latest in Stage II technology employs bootless nozzles. With these nozzles, refueling vapors are collected by the application of a slight vacuum to the vapor collection ports located near the tip of the nozzle. These ports draw the vapors into an annular space around the outside of the coaxial spout. Fillpipe emissions can also be controlled without Stage II nozzles through the installation of on-board refueling emissions controls on the vehicle.

The application of refueling emissions controls reduces fill-pipe emissions by 95 percent, from 4.54 to 0.23 g/gal dispensed (California Air Resources Board, 1983). In addition, the return of vapors to the underground storage tank eliminates the ingestion of air as gasoline is pumped from the tank. Further vapor generation is prevented, and breathing losses are reduced by 90 percent to 0.05 g/gal. Refueling emission controls also reduce spillage emissions by 61 percent, from 0.32 to 0.12 g/gal. With Stage I and Stage II or on-board systems in place, the refueling emissions from gasoline vehicles are as summarized in Table 4.

Table 4
Refueling Emissions From Gasoline Vehicles

	Grams/Gallon	Grams/Mile
Vapor Emissions	0.50	0.018
Spillage Emissions	0.12	0.004
Total Emissions	0.62	0.023

Source: Sierra Research Inc., 1988

In estimating refueling emissions for methanol, several corrections are required. For M100 vehicles, the difference in vapor-phase emissions (on a per-gallon-dispensed basis) must be adjusted to account for differences in the vapor pressure of methanol and gasoline. For M85 vehicles there are no significant differences in vapor pressure, but the relative contribution of gasoline and methanol vapor must be estimated. Spillage emissions for M85 and M100 can reasonably be assumed to equal those for gasoline on a g/gal dispensed basis. For both M85 and M100 vehicles, an adjustment to vapor and spillage emissions must be made to account for the difference in the fuel economy of gasoline- and methanol-fueled vehicles.

To compute the vapor displaced with M100 relative to gasoline, Sierra Research used the same approach used by Carnegie Mellon University during a recent of study for CARB:

$$[\text{Methanol Mass}] = [\text{Gasoline Mass}]$$
$$\times \ (\text{Mol.Wt. Methanol} \div \text{Mol.Wt. Gasoline})$$
$$\times \ (\text{RVP Methanol} \div \text{RVP Gasoline})$$

Sierra Research estimated the molecular weight of gasoline vapors at 72 (based on pentane) and the molecular weight of methanol is 32. Using 4 psi for the RVP of M100 and 9 psi for the RVP of gasoline, methanol emissions due to vapor displacement are estimated to be 20 percent of gasoline vapor emissions per gallon.

For M85, the vapor pressure of the fuel is expected to be about the same as for gasoline. However, the amount of methanol and gasoline vapor in the fuel vapors is clearly not in proportion to their concentration in the liquid because of the higher vapor pressure of the gasoline component of the blend. Using Raoult's Law, Carnegie Mellon calculated the vapors of M85 to consist of 19 percent methanol and 81 percent gasoline. Sierra Research estimated the gasoline vapor emissions per gallon of vapor displacement by taking 81 percent of the pure gasoline emissions rate. Methanol emissions were estimated by taking 19 percent of the pure gasoline rate and reducing the mass to account for the difference in the molecular weight of methanol and gasoline vapor. Fuel economy differences were computed based on flexible-fuel vehicle test results for M85 and an assumed 10 percent efficiency advantage for M100.[1] Table 5 summarizes the results of Sierra Research's refueling emissions estimates.

Although an entirely different methodology was used, Sierra Research's refueling emissions estimates for M85 and M100 turn out to be relatively close to EPA's. As in the case of other emission categories, Sierra Research's analysis indicates that EPA has

Table 5
Sierra Research's Refueling Emissions Estimates
for Gasoline and Methanol Vehicles

	Gasoline		M85		M100	
	HC	Meth.	HC	Meth.	HC	Meth.
Vapor (g/gal)	0.50	0.00	0.41	0.04	0.00	0.10
Spillage (g/gal)	0.12	0.00	0.02	0.10	0.00	0.12
Total (g/gal)	0.62	0.00	0.43	0.14	0.00	0.22
Total OMHCE/gal	0.62		0.49		0.095	
Miles/Gallon	27.5		16.2		15.3	
Total OMHCE/mile	0.023		0.031		0.006	

Source: Thomas Austin, Sierra Research Inc., 1989

overestimated gasoline vehicle refueling emissions significantly, even though Sierra included underground storage tank emissions in its estimate and EPA did not.

Summary

The benefits that EPA projects for methanol are largely due to pessimistic projections of future gasoline vehicle emission levels. In addition, the limited durability experience with methanol vehicles in actual use, and the scarcity of evaporative emissions data, make EPA's methanol vehicle forecasts highly uncertain. It is premature to conclude that any significant vehicle emissions reduction will be associated with a government mandate for the use of methanol-fueled vehicles.

Note

1. Thirty percent efficiency benefits for M100 vehicles have been projected by EPA. However, Sierra Research's analysis indicates that EPA included improvements from the use of "lean-burn"

engine calibrations that are incompatible with the low NO_x standards required in California and under the proposed Clean Air amendments.

References

California Air Resources Board (1983). A Report to the Legislature on Gasoline Vapor Recovery Systems for Vehicle Fueling at Service Stations.

California Air Resources Board, Mobile Source Division (1989). Definition of a Low-Emission Motor Vehicle in Compliance with the Mandates of Health and Safety Code Section 39037.05.

CARB (1989). Staff Report #89–17.

Gray, C.L., Jr., and J.A. Alson (1989). The Case for Methanol. *Scientific American*. November, 1989.

Russell, A., et al. (1989). Quantitative Estimate of the Air Quality Impacts of Methanol Fuel Use. Department of Mechanical Engineering, Carnegie Mellon University.

Sierra Research (1988). An Analysis of Stage II and On-board Refueling Emissions Control.

Stamper, K.R. (1980). Evaporative Emissions from Vehicles Operating on Methanol/Gasoline Blends. SAE Paper 801360.

U.S. Environmental Protection Agency, Emissions Control Technology Division (1988). Running Losses. Presented at MOBILE 4 workshop, Ann Arbor, Mich.

U.S. Environmental Protection Agency, Office of Mobile Sources (1989). Analysis of the Economic and Environmental Effects of Methanol as an Automotive Fuel.

Commentary

JERRY HORN

Philip Lorang of the U.S. Environmental Protection Agency (EPA) and Tom Austin of Sierra Research, Inc., have presented quite different comparisons of emissions from methanol and conventional fuels. The EPA analysis argues that in-use emissions from conventional-fueled vehicles will never meet government standards, regardless of what emissions regulations are enacted. The Sierra Research analysis assumes that future in-use conventional vehicles will meet future standards. Lorang presents data showing that current in-use vehicles do not meet federal standards, data that he believes support his contention that future emissions will also be high. If this assumption is made for gasoline-fueled vehicles, however, a similar assumption should be applied to M100 vehicles. The vehicle fuel systems (other than materials) and emissions-control systems will be similar for both fuel types and therefore subject to the same types of in-use deterioration that causes emissions to increase. EPA, however, did not appear to apply the same degradation factors to M100 vehicles as it applied to conventional-fueled vehicles. This appears to bias the EPA analysis in favor of M100.

The Sierra Research analysis argues that both future conventional-fueled and methanol-fueled vehicles could meet in-use standards. This is based on the assumption, which I believe to be reasonable, that recently enacted regulations in California (two on-board diagnostic regulations, more stringent standards, and 100,000-mile compliance), combined with an effective inspection and maintenance program, would significantly reduce in-use vehicle emissions.

The Sierra Research argument that the in-use performance of vehicles of the two fuel types would be similar is also, in my opinion, reasonable. This is essentially a "level playing field" approach to a comparison of emissions.

In evaluating the two approaches taken, it should be noted that EPA considered only M100 in its comparison. M85 (85 percent methanol, 15 percent gasoline) is currently considered the most viable alternative fuel. At this time M100 is not feasible. It is difficult, if not impossible, to start M100 vehicles at temperatures below 50 to 60 degrees F. There are also concerns about flame luminosity and fuel tank vapor space flammability. In time, with research and development, these problems can probably be overcome. However, this same level of research and development could also be applied to improving the emissions of conventional vehicles. In addition, the emissions results from the prototype M100 vehicles tested are not necessarily indicative of the emissions behavior one can expect from production vehicles.

EPA presented data showing low formaldehyde emissions from M100 vehicles. However, the vehicles had low mileage, with the highest (the Toyota Carina) having about 11,000 miles. Other data on M100 vehicles indicate that their formaldehyde emissions may be significantly higher. For example, EPA data on an M100 Sentra— one of EPA's "Good M100 Prototypes"—show formaldehyde levels of approximately 30 mg/mi at relatively low mileage. Toyota data on an M85 version of the Carina (differing from the EPA vehicle only in the computer programming used to control the engine, fuel, and emissions-control systems) show that formaldehyde emissions of 5 mg/mi at close to zero miles increased to more than 60 mg/mi at 30,000 miles. This is significantly greater than 15 mg/mi, which EPA assumes M100 vehicles will emit at 50,000 miles.

EPA concluded that M100 vehicles will easily meet future potential nitrogen oxide (NO_x) standards of 0.4 g/mi and even 0.2 g/mi, yet it offered no substantiating data. In fact, M100 vehicles using current technology produce levels of nitrogen oxide emissions that are not very encouraging. NO_x emissions data from the EPA's Good M100 Prototypes show the following:

1981 Volkswagon Rabbit — 0.68 g/mi
1986 Toyota Carina — 0.55 g/mi to 0.76 g/mi at under 2,000 miles, about 0.9 g/mi at 11,000 miles

California Air Resources Board data on an M85 version of the Carina show NO_x emissions up to and exceeding 1 g/mi. The Carina tested by Toyota (mentioned earlier) had NO_x emissions that ranged from about 0.4 g/mi at zero miles to 0.7 g/mi at 30,000 miles.

These data make it difficult to share EPA's optimism about meeting extremely low NO_x levels with M100, using "lean-burn" technology. Toyota echoed a similar concern in a recently published Society of Automotive Engineers paper, in which Toyota's second-generation methanol lean-burn system was described. This system produced NO_x emissions of 0.31 g/mi at low mileage. However, the authors concluded: "It is thought that it will be very difficult for a production methanol vehicle to achieve 0.4 g/mi NO_x, especially considering deterioration of part and production scattering."[1]

This information does not make M100 vehicles, given their current state of development, look as promising as EPA would like the public to believe. This is not to say that the technology will not improve in the future—most likely it will. However, it must be remembered that the time and effort needed to make M100 vehicles a practical alternative could also be spent improving conventional vehicles.

Note

1. Society of Automotive Engineers Paper 892060.

Commentary

HARRY SCHWOCHERT

Philip Lorang and Tom Austin have pointed out that the same engine emissions control technologies can be applied to both gasoline-fueled and methanol-fueled vehicles. Good engineering judgment and experience indicate that similar amounts of emissions (total organic emissions, carbon and nitrogen oxides) should be expected from engines using either fuel. It is the composition of the total organic emissions (since the amounts will be about the same) that will determine the effect of an alternative fuel on air quality. Both speakers described the progress in controlling emissions that has occurred during the past 25 years of regulating gasoline. Fundamental emissions-control components have been developed and perfected. Engine controls have improved, as have the emissions levels of in-use vehicles. All of these factors need to be considered when attempting to predict the future level of emissions of both methanol-fueled and gasoline-fueled engines.

Although a limited amount of emissions data are available from prototype methanol-fueled vehicles, the future performance of any fuel must be judged by projecting the emissions-control capability of a mature technology. This can be illustrated by reviewing some recent steps in limiting gasoline emissions. The emissions standard for hydrocarbons has been constant at 0.41 g/mi since the 1981 model year. The basic emissions-control technology has also been essentially the same during this interval; that is, the use of three-way catalyst systems with closed-loop fuel metering. However, due to improvements in fuel metering, engine management, and catalyst systems (the actual maturing of the total vehicle emissions-control system at constant emissions standards), there has been significant improvement in the level of emissions from in-use vehicles and also in the maintenance of these vehicles.

Hydrocarbon emissions, between the 1981 and 1987 models (with significantly higher mileage data for the latter year), have been lowered by a factor of approximately three for the average General Motors car, when projected to 50,000 miles. In fact, GM data indicate the average recent model GM car has projected hydrocarbon, carbon monoxide, and nitrogen oxides emissions at 50,000 miles that are well below the standards. Data on these emissions improvements were derived from in-use vehicles. There was little or no change, however, in the level of emissions demonstrated on certification (prototype) vehicles. Therefore, it is necessary to look beyond the specific emissions data of selected, highly experimental vehicles and use good engineering judgment to estimate the emissions level of a mature technology.

The recently established automobile and oil industry joint project will measure the speciated organic emissions from existing gasoline-fueled vehicles and from numerous prototypes of future gasoline-fueled vehicles, and compare them with the emissions levels of a number of prototype alternative-fuel vehicles. Results of these studies will be published in late 1990. Detailed speciated organic emissions data from alternative-fuel vehicles are scarce. However, General Motors has performed speciated emissions tests on a variable-fuel vehicle for different fuel mixtures ranging from pure gasoline to pure methanol. These results were reported to an Air and Waste Management Association meeting in California in June 1989.

The data supported certain trends that had already been perceived. Switching from gasoline to methanol fuel decreases combustion hydrocarbons (non-oxygenated) significantly, while methanol and formaldehyde emissions increase substantially. While it should be anticipated that unburned fuel and aldehyde emissions from the engine/emissions-control system will ultimately be reduced to low levels, some differences are likely to persist. Non-oxygenated

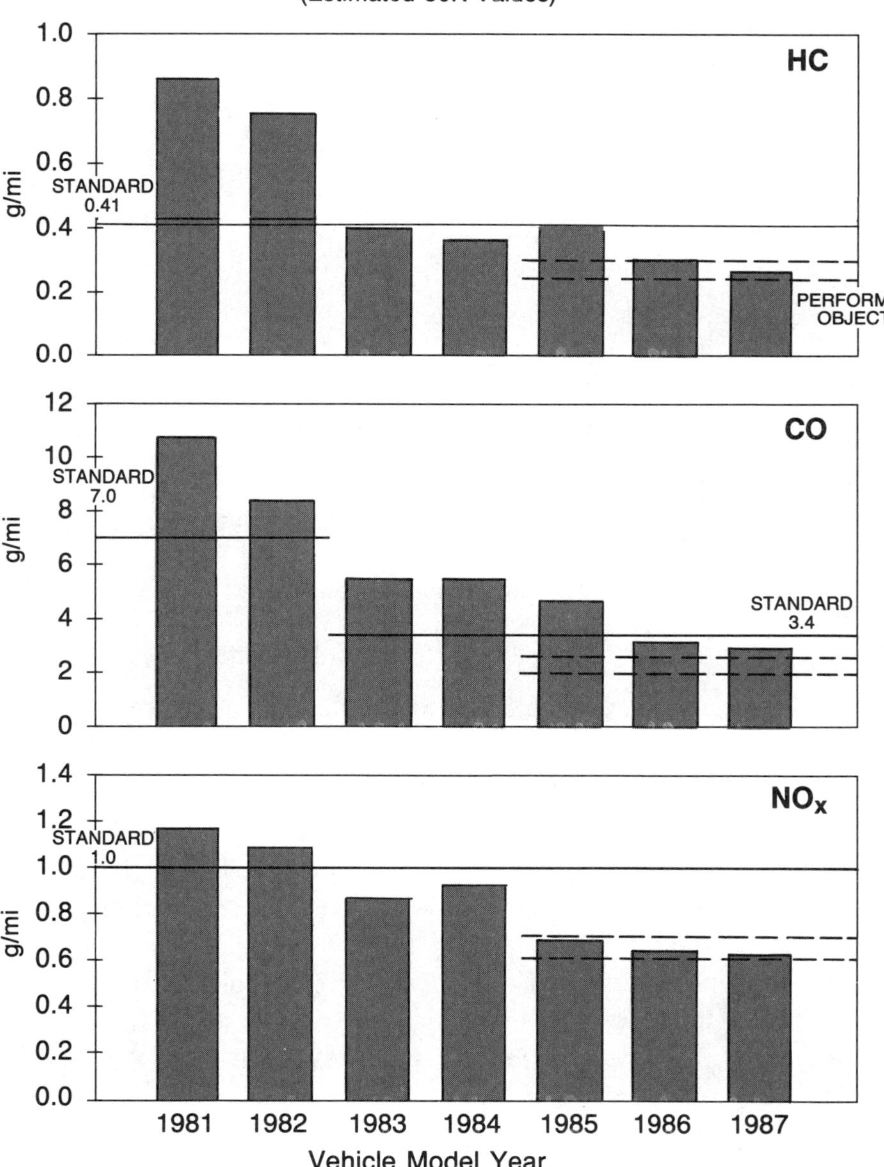

Source: General Motors Corporation, 1989

Figure 2
Variable-Fuel Vehicles Exhaust and Evaporative Emissions

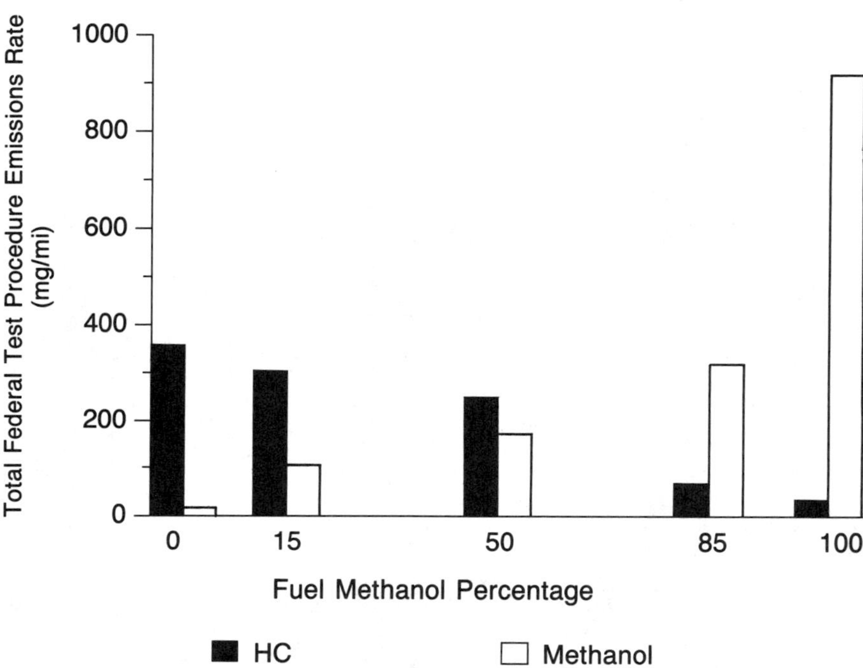

Source: General Motors Corporation, 1989

hydrocarbons from a gasoline-fueled vehicle will be higher, while methanol and aldehyde emissions will be relatively higher from a methanol-fueled vehicle.

An important use of this speciated emissions data is to predict the relative effects of emissions from gasoline-fueled and methanol-fueled vehicles on air quality. An air quality modeling study (also reported at the Air and Waste Management Association meeting) based on this data indicated that the amount of ozone formed from the emissions of gasoline-fueled or methanol-fueled vehicles is very sensitive to the non-methane organic carbon to

Figure 3
Variable-Fuel Vehicles Exhaust Emissions

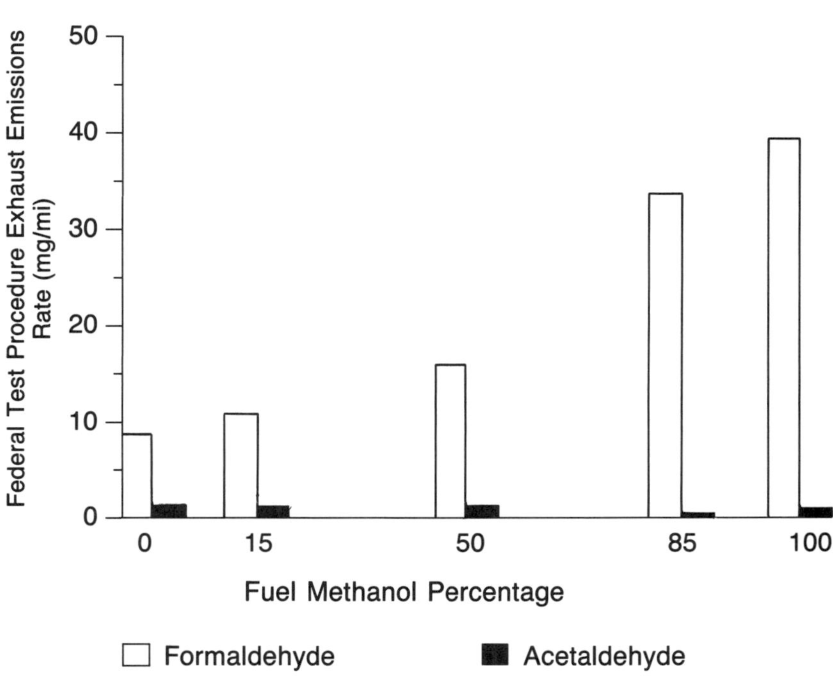

Source: General Motors Corporation, 1989

nitrogen oxides ($NMOC/NO_x$) ratio in urban atmospheres. This means that replacing gasoline-fueled vehicles with methanol-fueled vehicles may be appropriate as an ozone control strategy only in certain urban areas. The study showed that the greatest potential for ozone control occurred at very low $NMOC/NO_x$ ratios.

The amount of ozone formed from the emissions of methanol-fueled vehicles was also very sensitive to the formaldehyde fraction in the emissions. Therefore very low emissions levels (approaching current gasoline levels) are necessary to achieve a major reduction in ozone formation through a switch to methanol.

In conclusion, overall emissions from gasoline-fueled and methanol-fueled vehicles whose technology is at a similar level of maturity will be about equal. Therefore the composition of the exhaust, with respect to the organic emissions, must be used to estimate the relative air quality merits of these two vehicle and fuel systems. From the standpoint of ozone formation, the potential benefit of methanol-fueled vehicles is highly dependent on the level of formaldehyde in the exhaust as well as on the characteristics of the ambient environment in which the vehicles would operate.

Part II

Impacts on Urban Ozone and Other Air Toxics

CHAPTER 4

Methanol Fuel Use for Photochemical Smog Control

ARMISTEAD G. RUSSELL

Abstract

Using methanol as a fuel in motor vehicles is a potential strategy for improving air quality in the most polluted cities in the United States. Local and state agencies, as well as the Bush administration's clean air program, call for increasing the use of alternative fuels. However, the possible benefits of switching to methanol fuel have aroused considerable debate. This study used an advanced mathematical air quality model to simulate the probable air quality effects that methanol fuel use in California would produce if the administration's clean air proposals are implemented. It estimates the effect of introducing 300,000 M85 methanol-fueled vehicles per year, starting in 1997, on the concentrations of various pollutants, especially ozone, in the year 2010.

The results indicate that the low atmospheric chemical reactivity of methanol slows ozone formation and thus leads to lower ozone concentrations. Methanol is significantly less reactive than most of the organic compounds emitted by gasoline-fueled vehicles. Estimates of future peak ozone levels decrease by as much as 9 percent, and estimates of exposure to levels above the federal standard drop by as much as 19 percent if M85 methanol fuel is used in approximately half of the light-duty automobile fleet. Formaldehyde levels and exposure did not increase greatly in simulations of methanol use, and they decreased in some cases. Other photochemical pollutants, such as nitric acid, particulate matter, and NO_x (nitrogen oxides) decreased. Results of this study indicate that the long-range dispersion of methanol will not contribute greatly to ozone formation in downwind regions.

Introduction

More than 100 cities experience ozone concentrations above the federal limit of 0.12 ppm. Roughly two-thirds of the U.S. population reside in these cities. Reducing ozone concentration to comply with the standard has proved to be very difficult. Many urban areas believe that they have exhausted the conventional options that are technically, economically, and politically feasible. The search for a solution has led to consideration of such alternative fuels as methanol and natural gas for use in motor vehicles.

For example, at the state level, the recent Air Quality Management Plan for Los Angeles calls for increasing the use of alternative fuels (South Coast Air Quality Management District and Southern California Association of Governments, 1988), and California's state agencies are studying their use as well. Nationally, the Alternative Motor Fuels Act of 1988 and, more recently, the Bush administration, as part of its clean air program, called for the increased use of alternative fuels. In particular, the President's program calls for introducing one million alternative-fueled vehicles per year, starting in 1997. The alternative-fueled vehicles are to be made available in the most polluted cities, such as Los Angeles.

Of the fuels being considered, methanol appears to be one of the most feasible to replace gasoline and to improve air quality. Preliminary studies have shown that the use of methanol, because of its low atmospheric chemical reactivity, could reduce the formation of photochemical smog and ozone. However, methanol-fueled vehicles emit more formaldehyde than gasoline-fueled vehicles do, a fact that raises concern about possible increases in the ambient levels of formaldehyde. Recent studies have also questioned the effectiveness of switching to methanol during prolonged periods of heavy smog and in areas with high concentrations of reactive organic gases (ROG).

In view of these concerns, it is necessary to gain a better understanding of the effects of using methanol fuels before increasing their use. The present study uses an advanced, three-dimensional, Eulerian, photochemical air quality model to investigate how the use of methanol fuel would affect air quality. This paper attempts to determine whether the administration's program would help reduce ozone formation in the Los Angeles area, which experiences ozone concentrations up to three times the National Ambient Air Quality Standard. The study focuses on the following: the effect of methanol on chemical formation of pollutants in the atmosphere; whether methanol build-up during prolonged episodes would negate the benefits of its lower photochemical reactivity; whether emissions from methanol-fueled vehicles would severely increase ambient formaldehyde, thereby negating ozone reduction; and whether the long-range dispersion of methanol is apt to be a problem. Finally, this paper discusses the types of models and the inputs used in other studies and how the results of those studies compare with the results obtained with this model.

Previous Studies

Interest in using methanol fuels arose in the late 1970s and early 1980s in response to the oil crisis, when methanol was considered as a primary fuel or as a gasoline additive. Since then a number of studies have focused on methanol's effect on the environment. (For example, Pefley, Pullman, and Whitten, 1984; O'Toole et al., 1983; Bechtold and Pullman, 1980; Balentine et al., 1985; Carter et al., 1986; Whitten and Hogo, 1983; Whitten, Yonkow, and Myers, 1986; Nichols and Norbeck, 1985; Russell et al., 1989; Chang et al., 1989.) The results of these studies were based on chemical kinetic models, air quality models, environmental smog chambers, and smog chamber simulations. The earlier studies focused primarily on ozone formation, although Russell et al. (1989) considered other photochemical pollutants as well. These studies disagree about methanol's effect on the environment. In addition, the technology and assumptions involved in some of these studies have been found to have significant limitations, which are discussed below.

Previous Mathematical Modeling Studies

Photochemical smog formation results from a complex system of competing, nonlinear processes that occur simultaneously in the atmosphere. Since it is impossible to analyze air quality effects, numerical modeling analysis (i.e., photochemical air quality modeling) is commonly used. Previous studies have used various types of models to study the role that methanol could play in controlling ozone. Balentine et al. (1985) used a zero-dimensional, chemical kinetics modeling approach. Others used box, trajectory, and/or grid-based Eulerian models: O'Toole et al. (1983), Nichols and Norbeck (1985), Chang et al. (1989), Whitten and Hogo (1983), Whitten, Yonkow, and Myers (1986), and Russell et al. (1989).

The first modeling study by Systems Application Incorporated (Whitten and Hogo, 1983) investigated the potential air quality benefits of methanol in the Los Angeles region. An Empirical Kinetic Modeling Approach (EKMA) (level II) one-day trajectory model was used to study the effect on ozone in the South Coast Air Basin. This study considered various levels of formaldehyde in the exhaust of methanol-fueled sources, but changes in emissions inputs into the model were not treated in detail. It also directly tested sensitivity to initial conditions, methanol-fueled vehicle exhaust formaldehyde content, and methyl nitrite emissions. The study found that ozone was reduced by as much as 31 percent, from 0.273 ppm to 0.188 ppm, when using the forecast 1987 inventory as a base case. It also found a strong sensitivity to exhaust formaldehyde content. Predicted peak ozone was 0.188 ppm for no formaldehyde in the exhaust, 0.213 ppm for 10 percent formaldehyde in the exhaust, and 0.237 for 20 percent formaldehyde in the exhaust. However, tests showed a strong sensitivity to initial and boundary conditions (both of which would change with a large-scale conversion to methanol-fueled vehicles). When initial and upper-level organic gas concentrations were halved, ozone was reduced to 0.181 ppm.

A Jet Propulsion Laboratory study modeled a one-day trajectory traversing South Coast basin and considered the effect that methanol fuel use would have on air quality in the year 2000. A 1974 inventory for the basin was scaled forward, accounting for expected automotive emissions controls and economic growth. However, stationary source emissions were not changed, nor was the spatial distribution of emissions. The model did account for changes in fuel vapor pressure and the composition of evaporative emissions. The model employed was the Carnegie/California Institute of Technology (CIT) trajectory model. A single air parcel trajectory, less than 24 hours in extent, was used.

Some cases were evaluated as part of the Jet Propulsion Laboratory study, including the complete substitution of methanol vehicles for gasoline (not diesel) fueled vehicles in the year 2000, assuming: (1) the same reactive organic gases (ROG) and NO_x emissions rates as those of gasoline, (2) ROG and NO_x emissions rates at 50 percent those of gasoline, and (3) a 50 percent reduction of both ROG and NO_x. The composition of the methanol-fueled vehicle exhaust was 72.7 percent methanol, 21.2 percent formaldehyde, and 6.1 percent alkane by mass. With complete substitution and with methanol-fueled vehicles emitting at the same rate as conventional-fueled vehicles (case 1), it was found that ozone was reduced from 0.333 ppm to 0.285 ppm (14.4 percent). Greater reductions were found for cases 2 and 3, with 17.4 percent and 19.8 percent reductions, respectively. If no automotive emissions were used, the reduction was 25 percent. Thus, even with the rather large fraction of formaldehyde in the exhaust (21 percent versus 1 to 10 percent, as used here), methanol was found to be effective at reducing ozone—lowering the predicted ozone by about 60 percent of the amount calculated if all gasoline-fueled vehicle emissions were removed.

Although the Jet Propulsion Laboratory study was very detailed in its treatment of emissions inputs, it has been criticized. First, since the air parcel studied was tracked for less than 24 hours, the results were very sensitive to initial conditions. Second, trajectory models have limitations in their formulation. Also, the formaldehyde fraction of the exhaust was higher than current data would suggest, and much higher than the possible regulatory levels. On a reactivity basis, the formaldehyde fraction of the models' methanol-fueled vehicle exhaust overwhelms the methanol fraction. Finally, more recent and detailed data are now available by which to estimate future emissions from mobile and stationary sources.

Other modeling studies have concentrated on the effects of methanol fuel substitution outside the South Coast region. The chemical kinetic modeling by Balentine et al. (1985) was conducted to test the applicability of various mechanisms for inclusion in models. The results showed that methanol use decreased ozone. Ford Motor Company, in two studies, modeled 20 cities with an EKMA-type trajectory model (Nichols and Norbeck, 1985; Chang et al., 1989). These studies assumed immediate penetration of methanol-fueled vehicles into the fleet. Nichols and Norbeck found that ozone was reduced by 1 to 36 percent when the methanol-fueled vehicle exhaust was zero percent formaldehyde, and by zero to 13 percent when the exhaust was 10 percent formaldehyde. The results of Chang et al. were similiar. At low ROG/NO_x ratios, methanol substitution was effective; as the ratio increased, however, the effect of switching to methanol was diminished. In addition, the study found that at the higher ROG/NO_x levels, where methanol was less effective at decreasing ozone, reducing NO_x should be useful. Systems Application Incorporated (Whitten, Yonkow, and Myers, 1986) used three different models to study the effect in Philadelphia, but in each case results were sensitive to the initial and/or boundary conditions. That study indicated little or no benefit from methanol-fueled vehicle substitution.

Many techniques developed for the Russell et al. (1989) study have been used in this study. Three different types of models were used to study the effect of methanol use on the air quality in Los Angeles. Box and trajectory models were used to make preliminary estimates of the sensitivity of concentrations of pollutants to the introduction of methanol-fueled vehicles. Results from these different kinds of models were also used to extend the findings of the grid-based Eulerian airshed model. The grid-based model has the fewest limitations in its formulation.

The Russell et al. (1989) study found that ozone decreased by as much as 16 percent when the in-use vehicle fleet was converted to using pure methanol. This was nearly the same reduction as was found when in-use motor vehicle emissions were totally removed. Switching to a gasoline-methanol blend (M85) gave about half of the improvement. The study also found that formaldehyde concentrations and exposure did not increase severely and, in some cases, decreased when a switch to methanol was simulated. Concentrations of other photochemical pollutants, such as nitric acid and peroxyacetyl nitrate, a phytotoxin and eye irritant, also decreased. Ambient methanol and formaldehyde levels were not expected to be above levels of concern (Marnett, 1988; National Research Council, 1981).

The estimated emissions rates of formaldehyde and other pollutants in the Russell et al. (1989) study were criticized for being lower than would be experienced in actual use (Austin, 1988). Recent tests, however, found that formaldehyde levels similar to or lower than those estimated are being achieved (Hellman et al., 1989; CARB, 1988). The effect of increased formaldehyde emissions was tested. When formaldehyde emissions were increased from 15 mg/mi to 55 mg/mi, the ozone reductions decreased by one-half. Control of formaldehyde is critical if the full benefits of methanol fuel use are to be achieved. It was also found that the use of methanol in diesel trucks and stationary sources to reduce NO_x emissions was beneficial for reducing ozone and particulates. Many cities have a relatively high ROG/NO_x ratio (Baugues, 1986) and would respond favorably to NO_x emissions reductions (Russell and Cass, 1984; Milford, Russell, and McRae, 1989; Chameides et al., 1989).

An important finding of the Russell et al. (1989) study resulted from the use of three different levels of air quality models. By using only a trajectory model, the study would have greatly underestimated the potential air quality improvement from conversion to methanol. This point should be emphasized in interpreting recent studies that have relied primarily on trajectory modeling.

Several important points emerge from the results of these studies. The use of methanol fuels appears to be effective in reducing ozone under certain circumstances. The exhaust formaldehyde fraction is an important contributor to ozone formation. Modeling, as done in the Jet Propulsion Laboratory, Ford, and Systems Application Incorporated studies, can be very sensitive to initial conditions, and care must be taken in specifying trajectory paths and modeling regions. Finally, trajectory models can show a different effect from that of grid models. In the Russell et al. (1989) study the trajectory modeling portion showed less benefit from the use of methanol.

Experimental Studies

The University of California's Riverside Statewide Air Pollution Research Center performed a series of well-documented, methanol-related smog chamber experiments in a 6,400-liter indoor Teflon chamber and a 50,000-liter outdoor Teflon chamber (Carter et al., 1986). Pollutants, similar to those in a very dirty urban atmosphere, were introduced into a chamber, and the pollutant concentrations were followed for as many as three days. This was done for pollutant mixes corresponding to current emissions, and also for mixes corresponding to methanol-fueled engine emissions, replacing one-third of the base mixture. On the first day, peak ozone levels were significantly lower for the methanol emissions mix than for the conventional emissions mix. However, this difference decreased over time, and ozone levels were similar by the third day. This result raised a question about the effects of methanol during prolonged episodes of smog. It appeared possible that methanol could build up and that a large fraction would react within three days, thereby negating methanol's earlier benefits.

In a previous study the University of North Carolina (Jeffries et al., 1985) conducted 29 smog chamber experiments with a combination of methanol and various levels of formaldehyde. One-third of the "base" organic mixture was replaced with a methanol or a methanol-formaldehyde mixture, a CH_4O:formaldehyde ratio of 90:10 or 80:20. Ozone reductions varied widely, from zero to 80 percent, depending on the formaldehyde content and organic loading.

An earlier smog chamber experiment (Pefley, Pullman, and Whitten, 1984) also looked at methanol substitution. However, the conditions were atypical of urban conditions, and the experiment used isobutene (not formaldehyde) as part of the methanol-fueled vehicle exhaust composite. Again, the one-day experiments suggested that ozone would be reduced from methanol substitution.

The differences between smog chambers and the earth's atmosphere make it difficult to use smog chamber results for predicting basin-wide air quality changes. First, a smog chamber does not replicate atmospheric diffusion and transport of chemical species. Second, all pollutants were present from the outset of the smog chamber experiments, while in the atmosphere fresh pollutants are being emitted throughout the day. Consequently, the smog chambers had very low NO_x concentrations on the simulations' second and third days, making ozone levels relatively insensitive to changes in ROG. Third, even a very large smog chamber has a surface-area-to-volume ratio that is many orders of magnitude greater than that of the atmosphere. Therefore, surface reactions have a much greater effect on pollution formation in the smog chamber than they have in the atmosphere. While smog chamber tests are extremely valuable for investigating certain aspects of atmospheric chemistry, they cannot be used directly to predict the effects of various emissions-control strategies, especially for prolonged episodes of smog.

Study Description

Since the completion of the earlier Russell et al. study of 1989, the Bush administration announced plans to introduce one million cleaner fuel vehicles per year, starting in 1997 (with fewer vehicles to be introduced in 1995 and 1996). This plan raises the question of the probable effect on air quality in the more polluted U.S. cities. The present study attempts to answer that question. The CIT airshed model was used to predict how the introduction of 300,000 light-duty methanol-fueled vehicles per year (30 percent of the one million to be sold) would affect air quality in the South Coast basin in 2010. The year 2010 was chosen to allow for the widespread use of methanol in the vehicle fleet and because detailed source emissions projections were available. The basin, which surrounds Los Angeles, was the area chosen because it experiences the most severe photochemical smog in the United States and may be the first area to be targeted for the introduction of methanol-fueled vehicles. The South Coast basin is ideal for studying prolonged episodes of pollution, since sufficient input data are available to construct detailed models.

The CIT model mathematically describes the formation, dispersion, and effect of pollutants in the atmosphere (McRae, Goodin, and Seinfeld, 1982; McRae and Seinfeld, 1983). Mathematical models are the most scientifically reliable method of evaluating the effect of changes in emissions on air quality. The CIT airshed model has already demonstrated its capacity to follow accurately the dynamics of ozone and other photochemical pollutants (Russell, McCue, and Cass, 1988; Wagner and Ranzierri, 1984). It is the only model that has been thoroughly evaluated as a predictor of concentrations of trace nitrogenous compounds.

The modeling domain used was a 250-km-by-150-km region surrounding Los Angeles, resolved horizontally into 5-km-by-5-km computational cells and vertically into five layers ranging from 39 m to 430 m in height. Emissions and meteorological variables were similarly resolved. Meteorological conditions were set to those from August 30 to September 1, 1982, because those conditions were conducive to the formation of ozone and because supplemental, detailed data were available for model evaluation (Russell and Cass, 1984). The maximum ozone concentration recorded on September 1, 1982, was 0.35 ppm, approaching the highest levels recorded in recent years.

In addition to those cases analyzed in the earlier, 1989 Russell study, three more simulations of the model were conducted to test the effectiveness of the administration's program. One simulation was the base case, in which all the vehicles were conventionally fueled. The second case looked at the effect of introducing 300,000 M85-fueled vehicles per year, as the administration proposes. The third case was a blank simulation, used to compare the results of the first two.

In these simulations emissions forecasts for 2010 were used as inputs for the base (non-methanol) case. A major difference between the cases studied here and the 1989 study is that in the present study the stationary sources were expected to meet the South Coast Air Basin's Air Quality Management Plan Tier 1 emissions levels. Also, the inventory used here accounted for standards expected to be in place for conventionally fueled vehicles. Specifically, light-duty automobile emissions standards were taken to be: 0.25 g/mi ROG, 0.20 g/mi NO_x, and 3.4 g/mi carbon monoxide.

In contrast to the Russell et al. (1989) study, the present study assumed that the vehicles did not meet these levels in-use. Instead, light-duty automobile emissions were assumed to remain higher than the standard using the same ratio by which new vehicles currently are estimated to exceed existing standards, as prescribed in the

EMFAC7D documents (this is about a factor of two higher and does not include running losses that are not now included in the inventories). In the base case the ROG emissions were speciated using the profiles developed by the California Air Resources Board. Emissions from M85-fueled passenger cars were set equal to their gasoline-fueled counterparts on a carbon basis. Organic gas emissions were split two-thirds methanol and one-third non-methanol organic gases with a similar composition to that of gasoline-fueled vehicles (Snow et al., 1989). Of the non-methanol component, 34 percent was methane (Snow et al., 1989). In addition, formaldehyde emissions were set to 30 mg/mi (twice the California standard). Recent tests indicate that these levels, and possibly much lower ones, are achievable.

Results

Calculations show that the widespread use of M85-powered vehicles would decrease ozone levels throughout the South Coast basin. Two metrics are used to describe the degree of improvement in air quality (Table 1). First, the maximum predicted ozone concentration within the region is used to show how effective controls are in reducing ozone levels toward the national standard. However, the peak concentrations for the meteorological period simulated occurred in the less populated eastern portion of the basin. In an effort to show the effect on potential human exposure, population projections (for the year 2010) were combined with spatial and temporal variations in ozone concentrations. This population "exposure" is defined as:

$$E_{C_0} = \int_A \int_T P(\bar{x}) \, C(\bar{x}, t) \, H(C - C_0) \, dt \, d\bar{x}$$

where $P(\bar{x})$ is the population as a function of location, $\bar{x}$; $C(\bar{x}, t)$ is the predicted ozone concentration as a function of time and location; and $H(C - C_0)$ is the Heaviside function, which is one if the predicted concentration is above a cutoff concentration, C_0, and zero otherwise. In essence, this is the population-weighted ozone exposure above a prescribed ozone concentration. The threshold concentration (one-hr average), C_0, used was 0.12 ppm (the national standard).

Table 1
Predicted Peak Ozone and Ozone Exposure for September 1, 2010

	Peak [O_3]	O_3 Exposure [O_3] > 0.12
Base Case	0.21	3894
M85	0.19 (9%)	3154 (19%)
Blank	0.18 (13%)	2932 (25%)

Note: Peak ozone is given in ppm, and exposure to ozone and formaldehyde is given in person-ppm-hours, using the expected population distribution in 2010. Simulation descriptions and definition of exposure are given in the text. For ozone exposure and peak, the values in parentheses are the reductions relative to removing mobile source emissions of ROG from the vehicles that were modeled as being introduced as methanol-fueled vehicles. Percent reduction from the base case is given in parentheses.

Source: Russell, 1990

As seen in Table 1, converting to methanol-powered vehicles reduced the peak ozone concentration predicted at any station by 9 percent. The blank case resulted in a higher reduction. A more striking difference is in the exposure. The M85 calculation predicts a 19 percent decrease, compared with the blank case, in which a reduction of 25 percent is noted. These results can be used together to estimate that the per-vehicle added reactivity of the M85 vehicle, in terms of ozone formation, is about 25 percent of the conventional vehicle (as modeled). This corresponds to the fraction of gasoline-derived organic gases and the formaldehyde fraction of the exhaust. (The reactivity of the exhaust is extremely sensitive to the ROG and formaldehyde fraction of the exhaust. Recent reports suggest that the ROG fraction might be higher [Gabele, 1989; Williams et al., 1989].)

These results further suggest that the unburned methanol is not a major contributor to ozone formation in the central, populated regions of the basin and that control of formaldehyde emissions is important in reducing ozone. Exposure to ambient formaldehyde increased 9 percent in this simulation, and the peak increased 7 percent. Concentrations of other pollutants such as peroxyacetyl nitrate and nitric acid also decreased. This further suggests that progression to a pure methanol fuel (M100), with reduced formaldehyde emissions, has the potential of providing further air quality improvements if technological difficulties (cold start, luminosity, and higher formaldehyde emissions) can be overcome.

There has been some debate as to whether methanol would exacerbate ozone concentrations in downwind regions. Calculations showed this is not the case in the South Coast basin (the modeling region includes areas that are more than two days downwind of downtown Los Angeles). Predicted methanol concentrations one day downwind of the urban area were less than 0.1 ppm. At this level methanol would add little to the reactivity of the ambient organic gases. Further dispersion would significantly decrease methanol levels due to vertical and horizontal dispersion and chemical decay. The potential additional reactivity of the dispersed methanol can be compared with carbon monoxide emissions. CO emissions are more than 10 times those of methanol, and CO is only one-third as reactive. Thus, dispersion of CO would be of greater concern for ozone formation. Studies have also shown that the formation of ozone in downwind regions is primarily sensitive to NO_x levels, not to ROG, further minimizing the role that methanol could play.

Conclusions

This modeling study, and the one that preceded it, indicates that conversion to an alternative fuel, such as methanol, can reduce ozone formation, even over prolonged periods. The use of methanol-based fuels would improve urban air quality by decreasing the reactivity of the organic gases emitted from motor vehicles. Controlling formaldehyde and gasoline-derived emissions is critical for achieving the full benefits of conversion to methanol. The administration's program, it is predicted, would reduce peak ozone levels by as much as 10 percent, and ozone exposure above the federal standard would be reduced by 19 percent. A larger reduction would result if methanol-fueled vehicles were more widely used and if heavy-duty vehicles, especially diesel vehicles, were also converted to methanol.

Before these studies were conducted, there was concern that increased direct emissions of formaldehyde from methanol-fueled vehicles would drastically increase ambient concentrations of formaldehyde. However, it was found that reduced atmospheric formation of formaldehyde from the oxidation of ROG offsets the increased direct emissions, leading to minor increases or, as the earlier study revealed, possible reductions in ambient formaldehyde exposure. Predicted peak levels were less than current concentrations and significantly less than previous peaks. On the basis of these simulations, the prolonged buildup of methanol should not pose a problem in terms of ambient exposure. Concentrations of other photochemical pollutants, such as nitric acid and peroxyacetyl nitrate, also decreased.

The results of the earlier study show that trajectory modeling can underestimate the benefits of switching to methanol. This is, in part, due to the trajectories becoming NO_x deficient. A similar result would be found if the modeled emissions were richer in ROG than the inventory warrants. This study and earlier studies have found that regions that are NO_x rich benefit most from using methanol. Conversely, regions that do not respond as well from switching to a lower reactivity fuel would benefit from NO_x emissions reductions (Milford, Russell, and McRae, 1989). This provides regulators with the possibility of adopting a combined strategy: using methanol to reduce ozone in one region, and decreasing NO_x emissions to reduce ozone in other regions. However, to realize the potential benefits of such a strategy fully, formaldehyde emissions must be controlled.

This study demonstrates that use of methanol-fueled vehicles can decrease smog in urban areas. While this capability alone cannot solve the Los Angeles smog problem, methanol-fueled vehicles can make an important contribution to improving air quality there and in other cities as well.

References

Alson, J.A. (1988). The Emission Characteristic of Methanol and Compressed Natural Gas in Light Vehicles. Paper presented at the Eighty-first Annual Meeting of the Air Pollution Control Association, Dallas, TX, June 19–24, 1988.

Austin, T. (1988). Potential Emissions and Air Quality Effects of Alternative Fuels. Sierra Research Inc., Sacramento, Calif.

Balentine, H.C., (1984). Effects of Emissions from Methanol-Fueled Vehicles on Smog. Paper presented at U.S. Environmental Protection Agency Methanol Workshop, San Francisco, Calif.

Balentine, H.C., et al. (1985). An Analysis of Chemistry Mechanisms and Photochemical Dispersion Models for Use in Simulating Methanol Photochemistry. U.S. Environmental Protection Agency, Report No. EPA-460/3-85-008.

Baugues, K. (1986). A Review of NMOC, NO_x, and $NMOC/NO_x$ Ratios Measured in 1984 and 1985. U.S. Environmental Protection Agency, Report No. EPA-450/4-86-015, Research Triangle Park, NC.

Bechtold, R., and J.B. Pullman (1980). Driving Cycle Economy, Emissions, and Photochemical Reactivity Using Alcohol Fuels and Gasoline. SAE Paper 800260, Society of Automotive Engineers, Inc.

CARB (1986). Methodology to Calculate Emissions Factors for On-road Motor Vehicles. Technical Support Division, California Air Resources Board, Sacramento, Calif.

_____ (1988). Draft Supplement 1 to Methodology to Calculate Emissions Factors for On-road Motor Vehicles. Technical Support Division, California Air Resources Board, Sacramento, Calif.

_____ (1989). Alcohol Fuel Fleet Test Program. 8th Interim Report. Mobil Source Division, California Air Resource Board, El Monte, Calif.

Carter, W.P.L. (1988). Reactivity Issues Related to Methanol Use. From Proceedings of the APRAC Methanol Workshop, Orange, Calif., April 19–21, 1988. CRC Report No. 564, Coordinating Research Council, Inc., Atlanta, GA.

Carter, W.P.L., et al. (1986). Effects of Methanol Fuel Substitution on Multi-day Air Pollution Episodes. Final Report on Contract No. A3-125-32, California Air Resources Board, Sacramento, Calif.

Chang, T. (1988). Impact of Methanol-Fueled Vehicles on Ozone Air Quality. From Proceedings of the APRAC Methanol Workshop, Orange, Calif., April 19–21, 1988. CRC Report No. 564, Coordinating Research Council, Inc., Atlanta, GA.

Chang, T., et al. (1989). Impact of Methanol-Fueled Vehicles on Ozone Air Quality. *Atmospheric Environment*, 28:629.

Consensus Workshop on Formaldehyde (1984). Report on the Consensus Workshop on Formaldehyde. *Environmental Health Perspectives*, 58:323-381.

Finlayson-Pitts, B.J., and J.N. Pitts (1986). *Atmospheric Chemistry*, New York: John Wiley and Sons.

Gabele, P.A. (1989). Characterization of Emissions From a Variable Gasoline/Methanol Fueled Car (personal communication). U.S. Environmental Protection Agency, Research Triangle Park, N.C.

Health Effects Institute (1987). Automotive Methanol Vapors and Human Health: An Evaluation of Existing Scientific Information and Issues for Future Research. Report of the Health Effects Institute's Health Research Committee, Cambridge, MA.

Hellman, et al. (1989); Marnett, L.J. (1988). Health Effects of Aldehydes and Alcohols in Mobile Source Emissions. In *Air Pollution, the Automobile, and Public Health*, edited by A. Watson, R. Bates, and D. Kennedy. National Academy Press, Washington, DC. 579–603.

Jeffries, H.E., K.G. Sexton, and J.R. Arnold (1989). Validation Testing of New Mechanisms with Outdoor Chamber Data. Volume 2: Analysis of VOC Data for the CB4 and CAL Photochemical Mechanisms. U.S. Environmental Protection Agency, Cooperative Agreement CR-813107, Research Triangle Park, NC.

Jeffries, H.E., K.G. Sexton, and M.S. Halleman (1985). 1985 Outdoor Smog Chamber Experiments: Reactivity/Methanol Exhaust. US Environmental Protection Agency, EPA 68-03-3162.

McRae, G.J., W.R. Goodin, and J.H. Seinfeld (1982). Development of a Second Generation Mathematical Model for Urban Air Pollution I. Model formulation. *Atmospheric Environment*, 16:679–696.

McRae, G.J., and J.H. Seinfeld (1983). Development of a Second Generation Mathematical Model for Urban Air Pollution, II. Performance evaluation. *Atmospheric Environment*, 17:501–523.

Milford, J.B., A.G. Russell, and G.J. McRae (1989). A New Approach to Photochemical Pollution Control: Implications of Spatial Patterns in Pollutant Responses to Reductions in NO_x and Reactive Organic Gas Emissions. *Environmental Science and Technology*, 23:1290–1301.

National Research Council (1981). Formaldehyde and Other Aldehydes. Committee on Aldehydes, Board on Toxicology and Environmental Health Hazards, Assembly of Life Sciences, National Research Council. National Academy Press, Washington, DC.

Nichols, R.J., and J.M. Norbeck (1985). Assessment of Emissions from Methanol-Fueled Vehicles: Implications for Ozone Air Quality. Paper presented at the Seventy-eighth Annual Meeting of the Air Pollution Control Association, Detroit, MI, June 16–21, 1985.

O'Toole, R.P., et al. (1983). California Methanol Assessment, Vol. II, Technical Report. Air Quality Impact of Methanol Use in Vehicles, Jet Propulsion Laboratory Publication 83–18. Report for the California Energy Commission, California Institute of Technology, Pasadena, Calif.

Pefley, R.K., B. Pullman, and G. Whitten (1984). The Impact of Alcohol Fuels on Urban Air Pollution: Methanol Photochemistry Study. Final report. U.S. Department of Energy, DOE/CE/50036-1.

Rogozen, M.B., and R.A. Ziskind (1984). Formaldehyde: A Survey of Airborne Concentrations and Sources. Systems Application Incorporated, Publication No. 841624.

Russell, A.G., and G.R. Cass (1984). Acquisition of Regional Air Quality Model Validation Data for Aerosol Nitrate, Sulfate, Ammonium Ion and Their Precursors. *Atmospheric Environment*, 18:1815–1827.

Russell, A.G., K.F. McCue, and G.R. Cass (1988). Mathematical Modeling of the Formation of Nitrogen Containing Air Pollutants. 1. Evaluation of an Eulerian Photochemical Model. *Environmental Science and Technology*, 22:263–271.

Russell, A.G., et al. (1989). A Quantitative Estimate of the Air Quality Impacts of Methanol Fuel Use. Final report to the California Air Resources Board, Sacramento, Calif.

Snow, R., et al. (1989). Characterization of Emissions from a Methanol-Fueled Motor Vehicle, *Journal of the Air and Waste Management Association*, 39:48–54.

South Coast Air Quality Management District and Southern California Association of Governments (1988). Draft 1988 Air Quality Management Plan, September 1988.

Wagner, K.K., and A.J. Ranzierri (1984). Model Performance Evaluations for Regional Photochemical Models in California. Paper presented at the Seventy-seventh Annual Meeting of the Air Pollution Control Association, San Francisco, Calif., June 1984.

Whitten, G.Z., and H. Hogo (1983). Impact of Methanol on Smog: A Preliminary Estimate. Systems Application Incorporated, Publication No. 83044.

Whitten, G.Z., H. Yonkow, and T.C. Myers (1986). Photochemical Modeling of Methanol-Use Scenarios in Philadelphia. U.S. Environmental Protection Agency, Report No. EPA-406/3-86-001.

Williams, R.L., F. Lipari, and R.A. Potter (1989). Formaldehyde, Methanol and Hydrocarbon Emissions from Methanol-Fueled Cars. General Motors Research Laboratories, Warren Mich.

CHAPTER 5

Urban Air Quality Impact of Methanol-Fueled Vehicles Compared to Gasoline-Fueled Vehicles

TAI Y. CHANG

SARA J. RUDY

Abstract

The effect of methanol-fueled vehicles on urban ozone levels and other aspects of urban air quality was recently compared with that of gasoline-fueled vehicles. Photochemical modeling studies and an examination of nine urban areas showed that possible reductions in urban ozone levels, as a result of methanol-fueled vehicle usage, are modest. For example, a 1 to 5 percent reduction of peak urban ozone levels was recorded for large-scale usage of optimized methanol-fueled vehicles, compared with gasoline-fueled vehicles. However, high formaldehyde levels may result from methanol-fueled vehicle usage. Not only does formaldehyde produce ozone, but high levels of formaldehyde may develop in roadway tunnels and in underground parking garages. Formaldehyde emissions must be kept as low as possible to maximize the air quality benefit of methanol-fueled vehicles.

Introduction

Ozone is the most pervasive and persistent urban air quality problem in the United States. According to the U.S. Environmental Protection Agency (EPA), 101 urban areas failed to meet the National Ambient Air Quality Standard (NAAQS) for ozone from 1986 to 1988. Replacing gasoline-fueled vehicles with alternative-fueled vehicles (CARB, 1989; EPA, 1988 and 1989a) has been proposed as a strategy for reducing urban ozone. Vehicles using a mixture of 85 percent methanol and 15 percent gasoline (M85), or 100 percent methanol (M100) and other alternative fuels, are being considered (EPA, 1988; CARB, 1989).

The non-methane organic compound (NMOC) emissions from methanol-fueled vehicles are largely unburnt methanol. Methanol's photochemical reactivity for ozone formation is lower than that of NMOC emissions from gasoline-fueled vehicles. However, methanol-fueled vehicles emit more formaldehyde than do gasoline-fueled vehicles (EPA, 1986; Nichols et al., 1988). The potential benefits of methanol-fueled vehicles in reducing ozone formation are closely linked to their level of formaldehyde emissions. Not only does formaldehyde produce ozone, but it is also a health concern.

This paper examines some aspects of methanol-fueled vehicles, in order to evaluate their effect on urban ozone levels. Other air quality issues, including formaldehyde levels in microscale situations, are examined briefly.

Photochemical Modeling Studies: A Summary

Several photochemical modeling studies of the effect of methanol substitution on urban ozone levels in the South Coast Air Basin (SoCAB) of California, and in other urban areas, have been carried out. Most studies have been performed as one-day simulations and have used trajectory models. A multi-day three-dimensional airshed simulation study with emissions scenarios in the South Coast basin has been carried out by Russell et al. (1989). The studies assume that the NO_x (nitrogen oxides) emissions rate from methanol-fueled vehicles is the same as that from vehicles using conventional fuels, and they evaluate the effect on ozone levels of replacing hydrocarbon (HC) emissions from conventional vehicles with NMOC emissions from methanol-fueled vehicles. Table 1 summarizes various photochemical modeling studies. In general, results relate to cases of equivalent-carbon emissions substitution, that is, cases in which the base NMOC mixture has been replaced by the indicated NMOC substitute on an equivalent-carbon basis. Table 1 lists the base year, substituted sources, substituted NMOC compositions, and peak one-hour ozone reductions.

Clearly, ozone levels calculated by photochemical models are sensitive to such input variables as initial and boundary conditions, meteorological variables, emissions rates, and photochemical mechanism. Furthermore, a one-day simulation using the standard Empirical Kinetic Modeling Approach (EKMA) starts from an air parcel (or column) in the center-city area at 8:00 A.M. The high sensitivity of peak 1-hr ozone concentrations, to initial concentrations of NMOC, NO_x and the $NMOC/NO_x$ ratio can be used as an advantage if city-specific measured values for initial concentrations are used. Although extensive NMOC and $NMOC/NO_x$ data are now available (Baugues, 1986), values for morning (6:00 to 9:00) ambient NMOC and $NMOC/NO_x$ ratios vary widely from day to day at a given site and from site to site in an urban area. Consequently, the choice of NMOC and $NMOC/NO_x$ values for the standard EKMA simulation is not straightforward. Airshed (three-dimensional grid) models can incorporate many of these physicochemical aspects. However, because of the extensive data requirements, airshed models can be applied to only a limited number of days in a limited number of areas. Airshed models applied to a specific ozone episode also require day-specific emissions inventories, which are highly uncertain.

Table 1 shows that reductions in peak ozone levels are quite variable and sensitive to the percentage of formaldehyde in the substituted NMOC mixture. Mobile-source fractions of the total

Table 1
Methanol Photochemical Modeling
(Summary)[a]

Modeled Urban Area	Base[b] Year	Substituted Sources
Los Angeles	1987	All gas vehicles
Los Angeles	1987	All gas vehicles
Los Angeles	2000	All gas vehicles
20 Cities	1982	Light-duty vehicles
Philadelphia	2000	Mobile sources
Los Angeles[f]	2000 2010	Mobile sources[h]
Los Angeles	[i]	Light-duty vehicles
20 Cities	2000	Light-duty vehicles

[a] Unless noted otherwise, 1-day trajectory modeling with no change of NO_x and equivalent-carbon emissions.
[b] Year of emissions inventories used.
[c] Substituted NMOC expressed as percent methanol/percent formaldehyde/percent HC on a carbon basis.
[d] NO_x emissions reduced by 50 percent.
[e] Twenty-city average (range in parentheses).
[f] Three-day airshed modeling.
[g] HC emissions from conventional-fueled vehicles were replaced by equivalent-carbon methanol emissions from methanol-fueled vehicles. Formaldehyde

Table 1
(Continued)

Substituted[c] NMOC (%)	Peak 1-hr O_3 Reduction (%)	Reference
100/0/0	31	Whitten & Hogo
90/10/0	22	(1983)
80/20/0	13	
90/10/0	18	Pefley, Pullman, & Whitten (1984)
57/17/26	14	O'Toole et al.
28.5/8.5/13[d]	20	(1983)
100/0/0	13 (1-36)[e]	Nichols & Norbeck
90/10/0	3.5 (0-13)	(1985)
100/0/0	10	Whitten, Yonkow, &
110/6/0	7	Myers (1986)
105/12/0	3	
[g]	9	Russell et al.
	6	(1989)
62/8/30	0 (2)[j]	Dunker (1989)
92/4.5/3.5	2 (18)	
69/1/30	3 (24)	
95/1/4	4 (30)	
53/3/44	1.3 (0-3)[e]	Chang et al.
86/1/13	2.6 (1-5)	(1989)
90/10/0	1.9 (0-4)	
100/0/0	2.9 (1-6)	

emissions from methanol-fueled vehicles were assumed to be comparable with those from conventional-fueled vehicles.

[h] Full penetration of advanced-technology methanol-fueled vehicles into on-road advanced-technology conventional-fueled vehicles and methanol conversion of off-road mobile source.

[i] Light-duty vehicles are assumed to account for 30 percent of the emission inventory.

[j] The number given is for an $NMOC/NO_x$ ratio of 10, and the number in parentheses is for a $NMOC/NO_x$ ratio of 7.

Source: T.Y. Chang and S. Rudy, Ford Motor Co., 1990

emissions inventories in the year 2000 and beyond are projected to be less than those in recent years. Therefore, ozone reduction for methanol-substitution scenarios in and after 2000 is smaller than the ozone reduction in recent years. All studies shown in Table 1 suggest that conversion of gasoline-fueled vehicles to methanol-fueled vehicles will have some potential to reduce ozone levels in urban areas. However, the most recent studies indicate a much smaller improvement in ozone formation than estimated in earlier, overly optimistic studies.

Emissions Characteristics

Only limited data are available for emissions factors of methanol-fueled vehicles and especially for M100 vehicles. These data, moreover, have been obtained from development and demonstration vehicles, which are not expected to be representative of fully developed production vehicles.

Table 2 lists the exhaust and evaporative emissions factors used recently by EPA (1989a). The EPA data are in-use non-methane organic compound emissions factors. The emissions factors of current gasoline vehicles are based on the EPA (1989b) MOBILE 4 model, which assumes that formaldehyde emissions are 1 percent of the exhaust NMHC emissions by weight. Special changes have been made in the MOBILE 4 model to estimate emissions rates for proposed gasoline vehicles. The emissions rates for M85 and M100 vehicles are estimated in-use emissions factors.

Detailed NMOC emissions composition data are also very limited. Table 3 lists the currently available data, measured for vehicles run on the Federal Test Procedure (FTP). Emissions fractions of NMHC, methanol, formaldehyde, and acetaldehyde are listed, as well as carbon fractions of the NMHC classes used in the Lurmann-Carter-Coyner (LCC) chemical mechanism (Lurmann, Carter, Coyner, 1987). The data in tables 2 and 3 have been used to evaluate the ozone-forming potential of vehicle emissions (EPA, 1989a; Chang and Rudy, 1989).

Table 2
Vehicle Emission Rates

Fuel/Species	Exhaust (g/mi)	Evap. (g/mi)	Total C[a] (mole/mi)	Fractional C
Current Gasoline				
NMHC	0.700	1.030	0.12246	0.998
Formaldehyde	0.007		0.00023	0.002
	0.707	1.030	0.12269	1.000
Proposed Gasoline				
NMHC	0.530	0.410	0.06686	0.998
Formaldehyde	0.005		0.00016	0.002
	0.535	0.410	0.06702	1.000
Optimized M85				
NMHC	0.150	0.160	0.02200	0.472
Methanol	0.500	0.250	0.02341	0.503
Formaldehyde	0.035		0.00117	0.025
	0.685	0.410	0.04658	1.000
Optimized M100				
NMHC	0.050		0.00360	0.164
Methanol	0.500	0.072	0.01785	0.813
Formaldehyde	0.015		0.00050	0.023
	0.565	0.072	0.02195	1.000

[a] The average molecular weight per mole C in exhaust and evaporative NMHC emissions are assumed to be 13.88 and 14.30, respectively (EPA, 1986).

Source: EPA, 1989a

To assess properly the relative air quality benefits of gasoline-fueled vehicles and methanol-fueled vehicles, one must know both the quantity and nature of the emissions from fully developed vehicles. Improvements in the emissions-control systems for both gasoline-fueled vehicles and methanol-fueled vehicles continue to be made, and much effort is being directed at improving gasoline fuel in order to reduce the NMOC emissions from gasoline-fueled vehicles. But no satisfactory methanol-fueled vehicles (M100 vehicles,

Table 3
Vehicle NMOC Emissions (Exhaust and Evaporative) Fractions

Compound Class	Fraction of carbon in compound class					
	Gas[a]	M85[a]	M100[a]	Current Fleet[b]	M85[c]	M100[c]
NMHC	0.982	0.491	0.087	0.993	0.295	0.035
Methanol		0.450	0.861		0.623	0.920
Formaldehyde	0.009	0.051	0.050	0.004	0.081	0.045
Acetaldehyde	0.009	0.008	0.002	0.003	0.001	0.000
	1.000	1.000	1.000	1.000	1.000	1.000

Carbon fractions of the NMHC classes[d] used in the LCC chemical mechanism:

	Gas[a]	M85[a]	M100[a]	Current Fleet[b]	M85[c]	M100[c]
ALK4	0.182	0.249	0.115	0.240	0.329	0.200
ALK7	0.440	0.477	0.460	0.312	0.227	0.200
ETHE	0.030	0.016	0.012	0.048	0.098	0.086
PRPE	0.033	0.018	0.011	0.015	0.061	0.086
TBUT	0.000	0.000	0.000	0.067	0.014	0.028
TOLU	0.181	0.079	0.207	0.097	0.142	0.143
XYLE	0.056	0.067	0.115	0.101	0.037	0.028
TMBZ	0.019	0.039	0.034	0.066	0.007	0.000
NR	0.059	0.055	0.046	0.054	0.085	0.229
	1.000	1.000	1.000	1.000	1.000	1.000

[a] CARB (1989).
[b] Light-duty gasoline vehicles equipped with catalyst (Shareef et al., 1988; Sigsby et al., 1987).
[c] Derived from Dunker (1989) and Williams, Lipari, and Potter (1989).
[d] Defined in Lurmann, Carter, and Coyner (1987).

Source: T.Y. Chang and S. Rudy, Ford Motor Co., 1990

in particular) have yet been developed, and it is not possible to project gasoline-fueled vehicles and methanol-fueled vehicles emissions with confidence. Nevertheless, the vehicle emissions factors in Table 2 can be used to estimate a range of ozone reductions from methanol-fueled vehicles compared with reductions from gasoline-fueled vehicles: (1) reduced-carbon emissions (column 4) for optimistic ozone reductions and (2) equivalent-carbon emissions (column 5), where the total reactive-carbon emissions from advanced gasoline-fueled vehicles are assumed to be the same as those from advanced methanol-fueled vehicles, for the lower bound of ozone reductions.

Ozone-Forming Potential of Vehicular NMOC Emissions

To consider the ozone-forming potential of methanol-fueled vehicles versus gasoline-fueled vehicles, reactivities of formaldehyde and methanol relative to NMHC were introduced (Gold and Moulis, 1988; Chang et al., 1989). The ozone-forming potential (P) of vehicular NMOC emissions is defined as

$$P = \frac{z(\text{subst.}) - z(\text{blank})}{z(\text{base}) - z(\text{blank})} \tag{1}$$

where $z(\text{base})$ is the maximum ozone concentration for the base case calculated with the base NMOC mixture using a photochemical model, $z(\text{blank})$ is the maximum ozone concentration calculated for the blank case in which a portion of the total NMOC is removed from the base case, and $z(\text{subst.})$ is the maximum ozone concentration calculated for cases in which the substituted vehicular NMOC mixtures are added to the blank case.

Relative reactivities of methanol (R_m) and formaldehyde (R_f) are defined by

$$P = (C_{hc} + R_m C_m + R_f C_f) \, / \, C_c \tag{2}$$

where C_c is the total carbon (mole) in the NMOC mixture for the base scenario. C_{hc}, C_m, and C_f represent, respectively, the NMHC, methanol, and formaldehyde carbon (mole) of the NMOC mixture for the substitution scenarios. The coefficients, R_m and R_f, are relative reactivities of methanol and formaldehyde and can be obtained by using computed values of P (Equation 1) for a number of substitution scenarios.

Recently, Carter and Atkinson (1989) introduced incremental reactivity (IR) for a NMOC species or mixture defined as*

$$IR(a) = \frac{iz}{ia} = \frac{z(b + ia) - z(b)}{ia} \qquad (3)$$

where z(b) is the maximum ozone concentration calculated with the base NMOC mixture and z(b + ia) is the maximum ozone concentration calculated with the test NMOC species added to the base NMOC mixture and ia is the incremental concentration (in terms of carbon) of the test NMOC species that would be present at the end of the simulation if there were no chemical reactions. Incremental reactivities and relative reactivities of NMOC species have been compared (Rudy and Chang, 1989).

To compare ozone-forming potentials calculated by different methods, the relative reactivity method and the incremental reactivity method were evaluated (Chang and Rudy, 1989a) using the Ozone Isopleth Plotting Package with Optional Mechanism (OZIPM) (Hogo and Gery, 1988) and the LCC chemical mechanism (Lurmann et al., 1987). These calculations assume that the NMHC portion of NMOC emissions from methanol-fueled vehicles has the same reactivity as that from gasoline-fueled vehicles. Ozone-forming potentials were also calculated directly from Equation 1 by using the speciated emissions data given in Table 3 (direct method).

Figure 1 compares the ozone-forming potential per carbon of vehicle emissions calculated by the three different methods. Simulation conditions for Figure 1 were chosen to represent high-ozone days in the eastern United States, with initial NMOC concentration = 0.7 ppmC and NMOC/NO_x ratio = 8. In principle, the ozone-forming potentials based on the direct method should be more accurate; however, the procedure requires accurate NMOC emissions composition data. As seen in Table 3, the NMHC emissions composition of California Air Resources Board's gasoline (based on two measurements) is different from that of the current fleet (Shareef

---

*In this equation "i" means "change in."

Figure 1
Ozone-Forming Potential for California Air Resources Board Vehicles and Three Others Listed in Table 3

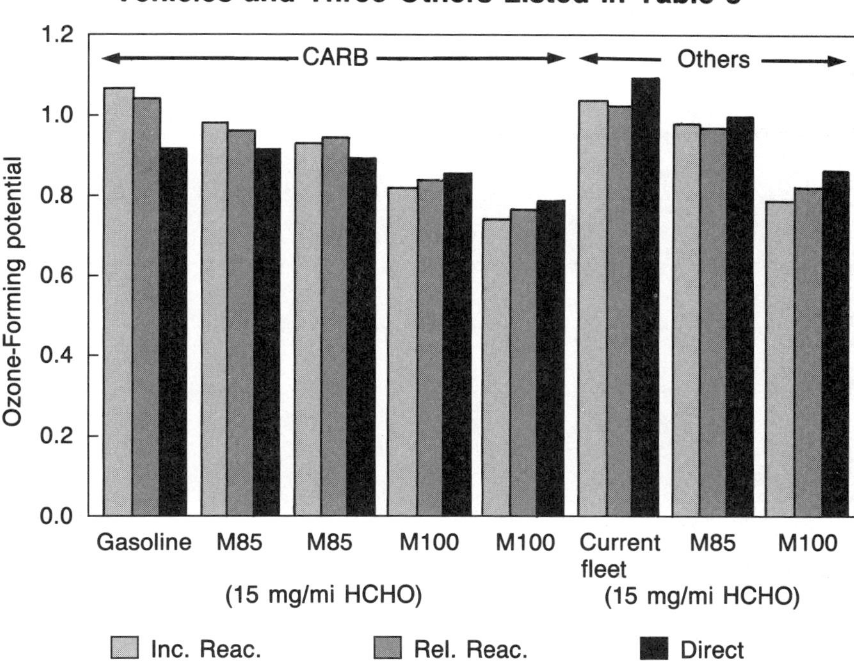

Source: T.Y. Chang and S. Rudy, Ford Motor Co., 1990

et al., 1988) based on a number of vehicle measurements (Sigsby et al., 1987). These differences are reflected in the ozone-forming potentials calculated by the direct method.

In Figure 2 ozone-forming potentials per carbon of vehicle emissions, based on the relative reactivity method, are plotted against the $NMOC/NO_x$ ratio. The ozone-forming potential of methanol-fueled vehicles relative to those of gasoline-fueled vehicles increases as the $NMOC/NO_x$ ratio increases. The ozone-forming potential of methanol-fueled vehicles compared with that of gasoline-fueled vehicles is smaller at low and intermediate $NMOC/NO_x$ ratios but larger at high $NMOC/NO_x$ ratios.

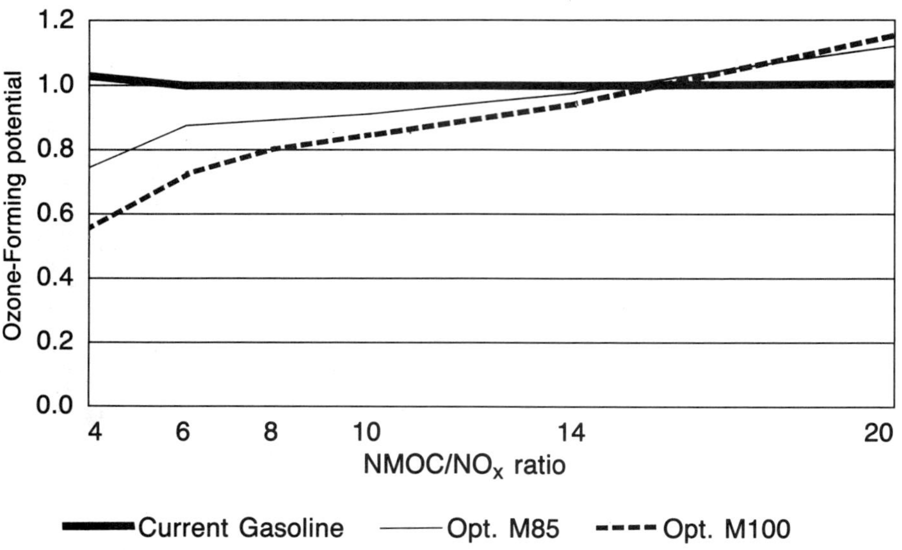

Figure 2
Ozone-Forming Potential Based on the Relative Reactivity Method and Relative to the Average Urban NMHC Mixture at each NMOC/NO$_x$ Ratio

Source: T.Y. Chang and S. Rudy, Ford Motor Co., 1990.

Potential Impact of Methanol-Fueled Vehicles on Urban Ozone Levels

The EPA (1989a) used the emissions factors of gasoline-fueled vehicles and methanol-fueled vehicles, given in Table 2, the relative reactivities of NMHC, methanol, and formaldehyde, and projected emissions inventories to estimate NMOC-equivalent reductions in nine urban areas (Los Angeles, Houston, New York, Milwaukee, Baltimore, Philadelphia, greater Connecticut, Chicago, and San Diego). These urban areas have 1986–88 ozone design values greater than or equal to 0.18 ppm. An additional examination of these nine areas is given in this paper.

The year 2015 is chosen to consider full penetration of proposed gasoline-fueled vehicles and methanol-fueled vehicles. In this paper we consider steady-state penetration scenarios where it is assumed that sufficient time will have elapsed to allow methanol-fueled vehicles to integrate completely with the in-use vehicle fleet. The NMHC emissions inventories for 2015 are projected for current emissions-control and moderate growth scenarios by using the EPA's data base and procedure (the MOBILE 4 model for mobile sources along with inventory rollback calculations is used, as described in EPA, 1987a).

Table 4 lists the percentages of the total NMHC emissions inventories attributed to mobile sources and light-duty vehicles (LDV) for the nine urban areas. The averages of mobile sources and LDV emissions are projected to be 21 percent and 11 percent, respectively, of the total NMHC emissions inventories. The NMOC concentrations, $NMOC/NO_x$ ratios, and temperatures used in the photochemical model are also listed. The NMOC concentrations

Table 4

Projected NMHC Emission Inventories and Some Input Variables

	2015 Emission Inventory				
City	**Mobile Sources (%)**	**LDV (%)**	**NMOC (ppmC)**	**$NMOC/NO_x$ ratio**	**Temp. (°K)**
Baltimore, MD	23.9	12.2	0.70	7.0	304
Chicago, IL	17.6	8.6	0.82	7.8	302
Greater CT	20.8	11.7	0.37	9.4	303
Houston, TX	14.0	4.8	0.82	10.9	307
Los Angeles, CA	25.0	11.6	0.85	7.1 (10.3)	301
Milwaukee, WI	22.3	14.5	0.70	8.0	300
New York, NY	24.1	9.7	0.71	8.0	303
Philadelphia, PA	19.3	12.3	0.57	7.3	303
San Diego, CA	25.5	13.0	0.31	7.4	301
AVERAGE	21.4	10.9			

Source: T.Y. Chang and S. Rudy, Ford Motor Co., 1990

and NMOC/NO$_x$ ratios are generally chosen based on EPA's 6:00 to 9:00 A.M. data. The temperatures are generally based on observed July maxima. The NMOC concentration and NMOC/NO$_x$ ratios in the Los Angeles basin are taken from Kelly and Gunst (1989). The base urban NMOC mixture is taken from Jeffries et al. (1989). The mixing heights are chosen to increase from 300 meters at 8 A.M. LDT to 600 meters at 3 P.M. LDT in Los Angeles and San Diego, and from 300 meters at 8 A.M. LDT to 1,500 meters at 3 P.M. LDT in other urban areas.

With input variables in Table 4 and emissions factors in Table 2, the OZIPM trajectory model with the LCC chemical mechanism is used to calculate maximum ozone concentrations for the base (current gasoline), substitution (proposed gasoline, M85 and M100), and blank (no light-duty vehicles) cases. Using these calculated ozone concentrations, relative ozone-forming potentials defined by Equation 1 are calculated for each city.

Table 5 lists the nine-city average ozone-forming potentials calculated in this paper, along with ozone-forming potentials based on relative reactivities. Ozone-forming potentials based on relative reactivities are calculated from Equation 2 by using emissions factors in Table 2 and three different sets of reactivities. Table 5 includes relative per-carbon ozone-forming potentials, or ozone-forming potentials for equivalent-carbon emissions (column 5, Table 2), where the total reactive carbon emissions are not changed, as well as relative per-mile ozone-forming potentials, or ozone-forming potentials for reduced-carbon emissions (column 4, Table 2). There are differences among the ozone-forming potentials; for example, M85 and M100 ozone-forming potentials, based on reactivities of EPA (1987b), are lower than others. Nevertheless, the ozone-forming potentials derived by somewhat different methods show consistent trends.

The EPA (1989a) used ozone-forming potentials to calculate NMOC-equivalent reductions. The NMOC-equivalent reduction (U) is calculated by

$$U = F L (1 - P) \tag{4}$$

Table 5
Relative Ozone-Forming Potential of Vehicular Emissions

Vehicle	Per-Carbon Potential (equivalent-carbon)				Per-Mile Potential (reduced-carbon)			
	a	b	c	d	a	b	c	d
Current Gasoline	1.00	1.00	1.00	1.00	1.00	1.00	1.00	1.00
Proposed Gasoline	1.00	1.00	1.00	1.00	0.55	0.55	0.55	0.57
Optimized M85	0.80	0.81	0.89	0.89	0.31	0.31	0.34	0.34
Optimized M100	0.62	0.68	0.79	0.76	0.11	0.12	0.14	0.14

[a] Relative reactivities: NMHC = 1; methanol = 0.43; formaldehyde = 4.83 (EPA, 1987b). Averages for a number of cities.
[b] Relative reactivities: NMHC = 1; methanol = 0.58; formaldehyde = 2.1 (Chang et al., 1989). Twenty-city averages.
[c] Relative reactivities: NMHC = 1; methanol = 0.69; formaldehyde = 3.03 (Rudy and Chang, 1989). Simulation with $NMOC/NO_x$ = 8.
[d] Nine-city averages calculated by use of Equation 1 and the OZIPM model.

Source: T.Y. Chang and S. Rudy, Ford Motor Co., 1990

where F is the percent of LDVs replaced by methanol-fueled vehicles, L is the LDV emission inventory fraction of the total NMHC inventories, P is the ozone-forming potential, and (1 − P) is the ozone reduction potential.

The ozone reduction is calculated by two methods. The first method is to calculate the ozone reduction (S) by

$$S = F\, S_0\, (1 - P) \tag{5}$$

where S_0 is the reduction in maximum ozone level when 100 percent of the LDV emissions are removed, and is given by

$$S_0 = 1 - z(\text{blank})/z(\text{base}) \tag{6}$$

Equation 5 was introduced to consider different percent penetrations by methanol-fueled vehicles (Chang et al., 1989). The second method is to compute ozone reductions directly by applying the OZIPM model.

The nine-city average NMOC-equivalent and peak ozone reductions for full penetration (F = 100 percent) of LDV are shown in Table 6. The reductions based on EPA's ozone-forming potentials (columns 2 and 6, Table 5), the nine-city average ozone-forming potentials (columns 5 and 9, Table 5), and the nine-city average ozone reductions calculated with the OZIPM model have been listed for cases of equivalent-carbon and reduced-carbon emissions. The nine-city average ozone reductions calculated with the OZIPM model agree with those from Equation 5. The nine-city average ozone

Table 6
NMOC-Equivalent and Ozone Reductions in 2015 for Full Penetration (Steady State) of LDV Relative to Current Gasoline

| | 9-City Average Reductions (%) | | | | |
| | NMOC-equiv. reductions | | Ozone reductions | | |
	Based on EPA O_3-form. pot.	Based on 9-city average O_3-form. pot.	Based on EPA O_3-form. pot.	Based on 9-city average O_3-form. pot.	Based on photo-chemical model calc.
Equivalent-carbon					
M85	2.1	1.2	1.1	0.6	0.6
M100	4.2	2.7	2.2	1.4	1.4
Reduced-carbon					
Proposed gasoline	4.9	4.7	2.5	2.4	2.4
M85	7.6	7.3	3.9	3.7	3.7
M100	9.7	9.5	5.0	4.9	5.0

Source: T.Y. Chang and S. Rudy, Ford Motor Co., 1990

reductions are approximately half of the NMOC-equivalent reductions based on Equation 4. The EPA (1989a) estimated the nine-city average NMOC-equivalent reductions to be 3.3 percent in 2015 for 30 percent penetration of LDV by M100, an estimate that is consistent with the NMOC-equivalent reductions shown in Table 6.

Ozone reductions in the Los Angeles basin are much more sensitive to $NMOC/NO_x$ ratios, compared with other areas. For an $NMOC/NO_x$ ratio of 7.3, ozone reductions in the Los Angeles basin are about three times the nine-city averages; however, for a $NMOC/NO_x$ ratio of 10.3, ozone reductions in the Los Angeles basin are about the same as the nine-city averages.

Formaldehyde Concentrations in Microscale Situations

The EPA (1989c) proposed that the level of concern for formaldehyde be 500 $\mu g/m^3$. Exposure at this level may cause irritation for most individuals. A fraction of the population may be sensitive to formaldehyde concentrations as low as 50 to 100 $\mu g/m^3$ (Gold and Moulis, 1988). Because of possible health concerns due to higher formaldehyde emissions from methanol-fueled vehicles compared to gasoline-fueled vehicles, the potential impact of methanol-fueled vehicles on formaldehyde levels in such microscale situations as tunnels, parking garages, and urban street canyons has been examined (EPA, 1986; Gold and Moulis, 1988). These studies are based on air quality models and physical variables suggested by Ingalls (1981). These studies showed that in such semi-enclosed spaces as tunnels or underground parking garages high levels of formaldehyde may result from methanol-fueled vehicles usage. Additional studies are under way, sponsored by the American Petroleum Institute, South Coast Air Quality Management District (SCAQMD) of California, and the Coordinating Research Council, Inc. (CRC).

Recently, quasi-steady-state, analytic air quality models for various types of roadway tunnels have been derived and applied to estimate formaldehyde concentrations resulting from methanol-fueled vehicles

(Chang and Rudy, 1989b). Roadway tunnels can be grouped by the most commonly used ventilation systems: natural (including the vehicle piston effect), longitudinal, semi-transverse, and balanced transverse. Maximum formaldehyde concentrations versus vehicle formaldehyde emissions factors have been calculated and are shown in Figure 3 for typical and severe conditions in a one-way, two-lane commuter tunnel.

Typical and severe conditions are assumed to have median vehicle speeds of 40 mph and 5 mph, respectively. Gold and Moulis (1988) reviewed emissions measurements of current-technology methanol-fueled vehicles and derived an average formaldehyde emission rate of 110 mg/mi for the FTP cycle. As advanced

Figure 3
Concentrations of Emissions in Tunnels of Varied Length with Varied Systems of Ventilation

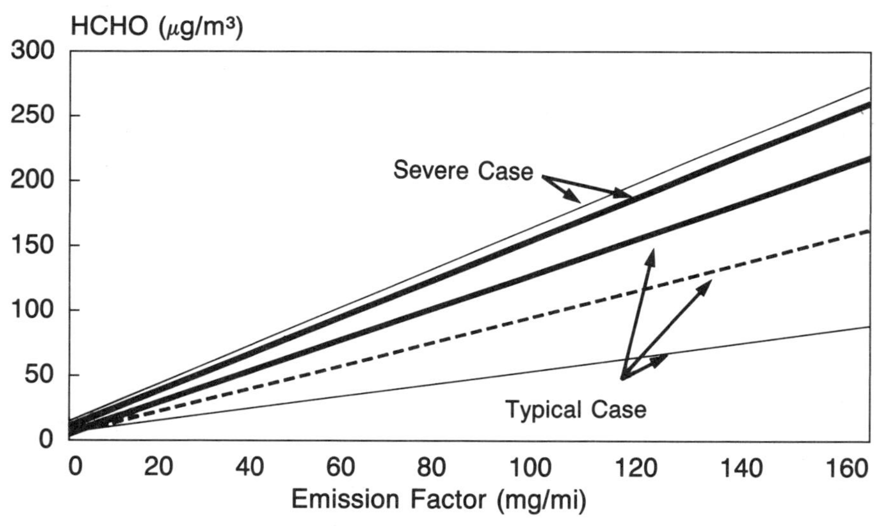

Source: T.Y. Chang and S. Rudy, Ford Motor Co., 1990

technologies are implemented, methanol-fueled vehicles will have much lower formaldehyde emissions rates than at present. If future formaldehyde emissions rates are assumed to be less than 20 mg/mi for typical conditions, and to be 60 to 100 mg/mi for severe conditions, maximum formaldehyde concentrations would be below 30 $\mu g/m^3$ for typical conditions and would be 100 to 200 $\mu g/m^3$ for severe conditions. It appears that formaldehyde concentrations in roadway tunnels resulting from methanol-fueled vehicles are not a health concern for the general public, but during heavily congested traffic conditions they may cause eye irritation for sensitive individuals.

Another possible concern is underground parking garages when a large number of cold-starting vehicles are in operation. Under these conditions high formaldehyde concentrations may result from methanol-fueled vehicles usage. (A study is in progress, sponsored by SCAQMD.) Formaldehyde emissions rates from cold-starting methanol-fueled vehicles are needed.

Other Air Quality Concerns

Little quantitative information is available concerning other toxic emissions from methanol-fueled vehicles, since data are very limited and fuel and oil specifications have not been completed. Ambient methanol and formaldehyde concentrations are expected to increase significantly. Ambient methanol concentrations, however, are expected to be well below the level of concern (Gold and Moulis, 1988; Russell et al., 1989). Ambient formaldehyde concentrations, except for those in semi-enclosed spaces, are also expected to be below the level of concern. No other air toxics have been identified that might increase as a result of methanol-fueled vehicles usage.

Benzene and polycyclic organic material (POM) emissions from methanol-fueled vehicles are expected to be much less than from gasoline-fueled vehicles, since vehicle benzene and POM emissions are approximately proportional to the aromatic content of the fuel.

Therefore, ambient benzene and POM concentrations are expected to decrease substantially through the use of methanol-fueled vehicles. A secondary pollutant, peroxyacetyl nitrate (PAN), is also expected to decrease as a result of methanol-fueled vehicles use, since methanol does not produce acetaldehyde, a precursor of PAN. Emissions of primary particulate matter, and such secondary particulate precursors as aromatics, from methanol-fueled vehicles are expected to be much lower than those from vehicles fueled by gasoline and diesel fuel. Therefore, ambient fine particle levels would decrease substantially, and urban visibility would improve significantly, especially if diesel-fueled trucks were replaced by methanol-fueled trucks (Russell et al., 1989).

Summary and Discussion

A summary of various photochemical modeling studies shows that percentage reductions in ozone levels through methanol substitution are quite variable. The ozone-forming potential of NMOC emissions, calculated by different methods, has been compared. The ozone-forming potential of light-duty vehicles (gasoline and M85 and M100) emissions has been evaluated, using vehicle emission rates derived by the EPA (1989a) and a simple trajectory model (the OZIPM model with the LCC chemical mechanism).

It was found that the ozone-forming potentials, calculated by various methods, differed but showed consistent trends. The impact of methanol-fueled vehicles on NMOC-equivalent and ozone levels for broad-scale usage of M85 and M100 in 2015 was calculated. It was found that average ozone reductions in nine urban areas were approximately one-half of the NMOC-equivalent reductions. The nine-city average reduction in ozone levels was modest, even with wide usage of M85 and M100. For example, average peak ozone reductions in nine urban areas were estimated to be 1 to 5 percent for broad-scale usage of optimized M100, compared with current gasoline. Ozone reduction in the Los Angeles basin was one to three times greater than the nine-city average reduction.

The effect of methanol-fueled vehicles on regional ozone levels needs further study. Since the relative reactivity of methanol increases as the NMOC/NO$_x$ ratio increases (see Figure 2), the effect of methanol-fueled vehicles on ozone levels in regional areas, where NMOC/NO$_x$ ratios are high, should be evaluated. (A study is planned, sponsored by CRC.)

The effects of increased formaldehyde and methanol concentrations on air quality also need to be studied. There are possible concerns about high formaldehyde levels in commuter tunnels during congested traffic conditions, and in underground parking garages when a large number of cold-starting vehicles are in operation. No other air quality effects that might increase as a result of methanol-fueled vehicles usage have been identified. Urban concentrations of benzene, polycyclic organic material, and fine particulate matter are expected to decrease significantly as a result of methanol-fueled vehicles usage.

References

Baugues, K. (1986). A Review of NMOC/NO$_x$ Ratios Measured in 1984 and 1985. U.S. Environmental Protection Agency, Report No. EPA-450/4-86-015, Research Triangle Park, NC.

CARB (1989). Definition of a Low-Emission Motor Vehicle in Compliance With the Mandates of Health and Safety Code Section 39037.05 (Assembly Bill 234 Leonard, 1987). Mobile Source Division, California Air Resources Board, El Monte, CA.

Carter, W.P.L., and R. Atkinson (1989). Computer Modeling Study of Incremental HC Reactivity. *Environmental Science and Technology*. 23:864-80.

Chang, T.Y., et al. (1989). Impact of Methanol Vehicles on Ozone Air Quality. *Atmospheric Environment*. 23:1629-44.

Chang, T.Y., and S.J. Rudy (1989). Ozone-Forming Potential of Organic Emissions from Alternative-Fueled Vehicles. Submitted to *Atmospheric Environment*.

_____ (1989b). Roadway Tunnel Air Quality Models. Accepted in *Environmental Science and Technology*.

Dunker, A.M. (1989). The Relative Reactivity of Emissions from Methanol-Fueled and Gasoline-Fueled Vehicles in Forming Ozone. AWMA 89-7.6, presented at the Air and Waste Management Association, Anaheim, CA, June 25–30, 1989.

Environmental Protection Agency (1986). Regulatory Support Document, Proposed Organic Emission Standards and Test Procedures for 1988 and Later Methanol Vehicles and Engines. Office of Mobile Sources, U.S. Environmental Protection Agency, Ann Arbor, MI.

_____ (1987a). Draft Regulatory Impact Analysis: Control of Gasoline Volatility and Evaporative HC Emissions from New Motor Vehicles. Office of Mobile Sources, U.S. Environmental Protection Agency, Ann Arbor, MI.

_____ (1987b). Air Quality Benefits of Alternative Fuel. Prepared for the Alternative Fuels Working Group of the President's Task Force on Regulatory Relief, Office of Mobile Sources, U.S. Environmental Protection Agency, Ann Arbor, MI.

_____ (1988). Guidance on Estimating Motor Vehicle Emission Reduction from the Use of Alternative Fuels and Fuel Blends. Office of Mobile Sources, U.S. Environmental Protection Agency, Report No. EPA-TSS-87-4, Ann Arbor, MI.

_____ (1989a). Analysis of the Economic and Environmental Effects of Methanol as an Automotive Fuel. Office of Mobile Sources, U.S. Environmental Protection Agency, Ann Arbor, MI.

_____ (1989b). *User's Guide to MOBILE 4 (Mobile Source Emission Factor Model)*. Office of Mobile Sources, U.S. Environmental Protection Agency, Ann Arbor, MI.

_____ (1989c). Standards for Emissions from Methanol-Fueled Motor Vehicles and Motor Vehicle Engines: Final Rule. Federal Register, 40 CFR Part 86, April 11, 1989.

Gold, M.D., and C.E. Moulis (1988). Effects of Emission Standards on Methanol Vehicles-Related Ozone, Formaldehyde, and Methanol Exposure. APCA Paper No. 88-41.4, presented at the Eighty-first

Annual Meeting of the Air Pollution Control Association, Dallas, TX, June 19–24, 1988.

Hogo, H., and M.W. Gery (1988). User's Guide for Executing OZIPM-4 with CBM-IV or Optional Mechanism. U.S. Environmental Protection Agency, Report No. EPA/600/8-88/073b, Research Triangle Park, NC.

Ingalls, M.N. (1981). Estimating Mobile Source Pollutants in Microscale Exposure Situations. U.S. Environmental Protection Agency, Report No. EPA-460/3-81-021, Ann Arbor, MI.

Ingalls, M.N., and R.J. Garbe (1982). Ambient Pollutant Concentrations from Mobile Sources in Microscale Situations. SAE Paper 820787, Society of Automotive Engineers, Inc., Warrendale, PA.

Jeffries, H.E., K.G. Sexton, and J.R. Arnold (1989). Validation Testing of New Mechanisms with Outdoor Chamber Data. Volume 2: Analysis of VOC Data for the CB4 and CAL Photochemical Mechanisms. U.S. Environmental Protection Agency, Cooperative Agreement CR-813107, Research Triangle Park, NC.

Kelly, N.A., and R.F. Gunst (1989). Response of Ozone to Changes in HC and NO_x Concentrations in Outdoor Smog Chambers Filled with Los Angeles Air. AWMA Paper 89-152.6, presented at the Air and Waste Management Association, Anaheim, CA, June 25–30, 1989.

Lurmann, F.W., W.P.L. Carter, and L.A. Coyner (1987). A Surrogate Species Chemical Reaction Mechanism for Urban-Scale Air Quality Simulation Models. Vol. 1—Adaptation of the Mechanism, vol. 2—Guidelines for Using the Mechanism. U.S. Environmental Protection Agency, Report Nos. EPA/600/3-87/014a and b, Research Triangle Park, NC.

Nichols, R.J., et al. (1988). A View of Flexible-fuel Vehicle Aldehyde Emissions. SAE Paper 881200, Society of Automotive Engineers, Inc., Warrendale, PA.

Nichols, R.J., and J.M. Norbeck (1985). Assessment of Emissions from Methanol-fueled Vehicles: Implications for Ozone Air Quality. APCA Paper No. 85-38.3, presented at the Seventy-eighth Annual Meeting of the Air Pollution Control Association, Detroit, MI, June 16–21, 1985.

O'Toole, R., et al. (1983). California Methanol Assessment. Jet Propulsion Laboratory, California Institute of Technology, Pasadena, CA.

Pefley, R.K., B. Pullman, and G. Whitten (1984). The Impact of Alcohol Fuels on Urban Air Pollution: Methanol Photochemistry. University of Santa Clara, Santa Clara, CA, under Contract No. DE-FG03-84CE50036, U.S. Department of Energy, Washington, DC.

Rudy, S.J., and T.Y. Chang (1989). Reactivity/Ozone-Forming Potential of Organic Gases. Submitted to *Atmospheric Environment.*

Russell, A.G., et al. (1989). A Quantitative Estimate of the Air Quality Impacts of Methanol Fuel Use. Prepared for the California Air Resources Board and the South Coast Air Quality Management District, under ARB Agreement No. A6-048-32.

Shareef, G.S., et al. (1988). Air Emissions Species Manual, Volume I. U.S. Environmental Protection Agency, Report No. EPA-450/2-88-003, Research Triangle Park, NC.

Sigsby, J.E., Jr., et al. (1987). Volatile Organic Compound Emissions from 46 In-Use Passenger Cars. *Environmental Science and Technology.* 21:466-75.

Whitten, G.Z., and H. Hogo (1983). Impact of Methanol on Smog: A Preliminary Estimate. Report 83044, prepared for Arco Petroleum Products Co. by Systems Applications, Inc., San Rafael, CA.

Whitten, G.Z., N. Yonkow, and T.C. Myers (1986). Photochemical Modeling of Methanol-Use Scenarios in Philadelphia. U.S. Environmental Protection Agency, Office of Mobile Sources, Report No. EPA 460/3-86-001, Ann Arbor, MI.

Williams, R.L., F. Lipari, and R.A. Potter (1989). Formaldehyde, Methanol, and HC Emissions from Methanol-Fueled Cars. GMR-6728, ENV No. 271, General Motors Advanced Engineering Staff, Warren, MI.

CHAPTER 6

Impact of Methanol-Fueled Vehicles on Rural and Urban Ozone Concentrations During a Region-wide Ozone Episode in the Midwest

SANFORD SILLMAN
PERRY J. SAMSON

Abstract

The use of methanol as a substitute for gasoline could affect both rural and urban air quality. Methanol-fueled vehicles do not emit as much volatile organic carbon (VOC) as gasoline vehicles, but they do emit greater amounts of certain kinds of methanol and formaldehyde. The effect of these emissions on air quality was tested for a case study of regional dispersion in the Midwest. A combined urban and regional-scale model was used to study an urban domain in the Detroit-Port Huron area and a regional domain extending from Michigan to Texas. Results indicate that the use of methanol would cause a 1 to 5 percent reduction in peak ozone concentrations in Detroit. Methanol would have virtually no impact on rural ozone concentrations or on regional dispersal, as these are sensitive to NO_x rather than to VOC. If NO_x emissions were also affected by the switch to methanol, there could be additional impacts on both urban and rural ozone.

Overview

During the next 10 years it appears that the United States will confront an increasing array of problems related to energy and the environment. A switch to methanol-fueled vehicles has been suggested as a possible solution to both problems (Gray and Alson, 1986). As a fuel derived from natural gas or coal, methanol would relieve pressure on domestic oil supplies. In addition, emissions of total VOC from methanol-fueled vehicles are likely to be lower than emissions from gasoline-fueled vehicles. To the extent that elevated ozone concentrations are caused by VOC emissions, methanol-fueled vehicles may alleviate the ozone problem.

This paper presents results of a preliminary investigation of the effect of methanol fuel on ozone in the eastern half of the United States, during a region-wide period of increased ozone concentration. A four-day period in August 1988, was selected for a case study of the effect of alternative emissions scenarios on air quality. Dynamical and photochemical processes during the four-day period were simulated, using a combined urban and regional-scale model (Sillman, Logan, and Wofsy, 1990; Sillman and Samson, 1989). The model results show excellent agreement with observed ozone concentrations, at both urban and rural locations (Sillman and Samson, 1989). Simulations for scenarios involving the use of vehicles fueled by a mix of 85 percent methanol and 15 percent gasoline (M85), and vehicles fueled by 100 percent methanol, along with changes in VOC and NO_x emissions from other sources, were carried out. Results indicate the probable changes in air quality, both in a specific urban area, Detroit, and throughout the rural Midwest, that might accrue as a result of changing emissions rates.

The Dual Nature of the Ozone Problem

Ozone is most commonly regarded as an urban problem associated with smog formation in major urban centers, such as Los Angeles. However, there has been increasing evidence during the last 15 years that ozone concentrations in rural locations throughout eastern North America are affected by anthropogenic emissions. Ozone concentrations of 0.08 ppm or higher have been observed over regions extending 1,000 miles or more (Samson and Ragland, 1977; Hov, Hesstvedt, and Isaksen, 1978; Wolff and Lioy, 1980; Logan, 1989). See Figure 1. Although rarely in violation of the National Ambient Air Quality Standard (NAAQS) for ozone, this regional "background" ozone is high enough to contribute significantly to violations of air quality standards in downwind urban centers (Sillman and Samson, 1989). In addition, rural ozone concentrations of 0.08 ppm have been identified as the cause of significant damage to crops and forests in the eastern United States

Figure 1
Observed Ozone Concentrations at Sites in
Eastern North America on July 16, 1977[1]

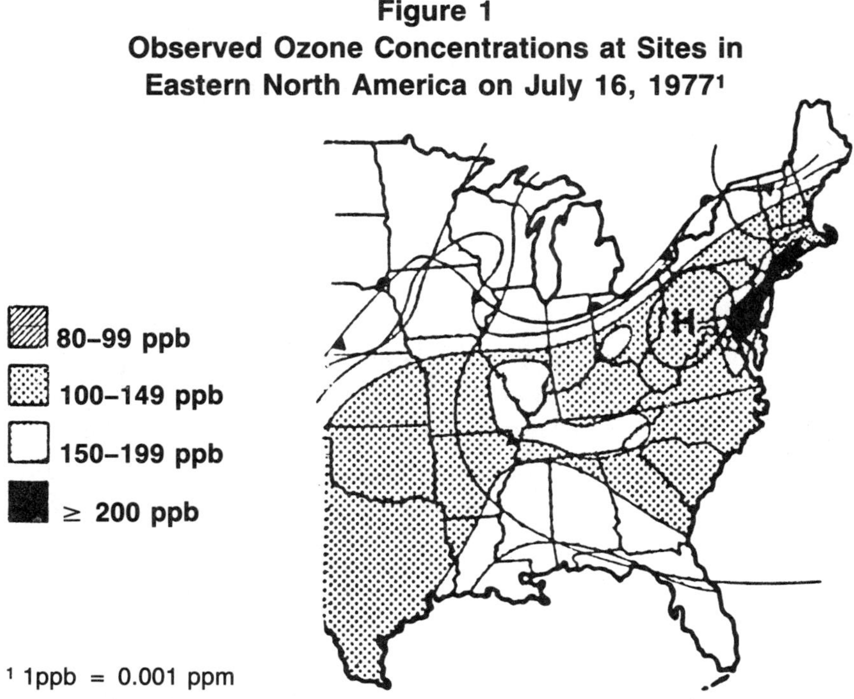

[1] 1ppb = 0.001 ppm

Source: G.T. Wolff and P.J. Lioy, "Development of an Ozone River Associated with Sinoptic Scale Episodes in the Estern United States," *Environmental Science and Technology,* 14

(Heck et al., 1982; Skarby and Selden, 1984; World Meteorological Organization, 1985). Wind speeds during these periods of regional ozone concentrations are lighter than average but strong enough to suggest that regional-scale transport may be a significant component (Samson and Ragland, 1977).

The ozone problem should therefore be thought of as a combination of two distinct problems, one involving urban-scale emissions and photochemistry, and the other involving regional-scale processes and large-scale dispersal. Although the urban and regional dimensions are interrelated—background regional ozone contributes to air quality violations in downwind urban centers, and exported urban ozone contributes to the regional background—it would be erroneous to assume that the sources of urban and rural ozone are identical. In particular, urban VOC and NO_x emissions may have a very different effect on ozone at far downwind locations than on ozone near the emissions source (Sillman, Logan, and Wofsy, 1990; Milford, Russell, and McRae, 1989). A control strategy that may be effective in reducing ozone concentrations near an urban center may have little or no impact on downwind ozone.

Simulation Methods

Analysis of the effect of methanol was based on a case study of an ozone episode in the Midwest and focused on air quality in southeastern Michigan. Violations of air quality standards were recorded downwind of Detroit on August 1 and 2, 1988, with the highest ozone concentration (0.144 ppm) observed at New Haven, Michigan, on August 2. There was also evidence of large-scale regional dispersal of ozone into southeastern Michigan during this period. Peak ozone concentrations at Saline, a rural location 40 miles southwest of Detroit, were 0.080 to 0.085 ppm, despite the lack of local emissions sources. Predominantly southwest winds precluded the possibility of transport of ozone to Saline from the Detroit area. Ozone concentrations of 0.07 to 0.08 ppm were also observed at rural sites in Indiana, Ohio, and Kentucky.

Meteorological conditions during the period were marked by clear skies, southwesterly winds, and extremely high temperatures (greater than 95 degrees F) throughout the Midwest. A back-trajectory analysis (Heffter, 1980) indicated that the air in southeastern Michigan on August 2 had originated in Texas on July 29, arrived in Missouri on July 30, and moved slowly across the Midwest from July 30 to August 2. Surface winds in Detroit on the morning of August 2 were extremely light (less than 5 mph), increasing to 10 to 15 mph during the afternoon. Vertical temperature profiles at Flint indicated that the mixed layer height was 1,850 meters on August 2.

The dynamical and photochemical processes that lead to increased ozone were simulated using a regional-scale model (the Plumes Model for Regional Air Quality, developed at Harvard University [Sillman, Logan, and Wofsy, 1990]) in combination with an urban-scale model developed at the University of Michigan. Both models used the photochemical mechanism developed by Lurmann, Lloyd, and Atkinson (1986) with photolysis rates from Logan et al. (1981). The photolysis rate calculation assumed an aerosol optical depth of 0.68 at 310 nm, based on turbidity data from the eastern United States (Flowers, McCormick, and Kurfis, 1969) and absorption cross sections from DeMore et al. (1985). Horizontal advection was by second-order moments (Prather, 1986) in the regional component and by the method of Smolarskiewicz (1983) in the urban component. The model included two vertical layers, a time-varying mixed layer, and an entrainment layer, with the diurnal variation of the mixed layer derived from van Ulden and Holtslag (1985). Deposition velocities were as follows: HNO_3, 2.5 cm/sec, O_3 and NO_2, 0.6 cm/sec; NO, 0.1 cm/sec; H_2O_2 and organic peroxides, 1.0 cm/sec; and PAN (peroxyacetyl nitrate), 0.25 cm/sec. Initial and boundary conditions were 0.040 ppm O_3, 0.005 ppmC VOC and 0.0002 ppm NO_x.

The combined model was exercised for a four-day period (July 30 to August 2) for a region bounded roughly by Detroit, Michigan; Atlanta, Georgia; Dallas, Texas; and Omaha, Nebraska. See Figure 2.

Figure 2
Domain for the Combined Urban/Regional Simulation

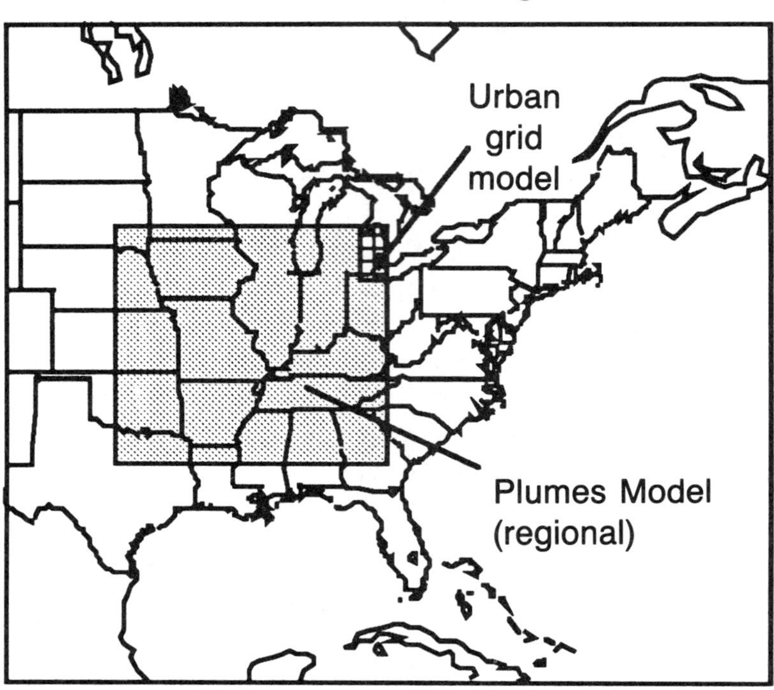

Source: Sillman and Samson, 1990

The urban component covered the corridor from Detroit to Port Huron. The extent of the urban and regional model domains included all emission sources likely to effect southeast Michigan on August 2. The resulting simulation showed excellent agreement with ozone concentrations observed at monitors throughout the Detroit area and at rural sites throughout the Midwest (Sillman and Samson, 1989). The simulation predicts a peak ozone concentration of 0.148 ppm at New Haven on August 2; the observed peak ozone at New Haven was 0.144 ppm. On August 1 the model predicted 0.136 ppm at Port Huron; the observed peak ozone at Port Huron was 0.126 ppm. The predicted rural background ozone was 0.078 ppm; observed peak ozone at upwind locations was 0.085 ppm at Saline and 0.080 ppm at Ann Arbor.

Emissions Scenarios

Anthropogenic emissions for the original simulation were derived from the National Acid Precipitation Assessment Program 5.2 inventory (1980) (EPA, 1986). This inventory has a VOC/NO$_x$ ratio of 6:1 for the city of Detroit, with strong geographical variations within the city; the 6:1 ratio appears broadly consistent with observed 6:00 to 9:00 A.M. VOC/NO$_x$ ratios (Sillman and Samson, 1989). Emissions scenarios were developed to represent 100 percent substitution of either pure methanol (M100) or a methanol-gasoline mix (M85) for gasoline in all urbanized areas (with more than 100,000 metropolitan area population). Vehicles in smaller towns and rural locations were assumed to run on gasoline.

Emissions rates for M85 and M100 vehicles were derived from recent EPA estimates (EPA, 1987). M85 vehicles are estimated to emit 19 percent less total VOC than conventional gasoline-fueled vehicles, and more than half of the M85 VOC emissions consist of methanol. Emissions of most other VOC species are reduced by 58 percent relative to conventional vehicles. However, M85 vehicles have a higher emissions rate for formaldehyde, equal to 3 percent of total VOC emissions. M100 vehicles emit 79 percent less total VOC than conventional vehicles, nearly all of it methanol. Emissions of VOC species other than methanol are reduced by 97 percent. The emissions scenarios assume that non-vehicular activities are unaffected by the switch to methanol. M85 and M100 emissions rates are substituted for 25 percent of total VOC emissions, based on the estimated fraction of VOC emissions attributed to light-duty vehicles (Chang et al., 1989).

In addition to methanol substitution, scenarios have been selected to reflect potential changes in future emissions of VOC and NO$_x$ from other sources. Estimates for future emissions suggest that NO$_x$ emissions could increase by 20 percent or more during the next 15 years (U.S. Congress, 1989). Emissions controls and energy conservation, however, could cause emissions rates for both VOC and NO$_x$ to decline. The authors have analyzed scenarios that include VOC reductions of as much as 40 percent from sources other than methanol-fueled vehicles, and changes in NO_x emissions ranging from a 40 percent reduction to a 20 percent increase. In the scenarios with reduced VOC emissions, the reduction has been applied to gasoline-fueled vehicles but not to methanol-fueled vehicles. However, in scenarios with changed NO_x emissions, the changed emissions rate has also been applied to methanol-fueled vehicles, as the actual NO_x emissions rate from those vehicles is uncertain (U.S. Congress, 1989).

Results of both urban and regional simulations are critically dependent on the assumed rate of emissions for biogenic hydrocarbons (isoprene) (Chameides et al., 1988; Trainer et al., 1987). Emissions rates for isoprene were derived from estimates by Lamb et al. (1985) with land-use data from Matthews (1983), in both the urban and regional simulations. Isoprene emissions in the Detroit-Port Huron corridor were derived by assuming 20 percent forest cover. The resulting emissions rate for isoprene was 3.15×10^{11} molecules cm²/sec (daytime average), a rate that reflects the abnormally warm temperatures during the period. As reported elsewhere (Sillman and Samson, 1989), the inclusion of biogenic emissions has a significant effect on predicted peak ozone in the Detroit area, but less effect on the predicted relation between ozone and precursors.

Results

Changed ozone concentrations, resulting from the use of methanol fuel in the original simulation, and VOC reduction scenarios are shown in Figure 3. The use of M85 vehicles was predicted to have a negligible effect (less than 2 percent) on peak ozone in the Detroit area case study. In contrast, M100 vehicles caused a reduction in peak ozone of 0.008 ppm (5 percent). These reductions were broadly consistent with the reductions attributed to M85 and M100 vehicles by Chang et al. (1989). The improvement gained from a switch to M100 vehicles has an impact on ozone equivalent to a 20 percent reduction in emissions of VOC from all sources. The gain

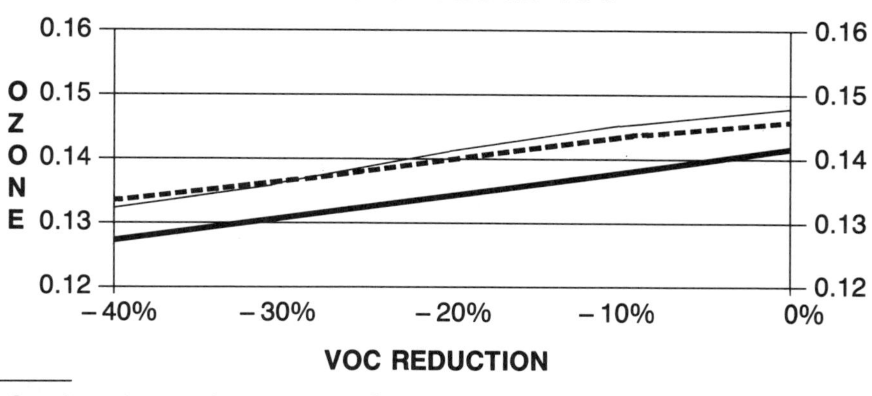

Figure 3
Predicted Peak Ozone in ppm in the Detroit-Port Huron Metropolitan Area in Response to Changed Emissions Rates for VOC

Continued use of gasoline-fueled vehicles
---- Switch to M85 vehicles
——— Switch to M100 vehicles

Source: Sillman and Samson, 1990

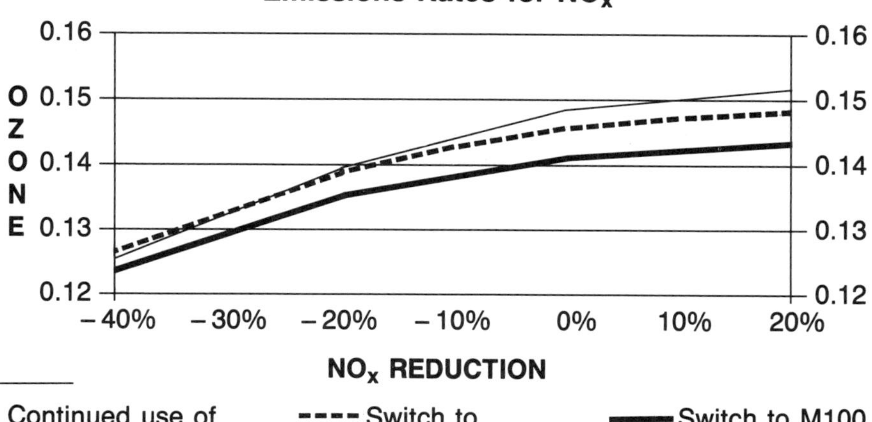

Figure 4
Predicted Peak Ozone in ppm in the Detroit-Port Huron Metropolitan Area in Response to Changed Emissions Rates for NO_x

Continued use of gasoline-fueled vehicles
---- Switch to M85 vehicles
———Switch to M100 vehicles

Source: Sillman and Samson, 1990

from a switch to M85 vehicles is only equivalent to a 7 percent reduction in overall VOC emissions, or to a 7 percent reduction in NO_x emissions. See Figure 4. The impact of the M85 scenario is also equivalent to the impact of gasoline-powered vehicles with a 33 percent reduction in VOC emissions, as evidenced by the crossover point in the original and M85 scenarios in Figure 3.

Simulations with changed NO_x emissions show that reduced NO_x, like reduced VOC, should lead to a reduction in peak ozone concentrations. A drastic (40 percent) reduction in NO_x has an even bigger effect than a similar reduction in VOC emissions. Methanol's capacity to achieve a major improvement in air quality is less if reductions in NO_x emissions are also made.

The behavior of rural "background" ozone in response to the various emission scenarios is shown in Figures 5 and 6. The curves representing the original, M85 and M100 scenarios in figures 5 and 6 are barely distinguishable from each other, because substitution of either M85 or M100 vehicles is found to have virtually no effect on rural ozone (or regional transport). Figure 5 also shows that anthropogenic VOC emissions from all sources have little effect on rural ozone (or regional transport). By contrast, changes in NO_x emission rates (Figure 6) have a major rural/regional impact. Prior results have shown that peak ozone in the Detroit area is also sensitive to NO_x emissions from distant upwind sources, which contribute to background ozone in southeastern Michigan.

The failure of VOC controls to reduce background ozone and the large impact of NO_x on background ozone can be attributed to two factors. First, the chemistry that governs production of ozone at downwind locations is different from the chemistry near urban centers (Sillman, Logan, and Wofsy, 1990). VOC/NO_x ratios are higher (Milford, Russell, and McRae, 1989; Fahey et al., 1986), and ozone formation is increasingly NO_x-limited, as one moves away from emissions sources (Fahey et al., 1986). Second, biogenic VOC emissions may equal or exceed anthropogenic sources on warm summer days over a region that includes extensive rural areas (Sillman, Logan, and Wofsy, 1990; Trainer et al., 1987).

Figure 5
Predicted Rural Ozone in ppm in Southeast Michigan in Response to Changed Emissions Rates for VOC

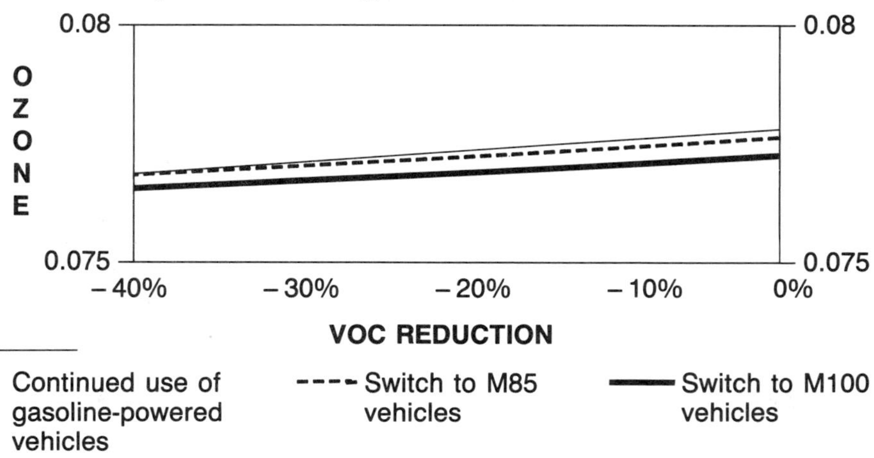

——— Continued use of gasoline-powered vehicles

---- Switch to M85 vehicles

——— Switch to M100 vehicles

Source: Sillman and Samson, 1990

Figure 6
Predicted Rural Ozone in ppm in Southeast Michigan in Response to Changed Emissions Rates for NO$_x$

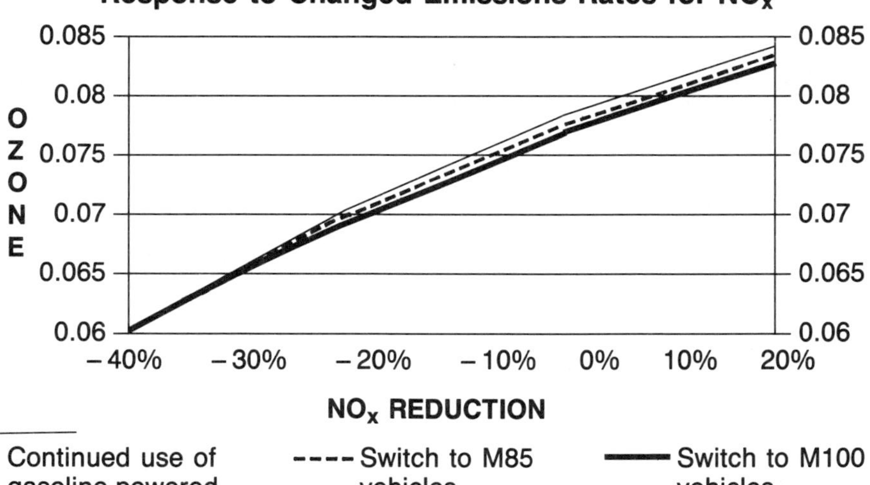

——— Continued use of gasoline-powered vehicles

---- Switch to M85 vehicles

——— Switch to M100 vehicles

Source: Sillman and Samson, 1990

Figure 7
Predicted Peak Concentration of PAN in ppm in the Detroit-Port Huron Metropolitan Area in Response to Changed Emissions Rates for VOC

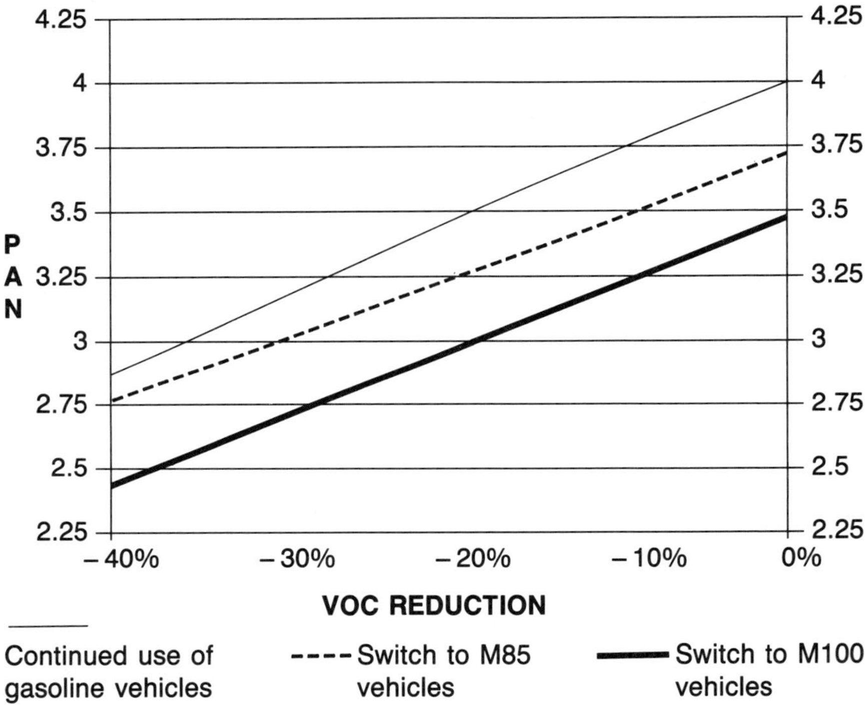

Source: Sillman and Samson, 1990

The effect of methanol fuel on peroxyacetyl nitrate is predicted to be greater than the effect on ozone. See Figure 7. PAN is predicted to reach 4 ppb (1ppb = 0.001ppm) in the Detroit area, and a switch to M85 and M100 vehicles causes reductions of 7 and 14 percent, respectively. M85 vehicles have a significant effect on PAN, because the methanol and formaldehyde emitted by M85 vehicles cannot cause PAN formation. Predicted peak methanol concentrations during morning rush hour are predicted to be 0.050 ppm in the M85 scenario, compared with a total VOC burden of 0.600 ppmC. Peak methanol in the M100 scenario is 0.020 ppm. Rural methanol concentrations are expected to be 0.001 ppm in the M85 scenario.

Discussion

A regional-scale simulation has been used to investigate potential effects of methanol-fueled vehicles on both urban and rural ozone in the Midwest. The model combined a detailed urban representation for an urban area (Detroit) with a regional simulation that accounted for precursor emissions from as far away as Texas. The combined model thus represents both the urban and regional components of the dual ozone problem in eastern North America.

Results suggest that a switch to M85 fuel will have little effect on ozone, even in urban areas. A switch to M100 has some effect (minus 5 percent) on urban ozone. A switch to either M85 or M100 vehicles has no effect on rural ozone concentrations or on regional dispersion, both of which are sensitive to emissions of NO_x rather than VOC. To the extent that methanol-fueled vehicles are associated with either increases or decreases in NO_x emissions, they may have a significant effect on rural ozone.

The poor results obtained with the M85 scenario here (a 1 percent reduction in peak urban ozone) contrasts with the 9 percent reduction predicted by Russell in the Los Angeles case study. The difference between the two M85 scenarios is apparently due to differing assumptions about M85 emissions. The emissions rate from M85 vehicles, used in this case study, would have an ozone formation potential approximately 70 percent that of current gasoline-fueled vehicles. In contrast, the ozone formation potential of Russell's M85 emissions (possibly reflecting a more recent estimate than that used here) is minus 33 percent, compared with that of gasoline. However, the M100 emissions rate, in this case study, has an ozone potential equal to 20 percent, compared with gasoline, and therefore provides a "maximum benefit" scenario.

The Los Angeles case study presented by Russell, and the Michigan case study presented here, represent two very different geographical and meteorological situations. The Los Angeles case represents an urban location under a severe thermal inversion with the highest ozone of the decade (0.350 ppm, nearly three times the NAAQ standard). In this situation VOC controls have their greatest effect. The Detroit analysis represents a more "typical" deterioration in air quality with lower ozone (0.144 ppm), lower precursor concentrations, and peak ozone occurring in a downwind suburb. The effect of VOC reductions (including reductions due to the use of methanol) is smaller in these cases. Finally, the rural situation in southeast Michigan represents low ozone (though still higher than "natural" background ozone levels) at locations far removed from precursor sources. Here, VOC effects are negligible and ozone concentrations are most sensitive to NO_x. The general effect of VOC controls is a shift in photochemical production of ozone away from the urban center and toward downwind locations. Consequently, VOC controls have little effect on rural sites or on regional-scale transport of ozone.

In summary, the effect of methanol vehicles on ozone appears to be 9 to 15 percent in the "Los Angeles ozone emergency" scenario, 5 percent in the "Detroit NAAQS violation" scenario, and zero in the "rural background/regional dispersal" scenario. From the viewpoint of regional-scale air quality, it is important that efforts to promote alternative fuels are not allowed to replace the need for tighter NO_x emissions standards.

References

Chameides, W.L., et al. (1988). The Role of Biogenic Hydrocarbons in Urban Photochemical Smog: Atlanta as a Case Study. *Science*, 241.

Chang, T.Y., et al. (1989). Impact of Methanol Vehicles on Ozone Air Quality. *Atmospheric Environment*, 23.

DeMore, W.B., et al. (1985). Chemical Kinetics and Photochemical Data for Use in Stratospheric Modeling. Jet Propulsion Laboratory Publication 85-37, California Institute of Technology, Pasadena, CA.

EPA (1986). Development of the 1980 NAPAP Emissions Inventory. U.S. Environmental Protection Agency, Report No. EPA-600/7-86-057a, Research Triangle Park, NC.

EPA (1987). Air Quality Benefits of Alternative Fuel. Prepared for the Alternative Fuels Working Group of the President's Task Force on Regulatory Relief, U.S. Environmental Protection Agency, Office of Mobile Sources, Ann Arbor, MI.

Fahey, D.W., et al. (1986). Reactive Nitrogen Species in the Troposphere: Measurements of NO, NO_2, HNO_3, Particulate Nitrate, Peroxyacetylnitrate (PAN), O_3 and Total Reactive Nitrogen at Niwot Ridge, Colorado. *Journal of Geophysical Research*, 91.

Flowers, E.C., R.A. McCormick, and K.R. Kurfis (1969). Atmospheric Turbidity Over the United States, 1961–1966. *Journal of Applied Meteorology*, 8.

Gray, C.L., and J.A. Alson (1986). *Moving America to Methanol: A Plan to Replace Oil Imports, Reduce Acid Rain and Revitalize Our Domestic Economy.* Ann Arbor, MI: University of Michigan Press.

Heck, W.W., et al. (1982). Assessment of Crop Loss from Ozone. *JAPCA*, 32.

Heffter, J.L. (1980). Air Resources Laboratories Atmospheric Transport and Dispersion Model (ARL-ATAD). NOAA Tech. Memo. ERL-ARL-81, Air Resources Laboratories, Silver Spring, MD.

Hov, O., E. Hesstvedt, and I.S.A. Isaksen (1978). Long-Range Transport of Tropospheric Ozone. *Nature*, 273.

Lamb, B., et al. (1985). Biogenic Hydrocarbon Emissions from Deciduous and Coniferous Trees in the United States. *Journal of Geophysical Research*, 90.

Logan, J.A. (1989). Ozone in Rural Areas of the United States. *Journal of Geophysical Research*, 94.

Logan, J.A., et al. (1981). Tropospheric Chemistry: A Global Perspective. *Journal of Geophysical Research*, 86.

Lurmann, F.W., A.C. Lloyd, and R. Atkinson (1986). A Chemical Mechanism for Use in Long-Range Transport/Acid Deposition Computer Modeling. *Journal of Geophysical Research*, 91.

Matthews, E. (1983). Global Vegetation and Land Use: New High-Resolution Data Bases for Climate Studies. *Journal of Climate and Applied Meteorology*, 22.

Milford, J., G.J. McRae, and A.G. Russell (1989). A New Approach to Photochemical Pollution Control: Implications of Spatial Patterns in Pollutant Responses to Reductions in Nitrogen Oxides and Reactive Organic Gas Emissions. *Environmental Science and Technology*, 23.

Prather, M.J. (1986). Numerical Advection by Conservation of Second-Order Moments. *Journal of Geophysical Research*, 91.

Samson, P.J., and K. Ragland (1977). Ozone and Visibility Reduction in the Midwest: Evidence for Large-Scale Transport. *Journal of Applied Meteorology*, 16.

Sillman, S., and P.J. Samson (1989). A Regional-Scale Analysis of the Sources of Elevated Ozone Concentrations in Southeastern Michigan. Submitted to *JAPCA*.

Sillman, S., J.A. Logan, and S.C. Wofsy (1990). A Regional-Scale Model for Ozone in the United States with a Sub-Grid Representation of Urban and Power Plant Plumes. *Journal of Geophysical Research*, in press.

Sillman, S, J.A. Logan, and S.C. Wofsy (1990). The Sensitivity of Ozone to Nitrogen Oxides and Hydrocarbons in Regional Ozone Episodes. *Journal of Geophysical Research*, in press.

Skarby, L., and G. Selden (1984). The Effects of Ozone on Crops and Forests. *Ambio*, 13.

Smolarskiewicz, P.K. (1983). A Simple Positive Definite Advection Scheme with Small Implicit Diffusion. *Monthly Weather Review*, 111.

Southeast Michigan Ozone Modeling Committee (1989). Assessment of Ozone Modeling Requirements for the State of Michigan. Report to Michigan Department of Natural Resources, Lansing, MI.

Trainer, M., et al. (1987). Models and Observations of the Impact of Natural Hydrocarbons on Rural Ozone. *Nature*, 329.

U.S. Congress (1989). *Catching our Breath: Next Steps for Reducing Urban Ozone*. Washington, DC: Office of Technology Assessment.

Van Ulden, A.P., and A.A.M. Holtslag (1985). Estimation of Atmospheric Boundary-Layer Parameters for Diffusion Applications. *Journal of Climate and Applied Meteorology*, 24.

Wolff, G.T., and P.J. Lioy (1980). Development of an Ozone River Associated with Synoptic Scale Episodes in the Eastern United States. *Environmental Science and Technology*, 14.

World Meteorological Organization (1985). Atmospheric Ozone 1985: Global Ozone Research and Monitoring Project Report #16. Washington, DC: NASA.

Commentary

ALAN C. LLOYD

D r. Chang and Dr. Russell have presented the results of two excellent studies, which use different modeling approaches to estimate the value of methanol-fueled vehicles as a strategy to improve urban air quality. It should be stressed that while models are necessary to demonstrate how motor vehicle emissions affect air quality, their usefulness depends on the quality of the input data and the assumptions used in the models. Clearly there are still significant uncertainties not only in our data base for alternative fuels but also about the durability of improved emissions-control technologies for gasoline-fueled cars and technologies for vehicles using alternative fuels.

Dr. Russell's study is based on an urban airshed model, previously used in the Los Angeles area, to assess the impact of methanol vehicles on air quality in the Los Angeles area in the year 2010. Dr. Chang's study uses a trajectory model, which concentrates on the nine cities mentioned in the Bush administration's alternative-fuels plan for the years 2010 and 2015. Both models indicate that the use of methanol improves air quality by reducing peak ozone concentrations. However, the models differ in their prediction of how much improvement in air quality would result from the use of methanol. Because the models differ in their approaches, it is difficult to compare their results.

Russell's study of Los Angeles shows that there is a decrease of as much as 9 percent for M85, and a decrease in ozone exposure of at least 19 percent, in addition to which air toxics are significantly decreased. This study by Dr. Russell uses a more realistic assumption for aldehydes (30 mg/mi), compared to his earlier one, but the study is limited, since it examines only one meteorological scenario.

Chang's results are difficult to apply to particular areas, since the use of city averages has very limited usefulness for planning purposes. This study shows that the use of methanol decreases peak ozone from 1 to 5 percent, but no data are presented for ozone exposure or toxic air contaminants. Chang emphasizes the importance of keeping formaldehyde emissions as low as possible, and that the hydrocarbon-to-nitrogen-oxides ratio has a critical effect on the ozone benefits. These are important points and should be emphasized in the national debate. In California, a strict formaldehyde emission standard of 15 mg/mi has already been enacted. This will minimize the effect of formaldehyde emissions on ozone formation and reduce the "hot spot" build-up of formaldehyde in garages, especially those underground. The South Coast Air Quality Management District is carrying out further research on formaldehyde build-up, the results of which will be available in late 1990.

Since the use of methanol fuel is a strategy to deal with hydrocarbon reactivity, it will not be effective in areas of higher hydrocarbon-(HC)-to-nitrogen-oxides-(NO_x) ratios, such as Houston. Nor can this strategy control rural ozone concentrations, where the HC-to-NO_x ratios are much higher, and where it is much more beneficial to control NO_x emissions. In areas where HC-to-NO_x ratios are higher, and where air toxics need to be controlled, an alternative-fuel strategy may be effective, but this should be evaluated on a case-by-case basis.

Both papers indicate that methanol will improve air quality, but one cannot assume from a nine-city average that a particular city will benefit. More extensive studies need to be carried out for each of the cities covered by Chang, since a more accurate understanding of the effect of methanol on both peak ozone and ozone exposure is needed. It is also important to recognize that in Los Angeles, where reducing ambient ozone has proved to be so difficult, even a few points' reduction in peak ozone can be significant.

Commentary

JANA B. MILFORD

The papers by Russell and by Chang and Rudy represent two different dimensions of the current ability to assess the effects of alternative fuels on air quality: depth versus breadth. Russell's modeling analysis examines in detail the effect of M85 fuel use on air quality in the Los Angeles area, under the meteorological conditions that existed for a three-day period in the summer of 1982. The three-dimensional Eulerian grid model used by Russell performs well, compared with detailed atmospheric concentration data for ozone and nitrogenous species (Russell, McCue, and Cass, 1988). The emissions inventory and meteorological inputs available for the Los Angeles area are more refined than those available anywhere else in the country. Apart from currently unavoidable uncertainty about emissions rates from future vehicles fueled by gasoline and methanol, the key limitation of Russell's work is its specificity to the area and meteorological conditions modeled. In contrast, Chang and Rudy examined the effect of methanol fuel use in the nine U.S. cities with the highest peak ozone levels. However, they used a simplified modeling approach that incorporates default values or approximations for many of the required inputs.

Ideally, any consideration of a federally mandated drive for alternative fuels would be based on data of both breadth and depth; that is, Los Angeles-level modeling efforts would be carried out for all of the cities involved. The best that can be done now, however, is to try to draw conclusions from detailed results for the Los Angeles area and comparatively rough results from other cities, bearing in mind that projections of emissions rates from future vehicles are uncertain.

Russell examined the effect of introducing 300,000 M85-fueled vehicles per year in the Los Angeles area, starting in 1997, on air quality in the year 2010. At this rate, by 2010, about half of the light-duty vehicles in the region would be fueled with M85. Organic emissions from M85-fueled vehicles were assumed to be equivalent to gasoline-fueled vehicle emissions in terms of carbon mass. One-third of the M85 emissions were assumed to have the same composition as gasoline-fueled vehicle emissions, with the remaining carbon emitted as methanol. Formaldehyde emissions from the M85 vehicles were assumed to be 30 mg/mi.

For the three-day period modeled, M85 use was predicted to reduce peak ozone by about 9 percent and to reduce exposure to concentrations above the standard by about 20 percent. Russell compared the results of fuel substitution with a case in which reactive organic gas (ROG) emissions from the affected fraction of vehicles were completely eliminated. Based on the peak ozone reduction of the two cases, the "reactivity" of M85-fueled vehicle emissions was roughly one-third that of emissions from conventional vehicles.

Chang and Rudy examined the effect of M85 use in all light-duty vehicles in nine cities, considering emissions projections for the year 2015. The calculations were performed with a simple, single-cell model. Initial concentration and emissions levels were based on morning concentrations of reactive organic gas and nitrogen oxides (NO_x) observed in the nine cities, but default meteorological conditions were used. Chang and Rudy assumed that M85 emissions were about one-half methanol and one-half emissions with the same composition as emissions from gasoline-fueled vehicles. They used a formaldehyde emissions rate of 35 mg/mi.

Like Russell, Chang and Rudy also examined a case in which the carbon mass in M85-fueled vehicle emissions was assumed to be equivalent to that of gasoline-fueled vehicle emissions. They found that full substitution of M85 in light-duty vehicles would yield a nine-city average reduction in peak ozone of 0.6 percent. With equivalent masses of carbon emitted, Chang and Rudy found that the reactivity of M85-fueled vehicle emissions ranged from about

80 to 100 percent of the reactivity of gasoline-fueled vehicle emissions.

Obviously, major differences exist between Russell's and Chang and Rudy's studies. Nevertheless, it is not obvious why they obtained such different estimates of the relative reactivity of emissions from M85-fueled vehicles, compared with those associated with gasoline use. Several factors might be contributing to the discrepancy: First, Chang and Rudy used a higher fraction of non-methane hydrocarbons emissions and higher formaldehyde emissions from M85-fueled vehicles. Second, the ROG-to-NO_x ratio of the inputs to Russell's simulations was lower than the ratios of the inputs used by Chang and Rudy. Third, differences might arise from artifacts of the methodology used in each study; for example, the model used by Chang and Rudy is very sensitive to initial conditions. Finally, real differences exist between chemical and meteorological conditions in the Los Angeles area and those in the other cities examined by Chang and Rudy.

In recent tests of low-mileage, flexible-fuel vehicles operated on M85 over the Federal Test Procedure driving cycle, on a moles-carbon basis, exhaust plus evaporative emissions of non-methane organic compounds ranged from 45 to 62 percent methanol, 6 to 9 percent aldehydes, and 30 to 50 percent non-methane hydrocarbons (CARB, 1989; Gabele, 1989; Snow et al., 1989; Williams et al., 1989). Russell's non-methane hydrocarbon fraction falls below the range of the test results; Chang and Rudy's falls at the high end. The formaldehyde fractions observed in the tests are about twice as high as the fractions used in the papers discussed here.

Figure 2 in Chang and Rudy's paper shows that, compared with emissions from gasoline-use, the predicted relative reactivity of M85 emissions increases with the ROG/NO_x ratio of the inputs. It is important to note that the ROG/NO_x ratios in Figure 2 refer to the ratio of total inputs, that is, emissions plus initial conditions. To set the initial conditions for each city in their study, Chang and Rudy used median values of 6 to 9 A.M. average ROG and NO_x concentration measurements reported by Baugues (1986). During the course of the day, emissions were assumed to contribute the same mass of ROG and NO_x as was present in the initial conditions. However, ROG and NO_x emissions inventories generally indicate that the ratios of these emissions in many cities are lower by a factor of two or more, than the concentration ratios reported by Baugues (NEDS, 1988; NAPAP, 1988). For the Los Angeles area, the ratio of ROG/NO_x emissions used by Russell was less than 4:1, below the range considered by Chang and Rudy (Russell et al., 1989). Extrapolating the trend in Chang and Rudy's Figure 2 suggests that the comparatively large benefit predicted by Russell for M85 is partially explained by the low ratio of ROG/NO_x emissions in the base 2010 inventory for the Los Angeles area.

The foregoing factors suggest that as estimates of the reactivity of emissions from an M85-fueled vehicle versus an equivalent amount of carbon from a gasoline-fueled vehicle, Russell's value of one-third is too low, and Chang and Rudy's range of 80 to 100 percent is somewhat high. However, the reactivity benefit of emissions from M85-fueled vehicles versus gasoline-fueled vehicles is very sensitive to several factors that are currently uncertain, such as the level to which formaldehyde emissions can be controlled.

Ozone concentrations depend on emissions of nitrogen oxides as well as reactive organic gases. A number of recent studies have indicated that in some places and on some days ozone production is limited by NO_x levels, whereas in other areas or at other times ROG are the limiting precursor (Sillman, 1987; Trainer et al., 1987; Chamiedes et al., 1988; Milford, Russell, and McRae, 1989). Although most efforts to reduce ozone have focused on controlling emissions of ROG, some areas may have to reduce NO_x emissions to meet the ozone standard. Russell and Chang and Rudy assumed that NO_x emission rates for methanol-fueled vehicles would be equal to those of gasoline-fueled vehicles. Emissions tests of methanol-fueled vehicles as well as theoretical considerations suggest that this assumption is reasonable (EPA, 1988).

However, a potential trade-off may exist between lowering NO_x emissions versus improving the fuel economy of methanol vehicles. One approach to improving fuel economy is to operate a motor vehicle "lean," that is, with a high air-to-fuel ratio. Higher air-to-fuel ratios can be used with methanol than with gasoline, because methanol's properties allow stable combustion across a wider range of ratios. Unfortunately, lean operation interferes with the use of a three-way reduction catalyst to lower NO_x emissions in vehicle exhaust, and thus may not be compatible with a NO_x standard of 0.4 g/mi or lower (DeLuchi, Johnston and Sperling, 1988).

Methanol reacts slowly and, therefore, has a long atmospheric lifetime, compared with most of the organic compounds emitted from gasoline-fueled vehicles. It is the low reactivity of methanol that is its beneficial property, from the standpoint of efforts to reduce urban ozone. However, as Russell points out in his paper, methanol's low reactivity has also raised concerns about the possibility that methanol could build up during multi-day smog episodes, and contribute to ozone formation downwind of the city where it was used.

Sillman and Samson analyzed the regional-scale effects of methanol use in Midwest cities for meteorological conditions from a four-day episode of elevated ozone concentrations in August 1988. Their modeling study indicates that substituting methanol (either M85 or M100) for gasoline would have a negligible effect on ozone concentrations outside the urban areas where it is used. In areas outside midwestern cities, according to their model, ozone formation is generally limited by the amount of existing nitrogen oxides and is, therefore, insensitive to small changes in the amount or composition of volatile organic compound emissions.

Carter et al. (1986) conducted three-day smog chamber experiments with methanol and observed a similar result: methanol substitution reduced ozone on the first day, but the benefit declined after that, until on the third day the effect was negligible. As in the areas outside cities in Sillman and Samson's model, ozone formation was apparently NO_x-limited on the second and third days of the smog chamber experiments.

The Carter et al. results and the Sillman and Samson study leave open the question of what effect methanol use would have on air quality in downwind areas that have excessive amounts of nitrogen oxides. This question is very important because use of methanol has been proposed for such cities as New York and Philadelphia, which are upwind of other urban and industrialized areas.

Sillman and Samson's results for the Detroit area appear to be comparable to the nine-city average results presented by Chang and Rudy. As shown in their Figure 3, switching to M85 is equivalent to a 6 to 7 percent reduction in total VOC emissions or a 25 to 30 percent reduction in mobile source emissions. Much of their predicted reduction with M85 is due to a decrease in the mass of carbon emitted per mile, as Sillman and Samson's M85 vehicles emitted about 20 percent less carbon per mile than the gasoline-fueled vehicles they replaced. Based on this commentator's calculations in comparing the reactivity of equal amounts of VOC from gasoline-fueled vehicles and from M85-fueled vehicles, the M85 emissions composition used by Sillman and Samson is about 90 percent as reactive as the gasoline emissions composition they used. Their result is thus comparable to the relative ozone-forming potential calculated by Chang and Rudy, and significantly higher than Russell's result for Los Angeles. Sillman and Samson's M85 emissions composition was similar to that used by Chang and Rudy; and the ratio of volatile organic compound emissions versus nitrogen oxide emissions in the Detroit area was relatively high, at 6:1. Based on a 6:1 ratio, Figure 2 in Chang and Rudy's paper shows a relative ozone-forming potential of 0.85. Sillman and Samson note that their M85 emissions composition differs significantly from the composition used by Russell. Moreover, emissions and meteorology in Los Angeles are very different from those modeled for Detroit.

References

Baugues, K. (1986). A Review of NMOC, Nitrogen Oxide and NMOC/Nitrogen Oxide Ratios Measured in 1984 and 1985. U.S. Environmental Protection Agency, EPA Report No. 450/4-86-015, Research Triangle Park, N.C.

CARB (1989). Definition of a Low-Emission Motor Vehicle in Compliance with the Mandates of Health and Safety Code Section 39037.05. California Air Resources Board, Mobile Source Division.

Carter, W. P. L., et al. (1986). Effects of Methanol Fuel Substitution on Multiday Air Pollution Episodes. California Air Resources Board, Sacramento, Calif.

Chameides, W. L., et al. (1988). *Science.* 241.

DeLuchi, M. A., R. A. Johnston, and D. Sperling (1988). Methanol vs. Natural Gas Vehicles: A Comparison of Resource Supply, Performance, Emissions, Fuel Storage, Safety, Costs and Transitions. SAE Technical Paper Series 881656, presented at the International Fuels and Lubricants Meeting and Exposition, Portland, Ore., October 10-13, 1988.

EPA (1988). Guidance on Estimating Motor Vehicle Emissions Reductions from the Use of Alternative Fuels and Fuel Blends. U.S. Environmental Protection Agency, Report No. EPA-AA-TSS-PA-87-4, Office of Mobile Sources, Ann Arbor, Mich.

Gabele, P. A. (1989). Characterization of Emissions From a Variable Gasoline/Methanol Fueled Car (personal communication). U.S. Environmental Protection Agency, Research Triangle Park, N.C.

Milford, J. B., A. G. Russell, and G. J. McRae (1989). *Environmental Science and Technology*, 23.

NAPAP (1988). 1985 Area Source Emissions Inventory. National Acid Precipitation Assessment Program.

NEDS (1988). 1985 National Emissions Data System. (Emissions inventory printout.) U.S. Environmental Protection Agency.

Russell, A. G., K. F. McCue, and G. R. Cass (1988). *Environmental Science and Technology*, 22.

Russell, A. G., et al. (1989). Quantitative Estimate of the Air Quality Impacts of Methanol Fuel Use, Final Report. California Air Resources Board, Sacramento, Calif.

Sillman, M. S. (1987). Models for Regional-Scale Photochemical Production of Ozone. Ph.D. thesis, Harvard University, Cambridge, Mass.

Snow, R., et al. (1989). *JAPCA*, 39.

Trainer, M., et al. (1987). *Nature.* 329.

Williams, R. L., F. Lipari, and R. A. Potter (1989). Formaldehyde, Methanol and Hydrocarbon Emissions from Methanol-Fueled Cars. General Motors Research Laboratories, Warren, Mich.

Part III

Sources of Methanol Supply, Energy Security, and Global Warming

CHAPTER 7

Methanol and Energy Security: Cost, Feedstock, and Source of Supply

CARMEN DIFIGLIO

Abstract

The use of methanol and other alternative transportation fuels is now under consideration as a strategy to reduce urban smog in the United States. This paper discusses the cost, feedstock, and possible sources of methanol. It concludes that, for the foreseeable future, methanol is likely to be produced from foreign sources of natural gas at prices that will become increasingly competitive with gasoline. Most of the methanol produced to satisfy new U.S. demand will probably come from producers in Central and South America, North Africa, and the Persian Gulf.[1] The diversity of methanol exporters will probably exceed that of oil exporters in the future.

The views expressed in this paper are those of the author and do not necessarily represent the views or policies of the U.S. Department of Energy.

The use of methanol as a transportation fuel would contribute to U.S. energy security for the following reasons: Methanol imports would be provided by a more diversified group of countries than the countries that would otherwise provide incremental oil imports to the United States. This diversity of suppliers should make the methanol market more competitive than the international oil market. Capital-intensive investments required to support methanol production would make it less likely that supplier countries could afford to limit production in an attempt to increase the price of methanol. Flexible-fuel vehicles could also operate on gasoline if methanol imports, for any reason, were curtailed. The use of methanol as a transportation fuel may reduce demand for oil, thereby reducing world oil prices.

Introduction

The use of methanol, and other alternative transportation fuels, has become an important energy and environmental issue. The administration's clean air proposals and the bipartisan commitment to amend the Clean Air Act, as well as growing concern about global climate change, have intensified environmental concerns.

The reduction of urban smog is a major focus of current environmental legislation. Among the measures now being considered, those involving motor vehicles are critical. Measures limiting gasoline vapor pressure and evaporative emissions, as well as stricter tailpipe standards, are likely to be adopted in new clean air legislation. In addition to these measures, the administration's program would require that, by 1997, alternative fuels, such as methanol, compressed natural gas, ethanol, propane, electricity, or reformulated gasoline, be used in about 30 percent of all new vehicles purchased in the nine urban areas unable to meet current air quality standards. The program would begin in 1995 and involve a half million new vehicles in its first year, growing to one million vehicles per year by 1997. Automobile manufacturers would have a major role in determining which fuel is used, subject to performance requirements set by the U.S. Environmental Protection Agency.

Methanol's use as a transportation fuel has become the focus of much of the public debate about the administration's proposal. This is due, in part, to the availability of flexible-fuel vehicles that can operate on gasoline, or a blend of 85 percent methanol and 15 percent gasoline (M85), or any combination of M85 and gasoline. Both Ford and General Motors have developed flexible-fuel vehicles, several of which are being tested by the state of California in its alternative-fuels demonstration program. Tests of these vehicles indicate that flexible-fuel, methanol vehicles could be ready for commercial-scale production, perhaps as early as the 1993 model year.

While most of the debate about the use of methanol-fueled vehicles focuses on its cost-effectiveness as a way to reduce urban smog, the use of methanol as a transportation fuel can have important energy security benefits as well. New supplies of methanol, at least for the next 20 years, are likely to be produced outside the United States. Therefore, the introduction of methanol fuels will not result in greater use of domestic energy supplies, at least in the foreseeable future. If methanol is imported, critics say, the United States is simply switching from its dependence on imported oil to a new form of imported energy. However, supporters of methanol argue that even though methanol is imported, it could provide a more secure supply of automobile fuel than imported oil. Moreover, foreign suppliers of methanol could be more competitive and reliable than those of oil.

Methanol: Likely Feedstocks and Cost

Methanol can be made from a wide variety of materials, including natural gas, naphtha, liquefied petroleum gas, coal, or biomass. Various processes can be used to produce oxides of carbon and hydrogen. These synthesis gases are then reacted over a catalyst to form methanol. The cost of methanol production varies considerably, depending on the feedstock used to produce the synthesis gas. Table 1 provides estimates of producing methanol from natural gas for the U.S. market, using current and advanced production technologies.

Table 1
Retail Methanol Prices (Natural Gas Feedstock)
(cents per gallon)

	Current Tech	Advanced Tech
Capital Recovery	0.21	0.12
Operating Cost (non-gas)	0.10	0.06
Operating Cost (gas)	0.11	0.10
Plant Gate Cost	0.42	0.28
Overseas Transportation Cost	0.06	0.06
Mark up to Retail (M100)	0.26	0.26
M100 Price	0.74	0.60
M85 Price	0.80	0.68
Gasoline-Equiv. Price (M85) (6% efficiency improvement)	1.30	1.11

Source: U.S. Department of Energy, Energy and Environmental Analysis, 1989

Estimates in Table 1 assume that (1) the methanol is produced in a remote location that has infrastructure, such as housing, roads, water, and utilities (for example, Venezuela or Saudi Arabia), (2) the price of natural gas is $1.10/mmBtu, (3) overseas one-way transportation is by dedicated methanol tankers that can use existing canals (40,000 deadweight tons), and (4) the one-way shipping distance is 6,000 miles. Since methanol has less energy per unit volume of fuel, gasoline-equivalent costs are provided and assume that a flexible-fuel vehicle would operate on methanol with a 6 percent efficiency improvement compared with gasoline (gasoline is assumed to cost $1.10/gal).[2]

The current-technology plant, typical of those built in recent years, produces 2,500 metric tons per day (20,000 b/d) and uses a steam-reforming process to produce synthesis gas. The advanced-technology plant uses a partial-oxidation process to produce synthesis gas. This plant is assumed to be four times as large (10,000 metric tons per day, or 80,000 b/d) and consists of a single train for production of synthesis gas and two trains for methanol synthesis. The large plant size results in significant economies of scale and reduces the cost of producing methanol from 42 cents/gal to 28 cents/gal. The fuel-equivalent cost to the consumer of using M85 is thereby reduced from $1.30/gal to $1.11/gal. This reduction is significant because methanol then costs no more than gasoline.

Concerns about global climate change have caused many to conclude that the use of methanol could exacerbate environmental problems. Critics argue that any benefits methanol may provide for reducing urban smog would be outweighed by an increase in CO_2 emissions if methanol is produced from coal. In the long term, it is argued, the need to improve U.S. energy security, and the rising costs of foreign natural gas, would lead to substantial production of methanol from U.S. coal.

Coal gasification is a well-understood technology that has been used in a few methanol plants, including one operated by Eastman Kodak, Inc. However, unless the price of coal is assumed to be abnormally low, or natural gas prices are assumed to be abnormally high, coal cannot compete with natural gas until the price structure of these feedstocks changes significantly or new technologies alter the economics of coal gasification. Coal requires more processing before it can be gasified, whereas natural gas requires little or no processing before it is reformed into synthesis gas. In addition, coal produces synthesis gas that is hydrogen-deficient, compared with the most efficient synthesis of methanol. Advanced technologies can mitigate but not eliminate these problems. Using advanced methanol synthesis technologies, and assuming that coal would cost $1.25/mmBtu, the cost of methanol from coal is estimated to be 75 cents/gal (DOE, 1989). This cost makes coal-derived methanol highly unlikely, unless a subsidy were provided to favor its production.

The cost of coal-derived methanol can be substantially reduced in the coproduction of coal and electricity, because hydrogen-deficient gas can be used, as produced in the coal gasifier, without having to provide hydrogen. The carbon-rich synthesis-gas-stream can be used in electricity generation after a substantial portion of the synthesis gas is converted to methanol. Assuming again that coal would cost $1.25/mmBtu, and also assuming that the plant would recover 5 cents/kwhr for the electricity generated, methanol could be produced at 55 cents/gal (DOE, 1989). Although still more expensive than methanol produced from natural gas, the cogeneration of methanol and electricity makes coal much more competitive than the production of methanol from coal. Should coal gasification become a competitive method of producing electricity, coal-derived methanol would be available only in relatively limited quantities, depending on the mix of electricity-generating capacity added in the United States. This methanol might also be used as a peaking fuel by the electricity generator, instead of being sold as automotive fuel.

Estimates of the cost of producing methanol from biomass are quite uncertain. The U.S. Department of Energy is funding research to develop biomass-to-alcohol technologies. Given the current state of this technology, it will be difficult to bring the cost of biomass-derived methanol to levels comparable with the cost of producing methanol from coal. DOE's current cost target is 60 cents/gal for bio-alcohol, but this is a goal for the year 2000, not a cost projection.

Timing of Methanol Supply

The likelihood of making methanol from natural gas during the next few decades increases when the time required for developing a methanol industry is considered. For example, if methanol is to displace as much as one million b/d of gasoline consumption (about one-seventh of total U.S. use), approximately two million b/d of methanol would be required. The current operating capacity of the U.S. methanol industry is about 400,000 b/d. Therefore, the existing methanol industry would have to be expanded by a factor of six in order to displace one-seventh of current U.S. gasoline demand. A single current-technology methanol plant takes about three years to construct. With an operating capacity of 2,500 metric tons per day, approximately 100 world-scale plants would have to be constructed to fulfill this demand.

If the methanol industry should expand at the maximum possible rate, considering only the technical limitations caused by availability of construction and engineering expertise, labor, and component parts, about 8 to 12 years would be required before the industry could supply two million barrels of methanol per day (Difiglio, 1988). If financial considerations are taken into account, such as the risks associated with such a rapid expansion of the methanol industry, the time required could be doubled.

Financial risk is an extremely important consideration. A current-technology plant, in a remote location with infrastructure in place, is estimated to cost about $300 million, while an advanced-technology plant would cost about $680 million (DOE, 1989). Therefore, the cumulative capital cost to provide two million b/d of additional methanol supply would be between $20 and $30 billion, depending on the mix of current- and advanced-technology plants built. Given the history of relatively low methanol prices in comparison with the cost of production, it is unclear whether investments of $1 to $2 billion per year in methanol plants would be continued. If the price of methanol continues to be volatile and frequently depressed, new methanol investments are likely to be postponed.

Therefore, it is not likely that more than two million b/d of methanol can be produced before the year 2010, due to limitations on the growth rate of methanol production capacity. It is also doubtful that there would be sufficient investment to expand production capacity at the maximum level that would be physically possible.

Sources and Availability of Natural Gas

As of the end of 1988, world proven gas reserves totaled more than 3,900 trillion cubic feet (tcf). The distribution of these reserves (as of January 1, 1988) is shown in Figure 1. Figure 1 indicates these reserves on a logarithmic scale along the Y axis and on the X axis displays the reserve-to-production ratio in each country identified. The country with the largest reserves, the USSR, is producing natural gas at exactly the world average of 1.8 percent of reserves. The horizontal dotted line drawn through the USSR, therefore, divides the countries listed into those with higher-than-average production (below the line) and those with lower-than-average production (above the line). The countries with the lowest average production rates are (in order) Iran, Qatar, Nigeria, the United Arab Emirates, Kuwait, Saudi Arabia, Libya, and Venezuela. Except for Kuwait and Libya, these countries all have very large reserves of about 100 tcf or more. The countries at the bottom of Figure 1 are either OECD (Organization for Economic Cooperation and Development) countries with well-developed gas distribution systems or South American countries with relatively low reserves (compared with Venezuela or Argentina).

Figure 1 is also divided by a vertical line, showing which countries are capable of producing 200,000 b/d of methanol at the world average rate of 1.8 percent production-to-reserves. The countries to the right of this line are all capable of producing what has been arbitrarily defined as a "significant" quantity of methanol (200,000 b/d, or half of current world methanol production). Ten such producers could provide the two million b/d that was estimated to be the maximum level of production achievable by 2010. Trinidad, Italy, and Egypt are the countries closest to this definition, although they have slightly smaller reserves than the definition implies. Of these countries, only Trinidad is producing at a rate that is lower than the world average.

Therefore, 11 Trinidads, producing natural gas at the world average rate of production, and devoting that production entirely

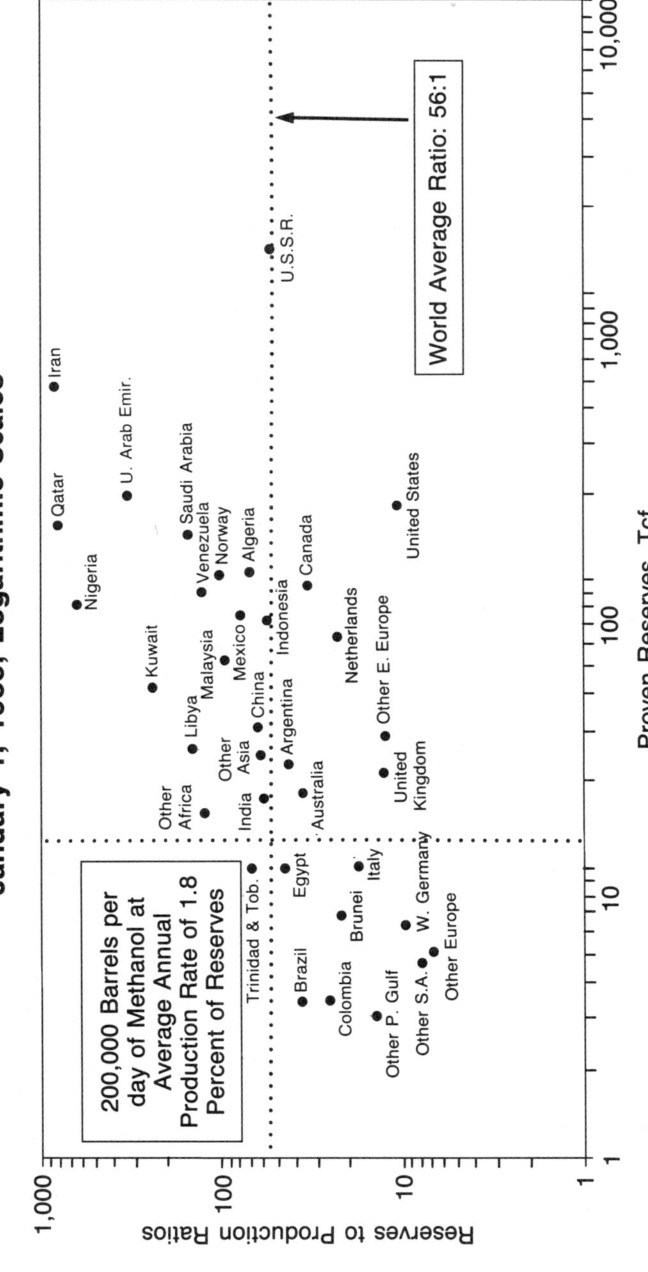

Source: U.S. Department of Energy, Energy Information Administration, *International Energy Annual 1987, 1988*

to methanol, could supply two million b/d of methanol. Alternatively, one Venezuela could also supply two million b/d of methanol, while Iran could supply more than four times that amount. There are, therefore, ample supplies of natural gas in the world to provide feedstocks for the maximum feasible production of methanol in and after 2010. It is unrealistic to assume that any country will devote its entire reserves of natural gas to the production of methanol. However, these statistics indicate that world supplies of natural gas could satisfy realistic estimates of production capacity for methanol that might occur by 2010, even when local uses of natural gas and the potential export of natural gas via liquefied natural gas are accounted for. For example, if the world average rate of gas production increased from 1.8 percent to 2.2 percent—the average depletion rate for petroleum production—world methanol production could increase to eight million b/d, enough to displace one-half of U.S. gasoline use. By the time enough methanol plants could be built to produce eight million b/d, it is likely that proven world natural gas reserves would be considerably larger and the required depletion rate would be lower than 2.2 percent.

Figure 1 indicates the countries that have sufficient gas reserves to produce methanol and those that might have low-cost gas. The OECD countries, with their high production-to-reserves levels, have exploited their low-cost reserves. Compared with the less developed countries with low rates of gas production, the OECD countries tend to have high production costs and high opportunity costs for natural gas.

Table 2 includes a list of countries that are potential low-cost suppliers of methanol and shows the proven gas reserves of each as of December 1988. Seventeen countries could produce 200,000 b/d or more of methanol, and seven countries could produce at least two million b/d. These countries have production-to-reserves ratios at or near the current world average.

The supply of low-cost gas available to produce methanol is large enough to support levels of methanol production that are not likely to occur until 2030 or later. The size of a country's gas reserves

Table 2
Areas with Potentially Low-Cost Natural Gas Reserves

	(Proven Reserves—TCF) (Year End—1988)
OECD	
Alaska	9
Canadian Arctic	56
Norwegian Arctic	19
Australia	17
New Zealand	5
PERSIAN GULF OPEC	
Iran	494
Saudi Arabia	152
Kuwait	49
U.A.E.	202
Qatar	157
OTHER OPEC	
Indonesia	84
Algeria	104
Libya	26
Nigeria	85
Venezuela	102
Ecuador	4
NON-OPEC, NON-OECD	
China	32
Thailand	4
Malaysia	52
Mexico	75
Trinidad, Tobago	10
Argentina	27
Chile	4
Soviet Union	1,500
WORLD RESERVES	3,936

Source: U.S. Department of Energy, Energy Information Administration, *International Oil and Gas Exploration and Development Activities.* Quarterly report, first quarter 1989; U.S. Department of Energy, Energy Information Administration, *U.S. Crude Oil, Natural Gas, and Natural Gas Liquids Reserves: 1988 Annual Report,* September 1989

is not a good indicator of which countries are likely to supply the near-term (to 2010) methanol market. Little or no activity is occurring in countries with the largest reserves—the USSR and Iran. However, many projects are under way in countries that have relatively small reserves but are, nonetheless, large enough to support significant quantities of methanol production, such as Chile. Several major methanol production projects using natural gas are under way in Venezuela. These projects are significant because, aside from the United States, Venezuela has the largest gas reserves in the Western Hemisphere. Argentina is also a possibility. Trinidad and Tobago are also possible candidates for future projects. Ecuador does not have current plans or projects, but its low-cost gas reserves make it a likely candidate for development. While Mexico has relatively large reserves, there is a large potential domestic demand for gas, and there are plans to expand the existing gas distribution system.

North Africa and the Middle East are other possible suppliers of methanol. The Persian Gulf country with the largest natural gas reserves, Iran, is unlikely to be a factor in the market in the foreseeable future. Iran probably lacks the technical wherewithal and capital to undertake such projects on its own, and it is currently viewed as an unattractive candidate for foreign investment. However, on the other side of the Persian Gulf, several countries could prove to be important. Saudi Arabia and Qatar both have large reserves of natural gas. Saudi Arabia has an existing petrochemical infrastructure and methanol production capacity, and additional plants could be built in Saudi Arabia at relatively low costs. Qatar, since it has relatively little oil, has a strong economic motive to export gas or products made from gas. Methanol production may provide an opportunity for expanding gas-derived exports.

The Middle East and North Africa could supply vast quantities of methanol—a significantly greater quantity than can be provided from the Western Hemisphere. However, it will take many years before demand for such large volumes of methanol can develop. South and Central America alone could provide more than two million b/d of methanol. In addition, countries in the Far East

(Indonesia, Malaysia, Thailand, Australia, and New Zealand) have large gas reserves and would be capable of providing methanol in large volumes.

The data show that virtually any part of the world has sufficient gas reserves to supply the U.S. market with methanol until and beyond the year 2010. Some countries in Central and South America, or in North Africa and the Persian Gulf, are likely to undertake new methanol projects if they are justified by U.S. demand for methanol fuel.

Energy Security Implications

The transportation sector's near-total dependence on oil plays a crucial role in U.S. oil-import dependence. Oil consumption by the transportation sector in 1989 was 10.8 million b/d, almost 20 percent higher than in 1973. Oil use, in the transportation sector alone, exceeded U.S. production of petroleum in 1989 by 900,000 b/d. Even with substantial energy conservation, fuel-switching, and cars that achieve fuel economy above 40 mi/gal by the year 2000, DOE estimates that U.S. oil use will remain at about the same level for the next 10 to 12 years (DOE, 1988). Since domestic oil production is expected to decline, even extraordinary conservation efforts will not prevent increased dependence on imported oil.

The energy security of the United States could be enhanced if flexible-fuel, methanol-powered vehicles were introduced in sufficient quantities between now and 2010 to reduce the nation's gasoline consumption by one-seventh. Although the two million b/d of methanol required would be imported, they would be replacing imported oil. The shift from one million b/d of imported oil, or oil products, to two million b/d of methanol would be positive for the following reasons:

Methanol imports would be provided by a more diverse group of countries than those from which the United States would import additional supplies of oil. Many countries in Central and South America may provide a substantial amount of increased U.S.

demand for methanol. As discussed earlier, these countries are capable of supplying all of the increase in U.S. methanol demand. They could not, however, export enough oil to satisfy increased demand for oil.

With this diversity of suppliers, the methanol market is likely to remain more competitive than the oil market. The world oil market is controlled by a core group of Persian Gulf producers and other oil-exporting nations within OPEC (Organization of Petroleum Exporting Countries). This cartel has enjoyed varying degrees of success in keeping oil prices high. As world oil demand increases, it is likely that the OPEC cartel will regain the dominant position it enjoyed in the 1970s. If oil demand is reduced, due to the increased use of methanol, OPEC's market power is also reduced. Since the suppliers of methanol are likely to be a more diversified group of countries than those that make up OPEC, it would be more difficult to coordinate production decisions of methanol suppliers.

The capital-intensive investments required for methanol production make it less likely that supplier countries would limit methanol production in an attempt to increase methanol prices. Countries that produce methanol are either investing their own capital, or allowing foreigners to invest. Therefore, these producing countries cannot afford to limit the production of methanol in an attempt to influence world market prices. In addition, the producing country is usually supplying natural gas under a long-term contract. This also reduces the incentive to reduce methanol output in an attempt to control prices.

Flexible-fuel vehicles could also operate on gasoline if methanol imports, for any reason, were curtailed. Even if methanol supplies were reduced, the price of methanol could not increase above the fuel-equivalent cost of gasoline, since flexible-fuel vehicles could switch to gasoline. Therefore, methanol demand would drop until the fuel-equivalent price of methanol was again competitive. Flexibility of fuel use provides greater price stability, even if supplies of one fuel or the other are interrupted.

Reduced oil demand, resulting from methanol use, is likely to reduce world oil prices. The DOE report, *Energy Security*, estimated that a one million b/d reduction in oil demand would reduce the world oil price by at least $2.00/b (DOE, 1987). According to this estimate, and the most recent Energy Information Administration projections, methanol use would reduce U.S. oil import expenditures by $10 billion in the year 2010.

In conclusion, the use of methanol would contribute significantly to increased U.S. energy security. Furthermore, if methanol use is justified by environmental considerations, the economic and security benefits it offers by reducing oil consumption are a bonus.

Notes

1. Use of remote, or lower quality U.S. natural gas, may provide opportunities to competitively expand U.S. methanol production. The U.S. Department of Energy is currently studying the relative economics of this source of domestic methanol compared to that available from the countries discussed in this paper. If economic, this source of methanol could substantially enhance the energy security benefits of methanol use.

2. The gasoline-equivalent cost is the retail cost for the volume of M85 required by a vehicle to traverse the same distance as it would on one gallon of gasoline.

References

Difiglio, C. (1988). Timing of Methanol Supply and Demand: Implications for Alternative Transportation Fuels Policies. *Transportation Research* 23A, No. 3.

DOE (1987). Energy Security: A Report to the President of the United States. U.S. Department of Energy, Report No. DOE/S-0057.

DOE (1988). Oil Conservation Potential in the U.S. Economy. Office of Policy, Planning and Analysis, U.S. Department of Energy, Report No. PE-0084.

DOE (1989). Assessment of Costs and Benefits of Flexible and Alternative Fuel Use in the U.S. Transportation Sector, Technical Report Three: Methanol Production and Transportation Costs. Office of Policy, Planning and Analysis, U.S. Department of Energy, Report No. DOE/PE-0093.

Energy and Environmental Analysis, Inc. (1989). Capital and Operating Costs of a Fuel Methanol Distribution System.

CHAPTER 8

Emissions of Greenhouse Gases From the Use of Gasoline, Methanol, and Other Alternative Transportation Fuels

MARK A. DeLUCHI

Abstract

This paper analyzes emissions of greenhouse gases (carbon dioxide, methane, and nitrous oxide) from the production, distribution, and use of petroleum and alternative transportation fuels. This study found that using coal as a feedstock to make any transportation fuel will increase emissions of greenhouse gases. The use of natural gas to produce methanol, compressed natural gas, liquefied natural gas, or electricity for electric vehicles was found to result in only slight reductions in CO_2-equivalent emissions, compared with gasoline and diesel fuel. The methanol-from-natural-gas process, for example, could emit as much as 30 percent more or less greenhouse gases per mile than the gasoline-from-crude-oil process, depending primarily on assumptions regarding the efficiency of methanol production and use. The main conclusion of the study is that if the transportation sector must reduce emissions of greenhouse gases, methanol from coal or natural gas must be no more than a transitional step to another fuel or feedstock.

The analysis shows that substantial reductions in emissions of greenhouse gases from the transportation sector can be accomplished only be greatly increasing the efficiency of vehicles, or using wood-based biofuels for natural gas and methanol vehicles or non-fossil electricity for hydrogen and electric vehicles. The author calls for more research and the development of environmentally-benign fuels, and policies for encouraging the production and use of these fuels.

Introduction

Long-term global environmental concerns will play an increasingly important role in energy policy. Foremost among these concerns is global warming, caused in part by carbon dioxide (CO_2) emissions from the production, distribution, and use of fossil fuels. In recent years there has been considerable interdisciplinary research on the changing atmospheric concentration of CO_2 and other greenhouse gases, the effects of anthropogenic releases of these gases on steady-state atmospheric concentrations, and the climatic effects of their increasing concentrations. Many scientists and energy analysts conclude that significant climatic change is likely and that actions should be taken to reduce the use of fossil fuels.

In the transportation field, concern about the greenhouse effect is coinciding with the need to reduce ozone formation in many urban areas. Policymakers, at the state and national level, are considering the relative potential of vehicles fueled by methanol, ethanol, compressed natural gas (CNG), or electricity to improve air quality and the energy security of the United States. By analyzing emissions of greenhouse gases from the production, distribution, and use of petroleum and alternative transportation fuels, with emphasis on methanol, this paper addresses the question of whether the use of these fuels will, compared with the use of gasoline and diesel fuel, mitigate or exacerbate global warming.

The earth is heated by shortwave radiation from the sun and radiates longer-wave infrared energy back. Carbon dioxide and other trace greenhouse gases, along with water vapor, absorb most of this energy from the earth and reradiate it back to earth. Although there are several important uncertainties in models of climate response to increases in emissions of greenhouse gases which leave room for differences of opinion about the causes and consequences of a greenhouse warming, researchers are in almost unanimous agreement that if the concentration of greenhouse gases doubled, a substantial warming would occur (MacCracken and Luther, 1985; Bolin et al., 1986, Ramanathan, 1988; Firor, 1989; MacCracken, 1989; Mahlman, 1989). This warming could shift global precipitation patterns, disrupt established crop-growing regions, raise global sea level, increase incidents of severe weather, change the distribution and abundance of biota and pathogens, and eventually melt portions of the polar ice caps.

Thus, continuing combustion of fossil fuels is likely to warm the planet and change the global climate, possibly with serious long-term consequences. But public policy can do much to delay and eventually halt the warming, by encouraging the use of fuels that contribute relatively little to the greenhouse effect. The U.S. highway transportation sector contributes about 25 to 30 percent of all carbon dioxide emitted from fossil fuel use. Hence, any policy aimed at reducing greenhouse gas emissions must include highway vehicles.

Analysis of Emissions of Greenhouse Gases from Highway Transportation Fuels

The most feasible alternatives to gasoline and diesel-fueled vehicles, in the near term, are methanol, ethanol, compressed natural gas, and electric vehicles. Hydrogen has been examined as a long-range option. Although several authors have estimated emissions of greenhouse gases from the production and use of these fuels (White, 1980; MacKenzie, 1987; Gushee, 1988; DeLuchi, Johnston, and Sperling, 1987a; Unnasch et al., 1989), there is no study that encompasses all the alternatives in a single analytical framework and that uses all the available data on emissions of methane and nitrous oxide. Interim results of a comprehensive, state-of-the art analysis of emissions of greenhouse gases from transportation fuels are the subject of this paper.

An energy use and emissions model was used to calculate emissions of carbon dioxide (CO_2), methane (CH_4), and nitrous oxide (N_2O), resulting from the end-use, distribution, and production of gasoline, diesel, and alternative fuels. The model also calculated the emissions from the recovery and transportation of fuel feedstocks (for example, coal, natural gas, crude oil, or biomass). The primary output of the model is grams of greenhouse gases emitted at each stage of the process, per mile of travel in the vehicle.

Emissions of Greenhouse Gases from End-Use

"End-use" refers to emissions from the combustion of transportation fuels on the highway. Here, estimates of end-use emissions began with a specification of the fuel economy and weight of base-case vehicles fueled by gasoline and diesel fuel. Alternative-fuel vehicles were then compared with this base case, by adjusting the base case for efficiency gains and losses due to the use of various alternative fuels (Table 1). Then, carbon dioxide emissions per mile, from the petroleum and alternative-fuel vehicles, were calculated from data on the amount of fuel consumed per mile, the carbon content of the fuel, and the amount of the fuel carbon that was fully oxidized. Emissions of methane and nitrous oxide are averages of several vehicle emissions tests. Calculated base-case emissions of greenhouse gases from vehicles are shown in Table 2.

Fuel Distribution

Fuel is distributed from fuel production facilities, such as petroleum refineries, to service stations by train, truck, ship, and pipeline. All of these modes are included in the greenhouse model.

Table 1
Base Case Fuel Economy Input Data

Base case gasoline vehicle efficiency: 32 mpg
Base case diesel vehicle efficiency: 5 mpg
Base case gasoline vehicle weight: 2,500 lbs
Base case diesel vehicle weight: 40,000 lbs
Efficiency of alternatives RELATIVE to gasoline and diesel:

	M100	NG	Hydrogen	Electric	Ethanol
Thermal efficiency, LDV's	1.15	1.10	1.25	5.90	1.10
Thermal efficiency, HDV's	0.90	0.90	1.05	—	0.90
Extra weight, lbs, LDV's	0.00	135.00	600.00	650.00	0.00
Extra weight, lbs, HDV's	0.00	450.00	300.00	—	0.00

Notes:
   The mi/mmbtu efficiency of an alternative-fuel vehicle is equal to the mi/mmbtu efficiency of the comparable petroleum-fuel vehicle multiplied by the thermal efficincy factor shown in the table, and then by a weight factor. The weight factor is calculated assuming an 0.70 percent decrease in fuel economy per 1.0 percent increase in vehicle weight. The weight factors were estimated assuming all vehicles have a 300 mile range, except hydride (hydrogen) and battery-powered vehicles, which have a 150 mile range. The EV (electric vehicle) uses a sodium-sulfur battery.

Source: Mark DeLuchi, Greenhouse Gas Emissions Model, 1990

In the model, tankers use residual fuel #6 (bunker fuel), trains and trucks use diesel fuel, natural gas pipelines use natural gas, and oil pipelines use electricity. The electricity used by the oil pipelines is assumed to be generated by a mix of fuels, which vary in the model. The amount of fuel used by each mode of distribution is based on data from the Energy Information Administration (EIA) (various titles and years), or is calculated as Btus required to move one ton of transportation fuels one mile, multiplied by the tons of product moved and miles of average haul. Tables 3 and 4 show the amount of energy and the mode breakdowns used to distribute fuels. The Btu/ton-mile data (which determine the amount of energy used to

Table 2
Lifetime Average Vehicle Tailpipe Emissions Data, GM/MI

	Gasoline	Diesel	LD-MeOH	HD-MeOH
CH_4	0.10	0.10	0.07	0.05
N_2O	0.07	0.06	0.07	0.06
CO_2 from engine oil	2.00	4.00	2.00	4.00
CO_2 from fuel	264.07	1963.93	213.59	1945.20
TOTAL CO_2 plus CO_2-equivalent of CH_4, N_2O	280.01	1983.59	229.12	1968.28

Notes: LD = light-duty application; HD = heavy-duty application; MeOH = methanol; H_2 = hydrogen; EtOH = ethanol.

Source: Mark DeLuchi, Greenhouse Gas Emissions Model, 1990

distribute transportation fuels) are derived from detailed estimates for each of the modes (Rose, 1979).

Carbon dioxide emissions for fuel distribution were calculated as they were for vehicles, as a function of the carbon content and energy density of the fuel. Methane and nitrous oxide emissions for trucks were based on direct measurements reported in the literature. Methane emissions from other modes were based on reports by the Environmental Protection Agency (EPA, 1985 and 1985a). Nitrous oxide emissions from distribution modes were assumed to be the same, on a g/mmBtu basis, as reported emissions of N_2O from power plants burning the same fuel.

Fuel Production

This stage includes emissions from petroleum refineries, ethanol fermentation facilities, wood gasification and synthesis plants, coal-to-methanol plants, natural-gas-to-methanol plants, natural gas compressors or liquefiers, hydrogen compressors or liquefiers, and electricity-generating plants. The amount and kind

Table 2
(Continued)

LD-NG	HD-NG	EV's	LD-H2	HD-H_2	LD-EtOH	HD-EtOH
1.40	3.00	0.00	0.005	0.005	0.07	0.05
0.07	0.06	0.00	0.000	0.060	0.07	0.06
1.00	2.00	0.00	2.000	4.00	2.00	4.00
189.51	1630.86	0.00	0.00	0.00	0.00	0.00
218.73	1679.36	0.00	2.14	14.64	14.11	15.79

of fuel used by each facility is specified in the model (Tables 3 and 4 show the base-case data). In the case of methanol production from natural gas and coal, where the product (methanol) contains less carbon by weight than the feedstock, the model tracks carbon through the conversion process. Carbon dioxide emissions are calculated as before. There are data on CH_4 and N_2O emissions from electricity generation, but not from other fuel production processes.

Feedstock Transportation

This stage includes emissions from the combustion of fuel used by the various modes to distribute the feedstocks to fuel production facilities: gas and oil pipeline compressors or pumps, ocean-going tankers, river barges, trains, and trucks. (See "Fuel Distribution," above, for further discussion, and tables 3 and 4.)

Feedstock Recovery

This stage includes emissions from fuel combustion at coal mines, oil and gas producing facilities, uranium mines, corn and

Table 3

MmBtu of Process Energy Consumed and NG Emitted per MmBtu of Net End-Use Energy Provided

	Gasoline	Diesel	Fuel oil	NG	Coal
Energy consumption:					
Byproduct credit					
Compress/liquefy				0.0530	
Fuel distribution	0.0099	0.0099	0.090		
Fuel production	0.0915	0.0915	0.0915		
Feedstock trans.	0.0108	0.0108	0.0108	0.0418	0.0100
Feedstock recovery	0.0300	0.0300	0.0300	0.0814	0.0063
NG emissions:					
Fuel transmission				0.005	
Feedstock recovery	0.024	0.024	0.024	0.010	0.030

Other data:
1.54 mmBtu of NG feedstock per mmBtu of methanol produced
1.80 mmBtu of coal feedstock per mmBtu of methanol produced

Notes:

Data are mmBtu of process energy consumed or NG emitted per mmBtu of net end-use energy (as defined in the text), except for MeOH/bio (methanol from biomass) and SNG/bio (synthetic natural gas from biomass) feedstock recovery and transportation and fuel production factors, which are energy consumed per unit of energy produced by the production plant, and fuel production for MeOH/coal and MeOH/NG, which are energy consumed per unit of energy produced. MeOH/bio and SNG/bio assume use of woody biomass.

tree farms, and fertilizer manufacturing facilities. Data on the amount and kind of energy used by fossil fuel and uranium recovery facilities are from the Bureau of the Census; biomass data are from sources in the technical literature. Data on the amount of feedstock produced by recovery operations are from the EIA (various titles and years) and other sources. Data on methane emissions from coal mines, natural gas systems, and oil production are from the EIA and the U.S. Bureau of Mines. Tables 3 and 4 present the base-case data.

Table 3
(Continued)

MeOH/ NG	MeOH/ coal	MeOH/ bio	SNG/bio bio	EtOH/ bio	Uranium	H2
				0.1400[a]		
			0.0530			0.0600
0.0464	0.0120	0.0300	0.0450	0.0273		0.1000
0.1200	0.3000	0.9000	0.3500	0.6000	0.0761[b]	
0.0200		0.0400	0.0400	0.0400	0.00007	
0.0814		0.1100	0.1100	0.5500	0.0197	
0.005						
0.010						

Electricity-based factors (compression, uranium enrichment) assume mmBtu of end-use power per mmBtu of product (CNG, CH_2, of kwh of nuclear power).

[a] The byproduct credit for ethanol is the amount of energy that would have been consumed to plant, fertilize, and harvest a food product with the same caloric and protein value as the ethanol byproduct.

[b] Uranium conversion and enrichment.

Source: Mark DeLuchi, Greenhouse Gas Emissions Model, 1990

Other Aspects of the Analysis

Several other aspects of the greenhouse model are worth mentioning. The fuel cycles in this analysis are "closed." Not only do the final emissions estimates include those from energy used to produce and transport primary feedstocks, they also include emissions from the use of energy used to recover, process, and transport the fuel used to recover and transport the primary feedstock, and so

Table 4
Use of Process Energy at Each Stage of the Fuel Cycle

Process Fuel or mode	Coal	Oil & Products	Uranium
Feed recovery			
Electricity[a]	0.40	0.20	0.40
Diesel[b]	0.30	0.70	0.35
NG	0.30	0.10	0.25
Biofuel truck			
Feed transmission			
Electric[a] pipeline	0.00	0.11	
Diesel tanker	0.20	0.87	
Diesel train	0.70	.00	0.00
Diesel truck	0.10	0.01	1.00
Biofuel truck			
Fuel production			
Electricity[a]		0.05	0.99
Diesel[b]		0.15	
NG		0.25	0.01
Refinery Gas		0.55	
Coal			
Wood			
Ethanol			
Fuel distribution			
Electric[a] pipeline		0.08	
Diesel truck		0.41	
Diesel tanker		0.49	
Alternative fuel truck[d]			
Diesel train		0.02	

Notes:

The numbers are: use of energy type E at stage S divided by total energy use at stage S. For example, the first number under the "coal" column, 0.4, says that 40% of the total energy used to recover (mine and prepare) coal is electricity.

All stages of the NG cycle (not shown here) use NG as the process fuel, except for a small amount of electricity used to run pipeline compressors.

Table 4
(Continued)

NG to Methanol	Coal to Methanol	Biomass Methanol	Biomass SNG	Ethanol
		0.04	0.04	0.04
		0.26	0.26	0.26
		0.60	0.60	0.60
		0.10	0.10	0.10
		0.20	0.20	0.20
		0.20	0.20	0.20
		0.60	0.60	0.60
0.002	0.010	0.03	0.01	0.10
		0.00	0.00	0.00
0.998		0.00	0.00	0.00
	0.990	0.00	0.00	0.90
		0.97	0.99	0.00
				0.00
			c	
0.02	0.15	0.05		0.01
0.25	0.35	0.00		0.68
0.53	0.20	0.05		0.05
0.20	0.20	0.80		0.00
0.00	0.10	0.10		0.26

[a] Electricity's share is counted on an end-use mmBtu basis: actual kwh consumed, converted to mmBtu, divided by total mmBtu of process fuel.
[b] "Diesel" represents diesel, residual, and crude oils.
[c] SNG from biomass is assumed to be distributed by SNG-fueled pipeline.
[d] Alternative fuel trucks use the fuel they are distributing.

Source: Mark DeLuchi, Greenhouse Gas Emissions Model, 1990

on—a complete accounting. This is accomplished not by iterative calculations ad infinitum, but by looking at steady-state energy flows in the United States.

Most estimates of emissions from fuel distribution, fuel production, feedstock transmission, and feedstock recovery are based on the entire system producing a *net* unit of energy available for transportation end-users (as in Table 3). That is, the relevant "output" of the process does *not* include fuel used by trucks delivering highway fuels, petroleum fuels used at petroleum refineries, electricity used to enrich uranium for nuclear power plants, natural gas used to generate electricity to compress natural gas, and so on. Rather, the net output is a unit of energy available to end-users, after energy used at all earlier stages has been deducted. This definition is significant because g/mi emissions of greenhouse gases is in most cases calculated as: grams of greenhouse gases emitted per mmBtu of energy produced or used by a process, multiplied by mmBtu produced or used by the process per mmBtu of net output to transportation, multiplied by mmBtu of net output to transportation per mile of travel provided.

Emissions of greenhouse gases from a particular process, such as petroleum refining, are directly related to the amount and kind of energy used. The quality of the entire analysis, then, is related directly to the quality of the data on the amount and kind of energy used at every stage for every fuel. For the fossil fuel energy cycles, national average data provided by, or derived from, reports of the EIA and the Bureau of the Census (various years) were used. The EIA collects data on energy production, disposition, and transportation from petroleum refineries, coal distributors, natural gas producers, electric utilities, and a host of other energy-related enterprises. The Bureau of Mines reports production and energy use data at coal mines, oil and gas wells, and uranium mines. Where possible, estimates based on aggregate EIA and Bureau of the Census data were checked against engineering estimates of the amount and type of energy required by a particular process (DOE, 1983).

The heating value, physical density, and carbon content of a particular fuel or feedstock varies slightly from batch to batch (in the case of coal, of course, there is considerable variation from one kind of coal to another). Generally, these variations do not significantly affect the final calculated emissions per unit of greenhouse gases. Values specified by the EIA, the American Society for Testing Materials, the American Petroleum Institute, and other standard references were used. All heating values are "gross" or "higher" and are consistent with values used by the EIA.

Efficiency—as in the efficiency of petroleum refineries, coal-to-methanol plants, and uranium enrichment, and the efficiency of natural gas vehicles, compared with gasoline-fueled vehicles—is probably the single most important determinant of greenhouse gas emissions. It is important to calculate or project, as precisely as possible, the efficiency of key processes in the production-and-use cycle of fuels. The efficiency estimates in this analysis are based on a fairly extensive review of each process in the recent literature. Base-case aggregate efficiencies for each step of the fuel cycles are shown in Table 3.

Converting Methane and Nitrous Oxide Emissions to Carbon Dioxide Emissions with the Same Temperature Effect

The combined effect on climate of all greenhouse gases emitted by the use of alternative fuels is simpler to understand if all of the emissions are represented by one measure. Therefore, CH_4 and N_2O mass emissions have been converted into a mass amount of CO_2 emissions, with the same temperature effect. The "same temperature effect" is defined here in terms of degree-years over a 125-year period (75 years of emissions, and then 50 years for the climate to equilibrate), where a degree-year is an increase in surface temperature of one degree C for one year. Methane emissions, for example, are converted to carbon dioxide-equivalent emissions, by calculating the annual mass emissions of carbon dioxide that would result in the number of degree years of warming produced by a given annual methane mass emissions rate.

In order to convert methane and nitrous oxide emissions into carbon dioxide emissions having the equivalent temperature effect, one needs to know, for both gases, the relationship between (1) equilibrium surface temperature and equilibrium atmospheric concentration and (2) the increase in yearly emissions and the increase in the equilibrium atmospheric concentration. These relationships are derived in DeLuchi, Johnston, and Sperling (1987a). The resulting conversion factors—11.6 for methane to carbon dioxide, and 175 for nitrous oxide to carbon dioxide, on a mass basis, are close to those derived by scientists at the Natural Resources Defense Council, using a more sophisticated approach (Lashof and Ahoja, 1989). The conversion factor for methane to carbon dioxide is sensitive to some uncertainties, including the number of years into the future being considered, and can range plausibly from 5 to 30 or higher, on a mass basis.

It is worth noting that the methane conversion factor used here, which accounts for the relatively short lifetime of methane, is much lower than the ratio of the radiative forcing of a molecule of methane divided by the radiative forcing of a molecule of carbon dioxide, which is typically around 25 to 30 on a part basis or 70 to 80 on a mass basis. The ratio of the radiative forcings is inappropriate for policy analyses because it does not account for the difference in lifetimes between carbon dioxide and methane. The former stays in the atmosphere, and warms it, much longer than the latter. Greenhouse strategies concerned with methane emissions should emphasize the short lifetime of methane, or policymakers may be led to believe that methane emissions are more important than they really are.

Base-Case Analysis and Results

With the input data and assumptions discussed above, and other, less important data not discussed here, the greenhouse-gas emissions model calculates grams of carbon-dioxide-equivalent greenhouse gases—actual carbon dioxide plus the carbon dioxide equivalent of methane and nitrous oxide—emitted per mile of travel by the vehicle, for each stage of the fuel production and use cycle, for petroleum fuels and alternative fuels, including biofuels. For all vehicles, the key base-case assumptions are shown in tables 1, 2, 3, and 4. Table 5 shows the results for the base-case data. Total fuel cycle emissions per mile, and percent changes relative to gasoline (light-duty vehicles) and diesel fuel (or heavy-duty vehicles) are shown at the bottom of the table.

The base case for methanol light- and heavy-duty vehicles assumes that methanol is made from remote natural gas. The results from making methanol from domestic gas and domestic coal, and of changing other base-case data for methanol, are shown in Table 6. The natural gas vehicle results shown are for compressed natural gas vehicles; the results for liquefied natural gas (LNG) vehicles are essentially the same. Electric vehicles in the base case are assumed to draw power in proportion to EIA's projected mix of fuels used by electric power plants in the year 2000: 52 percent coal, 16 percent nuclear, 16 percent natural gas, 8 percent oil, and the remainder, which is hydro, geothermal, wood, and renewables (with no emissions of greenhouse gases). It is assumed that hydrogen is made from water using nuclear power, but that it is compressed (the base case assumes hydride vehicles) using the projected average U.S. power mix (same fuel distribution as for electric vehicles). These base-case assumptions are varied in the scenario analysis discussed below.

Vehicles fueled by methanol and natural gas produce less carbon dioxide, in end-use (from the tailpipe), than do gasoline-fueled and diesel-fueled vehicles. This is due to the fact that methanol and natural gas contain less carbon per unit of energy than do gasoline and diesel fuel, and can be used more efficiently than gasoline (but not diesel); the net result is less carbon dioxide per mile. However, in the case of natural gas vehicles, higher emissions of methane eliminate some of the advantage of lower carbon dioxide

Table 5
Gm/mi of CO_2-Equivalent Emissions, Entire Fuel Production and Use Chain, Base Case

Gm/mi	Gasoline	Diesel	LD-MeOH	Hd-MeOH
Vehicles	280.0	1983.6	229.1	1968.3
with biofuels	0.0	0.0	14.4	17.4
Compress/liquefy	0.0	0.0	0.0	0.0
with biofuels	0.0	0.0	0.0	0.0
Fuel distribution	3.2	23.2	10.7	120.2
with biofuels	0.0	0.0	2.7	23.2
Fuel production	25.3	180.8	52.2	559.1
with biofuels	0.0	0.0	23.6	215.3
Feedstock trans.	3.8	27.3	0.0	0.0
with biofuels	0.0	0.0	4.5	43.4
Feedstock recovery	12.0	86.0	37.8[e]	424.2[e]
with biofuels	0.0	0.0	22.8	241.4
CH_4 leaks/flares	8.0	58.0	0.0	0.0
with biofuels	0.0	0.0	0.0	0.0
GRAND TOTALS, Gm/mi	335	2388	335	3131
with biofuels	—	—	67	516
% CHANGES	—	—	−0	30
with biofuels	—	—	−80	−77

Notes:

In all cases "with biofuels" shows the result if biomass is used as the feedstock rather rather than a fossil fuel. NG-biofuels case assumes that all NG, including any used at power plants, is made from biomass. CO_2 emissions from the combustion of biofuels are not counted here, because they are not net emissions to the atmosphere (the carbon in the fuel came originally from the atmosphere). Thus, the biofuel cycle produces less greenhouse-gas emissions wherever biomass replaces fossil fuel as the energy source.

Methanol case shown is methanol from remote natural gas.

Table 5
(Continued)

LD-NG	HD-NG	EV's	LD-H2	HD-H2	LD-EtOH	HD-EtOH
218.7	1679.4	0.0	2.1	14.6		
24.8	39.1	0.0	0.0	0.0	14.1	15.8
46.1	391.8	0.0	51.7	379.2		
39.7	337.1	0.0	0.0	0.0	0.0	0.0
0.0	0.0	0.0	0.0	0.3		
0.0	0.3	0.0	0.0	0.0	8.3	72.6
0.0	0.0	266.0[a]	110.0[b]	806.7[b]		
3.8	32.2	0.0	0.0	0.0	245.2	2145.8
8.1[c]	69.2[c]	0.0	0.03[d]	0.20[d]		
5.3	45.2	0.0	0.0	0.0	4.8	42.3
15.8	134.5	35.4[f]	13.2[g]	97.1		
29.3	248.6	0.0	0.0	0.0[g]	100.4	878.7
12.0	246.0	0.0	0.0	0.0		
0.0	0.0	0.0	0.0	0.0	0.0	0.0
293	2424	302	177	1299		
92	637	—	—	—	373	3155
−9	7	−9	−47	−45		
−69	−70	—	—	—	12	34

[a] Emissions from electricity-generation.
[b] Emissions from uranium conversion and enrichment.
[c] Includes all pipeline transportation of NG. For bio-NG, pipeline transmission is under fuel distribution. Pipeline compressors burn bio-NG.
[d] Emissions from highway transportation of uranium.
[e] Includes emissions from natural gas recovery and transmission. CH_4 leads are included here.
[f] Includes emissions from transportation of the fuel to the power plant.
[g] Emissions from uranium mining.

Source: Mark DeLuchi, Greenhouse Gas Emissions Model, 1990

Table 6

**Greenhouse-Gas Emission Scenario Analyses:
Percent Changes in Emissions Per Mile of CO_2-Equivalent
Gases Compared to Gasoline**

| | Percent change, relative to gasoline | |
| | With CH_4-CO_2 conversion factor = | |
Fuel and feedstock	**11.6**	**30**
EVs using solar power	− 100	
Hydrogen using solar power	− 99	
EVs using nuclear power	− 89	
EVs using coal-fire plants	+ 29	
EVs using gas-fired plants	− 21	− 17
CNG, favorable scenario	− 24	− 17
CNG, unfavorable scenario	+ 12	+ 38
Methanol scenarios:		
Methanol from RNG, favorable scenario	− 18	− 15
Methanol from domestic NG, favorable scenario	− 22	
Methanol from RNG, unfavorable scenario	+ 18	+ 31
Methanol from domestic NG, unfavorable scenario	+ 13	+ 26
Methanol from coal, base case	+ 80	
Methanol from coal, favorable scenario	+ 35	
OTM/LPM/IGCC	0	
Methanol from flared gas	− 90	
Methanol vs. diesel	+ 15 to + 30	

Source: Mark DeLuchi, Greenhouse Gas Emissions Model, 1990

emissions. And, as Table 5 shows, greater energy use at other stages in the fuel-production-and-use cycle eliminates methanol's advantage altogether.

Scenario Analyses (Table 6)

Electric and Hydrogen Vehicle Scenarios

Advanced electric vehicles drawing power from the projected (year 2000) electricity-generating mix in the United States will produce slightly less emissions of greenhouse gases than will gasoline-fueled vehicles, as shown in the base case (Table 5). (The result is quite sensitive to the assumed efficiency of electric vehicles, compared with gasoline vehicles, which is difficult to estimate. This analysis assumes an efficiency for an electric vehicle in the year 2000.) However, if electric vehicles draw their power entirely from coal-fired plants, there will be a moderate increase in emissions of greenhouse gases, compared with those of gasoline-fueled vehicles (Table 6). If they draw power only from natural gas plants, there will be a slight to moderate decrease. The use of nuclear power to recharge electric vehicles will nearly but not entirely eliminate emissions of greenhouse gases, since there will still be minor emissions from the use of electricity at gaseous-diffusion uranium-enrichment plants. The use of solar power by electric vehicles or hydrogen vehicles will totally eliminate emissions of greenhouse gases.

Compressed Natural Gas: Favorable Scenario

This scenario changes the base-case assumptions for compressed natural gas in the following ways: the efficiency advantage for compressed natural gas increases from 10 percent (in Table 1) to 20 percent; CNG compressors use 40 percent coal-derived power instead of 52 percent in the base case; CNG compression is 96 percent efficient instead of 94.7 percent efficient (in Table 3); natural gas leaks from the field and pipeline are 1 percent of output instead of 1.5 percent (in Table 3); CNG vehicles emit 1 g/mi methane instead of 1.4 (in Table 2); and venting and flaring of associated gas in oil production increases from 0.024 mmBtu (Table 3) to 0.030. These changes boost the reduction in greenhouse gases of compressed natural gas vehicles, compared with those of gasoline-fueled vehicles, above that of the base case; the amount depends on the magnitude of the methane-to-carbon-dioxide conversion factor. (Compare Table 5 result for compressed natural gas with Table 6 results.)

If the methane conversion is 5 (rather than 11.6 in the base case), and all the other favorable assumptions remain the same, the result is about the same (26 percent reduction, compared to gasoline, instead of 24 percent as shown in Table 6 under 11.6). This indicates that the methane-to-carbon-dioxide conversion factor is relatively unimportant if it is around 10 or less, especially if methane emissions are relatively low. (Other scenario analyses confirm this.) However, as will be demonstrated, the conversion factor becomes important if it is much larger than 10 and if emissions of methane are large.

Varying the N_2O-to-CO_2 conversion factor, here and elsewhere, did not change the results significantly.

Compressed Natural Gas: Unfavorable Scenario

The following parameters change in this scenario: the vehicular efficiency advantage for CNG is eliminated; CNG compressors draw 60 percent of their power from coal plants; compression is 94 percent efficient; gas leaks total 4 percent of output; CNG vehicles emit 1.8 g/mi methane; the efficiency of the petroleum refineries increases slightly; and venting and flaring declines to 0.02/mmBtu. The result is a slight to moderate increase in emissions of greenhouse gases, compared with gasoline, depending on the methane conversion factor. The relatively large spread between the results with the high and the base-case conversion factors, shows that the conversion factor can be important when it and methane emissions are relatively high.

Base-Case Methanol Scenario

For reference, the methanol base case assumes M85 fuel (a mixture of 15 percent gasoline and 85 percent methanol); a 15 percent efficiency advantage for methanol; an average haul of 5,500 miles in large tankers using 100 Btu/ton-mile; a 2 percent energy requirement for transporting gas from the field to the methanol plant; a 65 percent rate of efficiency in methanol production; a 1.5 percent gas leakage from field systems; and base-case petroleum assumptions. The base case assumes methanol is made from remote natural gas (RNG). Other base-case assumptions are shown in tables 3 and 4.

Methanol from Remote Natural Gas: Favorable Scenario

This scenario assumes 100 percent methanol (M100 instead of M85); a 25 percent efficiency advantage for methanol; a high carbon content in the gas used by petroleum refineries; high venting and flaring of associated gas; high coal use in electricity plants; an average methanol haul of 3,000 miles at 80 Btu/ton-mile; a 70 percent efficiency in the conversion of natural gas to methanol in stand-alone plants (no purchased electricity); low gas leaks from fields and pipelines; and low energy use in gas-recovery field operations.

Methanol from Domestic Natural Gas: Favorable Scenario

This scenario assumes M100 instead of M85; 25 percent efficiency advantage for methanol; a 71 percent efficiency for natural gas conversion to methanol in stand-alone plants (no purchased electricity); high carbon content in refinery gas used by petroleum refineries; high venting and flaring of associated gas; high coal use in electricity plants; low gas leaks from fields and pipelines; and low energy use in gas-recovery field operations.

Methanol from Remote Natural Gas: Unfavorable Scenario

This scenario assumes an 8 percent efficiency advantage over gasoline; low carbon content of refinery gas; a change in refinery energy consumption, from 9 to 7 percent of output; less use of coal in electricity generation; low venting and flaring of associated gas; a 63 percent efficiency in natural-gas-to-methanol conversion; a 6,000-mile average methanol haul by tankers at 120 Btu/ton-mile; and 2.5 percent total leakage in natural gas operations.

Methanol from Domestic Natural Gas: Unfavorable Scenario

This scenario assumes an 8 percent efficiency advantage for methanol; low carbon content of refinery gas; refinery energy use is 7 percent of output; electricity generation uses less coal; low venting and flaring of associated gas; a 63 percent efficiency in natural-gas-to-methanol conversion; 2.5 percent total leakage in natural gas operations.

Methanol from Coal: Base Case

The important assumptions in this scenario are 15 percent efficiency advantage for methanol vehicles; M85; 55 percent efficiency in coal-to-methanol conversion; base-case parameters for gasoline (see tables 1 to 4). Other base-case assumptions for methanol from coal are shown in tables 3 and 4.

Methanol from Coal: Favorable Scenario

This scenario assumes a 25 percent efficiency advantage for methanol; M100 instead of M85; 65 percent efficiency in coal-to-methanol conversion in a stand-alone plant (no purchased power); high carbon content in refinery gas used by petroleum refineries;

high venting and flaring of associated gas; high coal use in electricity plants; low methane emissions from coal mines.

OTM/LPM/IGCC: Coproduction of Methanol and Electricity

OTM/LPM/IGCC is once-through methanol, liquid-phase methanol catalyst, integrated-gasification combined-cycle power plant. "Once through" means that the synthesis gas from coal gasification passes over the methanol synthesis catalyst only once. (In conventional methanol production the synthesis gas is passed over the catalyst many times.) It is possible to use only one pass because the liquid-phase catalyst converts more synthesis gas to methanol in one pass than does an ordinary catalyst, and because the unconverted synthesis gas is not wasted but is sent to the combined-cycle power plant to produce electricity. Methanol production, therefore, also produces electricity.

Emissions of greenhouse gases from coproduced methanol depend on how the electricity generation credit is estimated. If one assumes that the coproduced electricity displaces electricity-generating fuels in proportion to national average inputs (about half coal), then the use of coproduced methanol from coal would result in about the same emissions of greenhouse gases as does the use of gasoline.

However, for several reasons only a small amount of methanol will be produced in this way. First, there must be a major market for the electricity for the integrated-gasification combined-cycle (IGCC) plant to be built, since electricity is the major output. Only a small portion of the projected need for new electricity-generating capacity in the United States will be satisfied by coal IGCC power plants; nuclear, gas turbine, conventional coal, solar technologies, and other sources will also contribute. Second, only a portion of the methanol produced (by the relatively small number of IGCC coproducing plants to be built) will be available for use as a transportation fuel, as these plants are supposed to produce methanol, during off-peak hours, for use as a peaking fuel by the plant itself.

Use of Flared Gas to Produce Methanol

A small and declining proportion of associated natural gas is vented and flared. Use of this wasted gas would result in virtually no emissions of greenhouse gases, compared with the continued venting or flaring of the gas. If all the carbon dioxide originally contained in flared natural gas is subtracted from the life-cycle CO_2-equivalent g/mi emissions from methanol vehicles on the grounds that it would have been emitted anyway, only 30 g/mi CO_2-equivalent emissions are left. These are from the use of diesel fuel to ship methanol in tankers and trucks, and from methane and N_2O emissions. This is more than a 90 percent reduction in life-cycle greenhouse gases, compared with gasoline from petroleum.

However, one cannot argue that this conclusion justifies favoring methanol over other fuels. The "wasted" natural gas could be made into liquefied natural gas or gasoline, with little or no energy penalty, compared with making it into methanol. More important, the amount of vented and flared gas that will be available to make fuel for the U.S. transportation sector will be quite small. This amount, as a percentage of highway fuel use in the U.S., can be calculated as:

$$P_y = 100 \cdot T_y \cdot E \cdot US \cdot C / H_y$$

where:

P_y = The percentage of highway fuel demand satisfied in year Y by methanol from vented and flared gas.

T_y = The total amount of vented and flared gas in year Y, in quads, in the absence of a policy to make methanol from remote gas.

E = The fraction of total vented and flared gas of economic interest to anyone wishing to make methanol.

US = The fraction of E that the United States will be able to capture.

C = The ratio of methanol energy delivered to the U.S. to methanol energy in the feedstock.

H_y = Highway fuel use in year Y.

What is the highest that P could be? Imagine an aggressive U.S. program to convert the highway fleet to methanol, and to develop methanol from remote natural gas. It would take 10 to 20 years to get a significant number of plants built and running, worldwide, and enough cars built, and in the hands of consumers, to use the methanol. What is T likely to be in 10 to 20 years? In 1987 less than three trillion cubic feet were vented and flared worldwide (EIA, 1989). This quantity has been declining steadily as natural gas has become more valuable, pipeline export markets have opened up, liquefied natural gas export markets have developed, and developing countries have constructed the infrastructure to use the gas locally. In 15 years the amount of wasted gas could easily be less than two quads, but 2.5 quads will be assumed to give the benefit of the doubt to policies that emphasize the conversion of wasted gas to transportation fuels.

Not all of the gas will be economically attractive. After all, gas is vented and flared largely because it is not worthwhile to use it. That is to say, there is an insufficient amount of it or an insufficient amount of capital to develop infrastructure for its use. Suppose, though, that 75 percent (= variable E) of the gas can be converted to methanol economically in small-scale plants and that the gas is located in countries willing to sell it and from which the United States is willing to purchase it. (Note that at various times in the recent past this last criterion would have eliminated the USSR and certain countries in the Middle East, all of which vent and flare much gas.)

The United States will not be the only country interested in methanol as a transportation fuel. Other countries will compete for the methanol, and this demand will be in addition to the factors that have reduced the quantity of wasted gas in the 1980s. Assume, generously, that the United States gets 75 percent (= variable US) of the wasted gas available for methanol. Counting gas used to process and pipe the gas to methanol plants, and allowing for relatively efficient methanol production technologies (using natural gas as a process fuel, as they all will), C will be approximately 65 percent.

As for H_y, assume generously that improvements in vehicle efficiency more than keep pace with increasing vehicle miles traveled and the desire for larger, heavier, more powerful cars, so that overall highway fuel consumption *falls* from more than 16 quads in 1987 to 15 quads in year Y. With these quite favorable assumptions, P = about 6 percent. With more realistic assumptions, P would probably be more on the order of 1 percent.

General Note on Methanol Compared with Diesel

All the preceding scenarios compare methanol with gasoline in light-duty applications. However, methanol also is being considered as a replacement for heavy-duty diesel-fuel applications. Generally, methanol heavy-duty vehicles will emit about 30 percent more greenhouse gases per mile than do diesel heavy-duty vehicles, given the current state of the art in methanol heavy-duty vehicles technology. If methanol heavy-duty vehicles attain mi/mmBtu fuel economy equivalency with diesel heavy-duty vehicles (they currently are not close), this increase will decline to about 15 percent.

Conclusions

The use of fossil-fuel feedstocks—coal, natural gas, and petroleum—to make transportation fuels will not significantly reduce emissions of greenhouse gases, compared with current gasoline and diesel fuel use, regardless of the feedstock, the fuel production process, and the final transportation fuel. Emissions of greenhouse gases from the transportation sector can be reduced significantly only by using non-fossil fuel feedstocks, or by significantly improving fleet efficiency.

The use of coal to produce methanol, electricity for electric vehicles, hydrogen, or synthetic natural gas (see DeLuchi, Johnston, and Sperling, 1987a, for analyses of hydrogen from coal, and discussion of synthetic natural gas from coal) will increase CO_2-equivalent emissions from the transportation sector. Coproduction of methanol and electricity from coal will result in about the same fuel cycle emissions of greenhouse gases as from the use of petroleum transportation fuels, but nationally this process can supply only a small amount of methanol. If coal is used to provide the energy needed to convert corn to ethanol, then the ethanol cycle would emit about the same amount of greenhouse gases as the gasoline cycle.

The use of natural gas to produce methanol, compressed natural gas, liquefied natural gas, or electricity for vehicles will result in, at the most, only slight reductions in CO_2-equivalent emissions, compared with gasoline and diesel fuel. Improving average fleet efficiency would reduce carbon dioxide emissions (but not necessarily CH_4 and N_2O) proportionately and is a good strategy for reducing emissions of greenhouse gases in the transportation sector. A doubling of the average efficiency of the whole U.S. vehicle fleet would reduce carbon dioxide emissions by 50 percent.

The production of alcohol fuels or synthetic natural gas from woody biomass can greatly reduce emissions of greenhouse gases, compared with the use of gasoline and diesel fuel. Small amounts of greenhouse gases are emitted from the use of energy and fertilizer in silviculture. Electricity from solar, hydro, or geothermal power, used to power electric vehicles or split water to make hydrogen fuel, results in essentially no emissions of greenhouse gases. The use of nuclear power results in a large decrease (but not total elimination) of emissions of greenhouse gases, given current gaseous-diffusion enrichment technology.

Focus on Methanol

The methanol-from-natural-gas cycle could emit as much as 30 percent more or less greenhouse gases per mile than the gasoline-from-crude-oil cycle, depending on assumptions. The most important unknowns are the efficiency of methanol vehicles compared with gasoline vehicles (this analysis considered efficiency advantages from 8 to 25 percent); the efficiency of conversion of natural gas to methanol (future plants may be more efficient than current state-of-the-art plants, but the driving forces will be economics and risk, not environmental concerns); the magnitude of gas leaks from fields and pipelines supplying methanol plants (which could vary between 0.5 and 4.0 percent); the average distance of methanol shipments by sea and the energy intensity of the tankers (which depends on their size, speed, and engine technology); the "true" methane-to-carbon-dioxide conversion factor (which depends in part on the extent to which one looks into the future); the efficiency of natural gas recovery and transmission operations; the composition of petroleum refinery gas (the main fuel used at petroleum refineries); the efficiency of future petroleum refineries; the amount of associated gas vented and flared (which in recent years has been declining); emissions of N_2O and CH_4 from tankers, refineries, and methanol production plants; and the mix of fuels used to generate electricity.

To sum up qualitatively, methanol from natural gas provides no greenhouse benefit. A series of optimistic but not unreasonable assumptions results in methanol's providing a moderate reduction in emissions of greenhouse gases; a series of pessimistic assumptions results in an equally moderate increase.

This analysis, then, supports the following statements about methanol and the greenhouse effect: If the policy objective is to hold the line on emissions of greenhouse gases from the transportation sector, methanol from natural gas is acceptable except in the worst cases. If the goal is to make moderate reductions in emissions of greenhouse gases, methanol from natural gas is acceptable only in the most optimistic cases. In all other cases it is not.

If the goal is to make large reductions in emissions of greenhouse gases from the transportation sector, and eventually eliminate such emissions, then methanol—from any fossil feedstock and using any production technology—is not acceptable under any condition. Methanol made from woody biomass is acceptable. Methanol from coal is acceptable only if one does *not* wish to reduce emissions of greenhouse gases. Therefore, this author concludes that unless we expect to be indefinitely indifferent about the greenhouse effect, fossil methanol should be no more than a transitional fuel to non-fossil energy sources.

Recommendations

This analysis has shown that substantial, long-term reductions in CO_2-equivalent emissions from the transportation sector can be accomplished only by greatly increasing the efficiency of vehicles, by using wood-based biofuels for natural gas or methanol vehicles, or by using non-fossil electricity for hydrogen and electric vehicles. From a greenhouse perspective, the policy question is: What is holding these options back, and what can be done to encourage their adoption?

Further improvements in fleet average efficiency are being stalled primarily by low world oil prices and consumer demand for larger and more powerful cars. Biofuels and solar energy are expensive (on a private cost basis), and nuclear energy is expensive and politically unpopular. Electric vehicles do not perform well enough to be used in all highway applications; hydrogen vehicles do, but are very expensive (on a private cost basis), and fuel storage is still a problem. In a nutshell, current petroleum fuels are relatively cheap, alternative fuels that do not produce carbon dioxide are relatively costly (on a private cost basis), and some alternative vehicle technologies have performance drawbacks. Two kinds of policies are needed to address these problems.

First, fuels and technologies should be priced at their social (full economic) cost, not at their private cost. Gasoline is currently the cheapest transportation fuel on a private cost basis, but the external costs of gasoline use, from air pollution to the cost of defending oil fields in the Middle East, are probably very large. It is possible that if the external costs of gasoline and diesel fuel use are at the high end of a plausible range, and the private costs of hydrogen or electric vehicles using solar power (with essentially no external costs) are at the low end of a plausible range, the use of hydrogen and electric vehicles, where their performance is acceptable, is more economically efficient, in the broadest sense.

Second, research and development should be directed at fuel and vehicle combinations with low external costs, especially those that do not produce carbon dioxide or exacerbate global tensions, since these external costs are difficult to estimate but are probably very large. For electric vehicles, this means research and development aimed at increasing the energy density and power of batteries, reducing battery cost, and reducing recharging time, without sacrificing battery performance and life. For hydrogen vehicles, research and development should focus on increasing the mass-energy density of hydrides, reducing the desorption temperature of hydrides, increasing the no-vent period for liquid-hydrogen (LH_2) vehicles, making the handling of LH_2 boil-off safe, and reducing storage costs for both hydride and LH_2 vehicles. Work on hydrogen-powered fuel-cell vehicles, which combine the best attributes of hydrogen and electric vehicles, should commence.

Today, the United States provides only modest support for the development of solar technologies and electric vehicles, almost no support for the development of hydrogen technologies, and no support at all for the development of hydrogen vehicles. Considering that these technologies are extremely benign environmentally, and technically promising, this lack of support is shortsighted.

Proper pricing of petroleum fuels will encourage efficiency improvements and reduce carbon dioxide emissions proportionately and will increase national efficiency of resource use. Proper pricing, combined with increased research and development on solar energy production and hydrogen and electric vehicles, will hasten the efficient adoption of sustainable, environmentally sound options for transportation fuels that do not produce carbon dioxide. We should start moving on this path today.

References

Bolin, B., et al., eds. (1986). *SCOPE 29: The Greenhouse Effect, Climatic Change, and Ecosystems.* N.Y.: John Wiley and Sons.

Bureau of the Census (1989). 1987 Census of Mineral Industries, Crude Petroleum and Natural Gas (Preliminary Report, Industry 1311). U.S. Department of Commerce, Washington, D.C., MIC87-I-13A(P).

_____ (1989a). 1987 Census of Mineral Industries, Coal Mining (Preliminary Report, Industries 1221, 1222, 1231, and 1241). U.S. Department of Commerce, Washington, D.C., MIC87-I-12A(P).

_____ (1989b). 1987 Census of Mineral Industries, Ferroalloy Ores, Metal Mining Services, and Miscellaneous Metal Ores (Preliminary Report, Industries 1061, 1081, 1092, 1094, and 1099). U.S. Department of Commerce, Washington, D.C., MIC87-I-10C(P).

_____ (1989c). 1987 Census of Mineral Industries, Oil and Gas Field Services (Preliminary Report, Industries 1381, 1382, and 1389). U.S Department of Commerce, Washington, D.C., MIC87-I-13C(P).

DeLuchi, M.A., D. Sperling, and R.A. Johnston (1987). A Comparative Analysis of Future Transportation Fuels. Berkeley, Calif.: University of California, Institute of Transportation Studies, UCB-ITS-RR-87-13.

DeLuchi, M.A., R.A. Johnston, and D. Sperling (1987a). Transportation Fuels and the Greenhouse Effect. Berkeley, Calif.: University of California, University Energy Research Group, UER-182.

Dickinson, R.E., and R.J. Cicerone (1986). Future Global Warming from Atmospheric Trace Gases. *Nature* 319, pp. 104–119.

DOE, Assistant Secretary for Environmental Protection, Safety, and Emergency Preparedness (1983). Energy Technology Characterizations Handbook, Environmental Pollution and Control Factors, 3rd ed. U.S. Department of Energy, Washington, D.C.

EIA (1988). Coal Distribution January–December 1987. U.S. Department of Energy, Energy Information Administration, Report No. DOE/EIA-0125(87/4Q), Washington, D.C. (also earlier years).

_____ (1988a). Natural Gas Annual 1987, Vol. 1. U.S. Department of Energy, Energy Information Administration, Report No. DOE/EIA-0131(87)/1, Washington, D.C. (also earlier years).

_____ (1988b). Uranium Industry Annual 1987. U.S. Department of Energy, Energy Information Administration, Report No. DOE/EIA-0478(87), Washington, D.C. (also earlier years).

_____ (1989). Petroleum Supply Annual 1988. Vol. 1. U.S. Department of Energy, Energy Information Administration, Report No. DOE/EIA-0340(88)/1, Washington, D.C. (also earlier years).

_____ (1989a). International Energy Annual 1988. U.S. Department of Energy, Energy Information Administration, Report No. DOE/EIA-0219(88), Washington, D.C. (also earlier years).

EPA (1985). Compilation of Air Pollutant Emission Factors. Vol. 2, Mobile Sources, 4th ed. U.S. Environmental Protection Agency, Office of Mobile Sources, Report No. AP-42, Research Triangle Park, N.C.

_____ (1985a). Compilation of Air Pollutant Emission Factors. Vol. 2, Stationary Sources, 4th ed. U.S. Environmental Protection Agency, Office of Air and Radiation, Report No. AP-42, Research Triangle Park, N.C. (with updates to September 1988).

Firor, J. (1989). The Straight Story About the Greenhouse Effect. Paper presented at the Annual WEA International Meeting, Lake Tahoe, June 18–22, 1989.

Gushee, D. (1988). Carbon Dioxide Emissions From Methanol as a Vehicle Fuel, 88-14S. Washington, D.C.: Congressional Research Service.

Lashof, D.A., and D.R. Ahoja (1989). Relative Global Warming Potentials of Greenhouse Gas Emissions. Submitted to _Nature_.

MacCracken, M.C. (1989). Greenhouse Gases: Changing the Nature of Our Environment. Paper presented at Winter Meeting of the American Nuclear Society, San Francisco, Calif., November 29, 1989.

MacCracken, M.C., and F.M. Luther, eds. (1985). Projecting the Climatic Effects of Increasing Carbon Dioxide. U.S. Department of Energy, Carbon Dioxide Research Division, Report No. DOE/ER-0237, Washington, D.C.

MacKenzie, J. (1987). Relative Releases of Carbon Dioxide from Several Fuels. Washington, D.C.: World Resources Institute.

Mahlman, J.D. (1989). Mathematical Modeling of the Greenhouse Warming: How Much Do We Know? In *Global Change and Our Common Future*, edited by Ruth S. Defries and Thomas F. Malone. Washington, D.C.: National Academy Press.

Ramanathan, V. (1988). The Greenhouse Theory of Climate Change: A Test by an Inadvertent Global Experiment. *Science* 240, pp. 293–99.

Ramanathan, V., et al. (1985). Trace Gas Trends and Their Potential Role in Climate Change. *Journal of Geophysical Research* 90, pp. 5547–66.

Rose, A.B. (1979). Energy Intensity and Related Parameters of Selected Transportation Modes: Freight Movements. Tenn.: Oak Ridge National Laboratory, ORNL-5554.

Unnasch, S., et al. (1989). Comparing the Impacts of Different Transportation Fuels on the Greenhouse Effect. California Energy Commission, Sacramento, Calif., P500-89-001.

Wang, W., and G. Molnar (1985). A Model Study of Greenhouse Effects Due to Increasing Atmospheric CH_4, N_2O, CF_2Cl_2, and $CFCl_3$. *Journal of Geophysical Research* 90, pp. 12791–80.

White, H.M. (1980). Assessment of Alternative Automotive Fuels Impact on CO_2 Emissions. The Aerospace Corporation, Mobile Systems Directorate, Los Angeles, Calif., for the U.S. Department of Energy.

Commentary

MICHAEL KELLY

Carmen Difiglio acknowledges in his paper that methanol production, in the short and long term, will be based on natural gas feedstocks and that the additional production of methanol fuel will come from overseas. Difiglio then presents production cost estimates for a methanol plant of 2,500-ton/day production capacity, using current steam-reforming technology, and estimates for a 10,000 ton/day plant based on a more advanced partial- oxidation process. In both cases the figures assume that there is $1.10/mmBtu cost of natural gas feedstock and that the plant is situated in a remote location with existing infrastructure. Difiglio arrives at a plant gate cost of methanol in the 10,000 ton/day case (including capital recovery, operating, and gas feedstock costs) of 28 cents/gal. However, only 12 cents of the total represents capital recovery which is substantially lower than other estimates.

In submissions to the California Energy Commission (CEC) earlier this year Bechtel Financing Services estimated the capital recovery cost per gallon of methanol for a 10,000 ton/day plant in six locations as follows:

Texas	14¢
Canada	16¢
Trinidad	23¢
Alaska	26¢
Saudi Arabia	29¢
Australia	31¢

These estimates were based on a nominal return on investment of 17 to 20 percent, depending on location and country-specific tax rates for corporations.

In its draft report (AB234) the CEC used lower rates of return to estimate the following capital recovery costs:

Texas	13¢
Canada	14¢
Trinidad	15¢
Alaska	26¢
Saudi Arabia	20¢
Australia	28¢

Thus Difiglio's figure of 12 cents/gal of methanol is lower than either the Bechtel or CEC figures. He appears to rely on economies of scale and advanced technology to achieve this reduction. However, as the technology is as yet unproven, it remains to be seen whether the lower capital costs can be achieved. For remote locations, "reliability of operation will be of paramount importance, so the plant must be built with proven technology and equipment," according to the Chem Systems report that Difiglio cited.

According to Difiglio, the current capacity of the methanol fuel industry is about 400,000 b/d. Therefore, the existing industry must expand by a factor of six in order to displace one million b/d of gasoline—about one-seventh of total U.S. use—with approximately two million b/d of methanol. This would require the construction of 25 advanced technology (10,000 ton/day) plants.

In regard to Difiglio's outlook for the timing of the development of a methanol fuel industry, the following points should be considered. Ultimately, how quickly methanol fuel plants will be built will depend on when a market for methanol develops. It is worth drawing some comparisons to the liquefied natural gas (LNG) market. In terms of gas volume, a 10,000-ton/day plant producing 80,000 b/d of methanol fuel is broadly equivalent to a single two-million-ton/yr LNG train requiring more than 100 bcf/yr of gas feedstock. Therefore, 25 such methanol plants would require roughly

2.5 tcf/yr of gas. In 1988, the total world trade in LNG amounted to 60.5 bcm (billion cubic meters gaseous volume), which is equivalent to 2.1 tcf. Consequently, Difiglio is postulating a trade in methanol fuel, for the U.S. market alone, that exceeds the size of the current world trade in LNG.

The first experimental shipment of LNG occurred in February 1959. Since then it has taken 30 years for the world LNG trade to reach its present size. In light of this experience, and the fact that methanol-fueled vehicles are not expected to be produced in quantity until the mid-1990s, it seems very unlikely that two million b/d of methanol fuel will be produced even by the year 2010.

With regard to the availability and cost of gas, Difiglio asserts that ample gas reserves exist in a variety of countries, at a cost equivalent to $1.10/mmBtu today, to support the development of 25 advanced-technology methanol fuel plants. In my opinion, the use of proven reserves and production numbers paints a rosier picture of gas availability than actually exists (Figure 1). Many gas reserves

Figure 1
World Exportable Gas Surplus
December 31, 1987

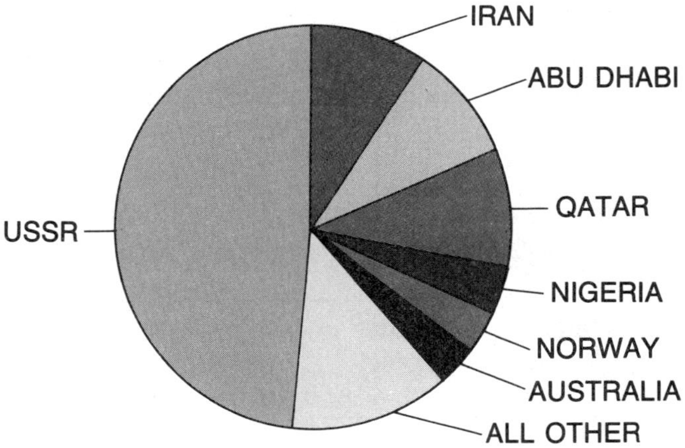

Total world exportable surplus 1,666 tcf

Source: Jensen Associates, Inc., 1988

are already committed to markets, and others are associated with oil, the production of which will dictate the rate at which the gas is produced.

In addition to identifying gas reserves as associated or non-associated with oil production, Jensen Associates, Inc., defines them according to three principal development categories, each of which has two subcategories. Gas reserves already committed to markets include those needed to meet domestic consumption requirements and those under contract for export by pipeline or as LNG. The second principal category comprises gas reserves that will be delayed in their development. These may be deferred, such as gas cap and reinjected gas in oil reservoirs, or frontier reserves, such as those in the arctic areas of the United States and Canada, whose development must await the provision of expensive transportation infrastructure. The third category of reserves represents surplus gas that may be marginally economic to develop or export.

Applying this methodology to Venezuela, for example, it can be shown that of 95 tcf of proven gas reserves at the end of 1987, 80.8 tcf were associated with oil production (Table 1). Similarly, 48.1 of the 95 tcf total were already committed to domestic markets, 28.4 tcf were in the deferred development category, and 4.3 tcf were marginally economic. Thus, only 14.2 tcf remain as an exportable surplus. However, this amount is still sufficient to support several advanced-technology methanol fuel plants.

Table 1
Venezuelan Gas Reserves, Year End 1987 (Tcf)

Proven Resources			Committed to Market		Delayed Development		Surplus	
Total	Assoc.	Non-Assoc.	Domest.	Export	Deferred	Frontier	Marginal	Export
95.0	80.8	14.2	48.1	0	28.4	0	4.3	14.2

Source: Jensen Associates, Inc., 1987

Figure 1 illustrates exportable surpluses by country at the end of 1987 and demonstrates that more than half the total exportable surplus is in the USSR and Iran. Special mention should be made of Saudi Arabia, for which no exportable surplus is shown. In the past, Saudi government policy has opposed the export of gas as fuel. Indeed, given that most of its reserves are associated with oil, the country suffered electricity shortages when oil production fell earlier in the 1980s. An attempt has been made to find non-associated gas reserves and has met with some successes. However, in the judgment of Jensen Associates, the Saudis are still not amenable to exporting gas. For methanol, the case is somewhat different. Although Saudi Arabia canceled a methanol project sponsored by Houston Natural Gas in the early 1970s, the Saudis have since gone ahead to develop a petrochemical industry. Consequently, it is likely that gas would be provided for a methanol fuel plant there.

Turning to the issue of gas feedstock prices for methanol fuel production, $1.10/mmBtu is not an unreasonable figure to assume, given current supplies, for a 10,000-ton/day plant. In some locations it may be available for less. Figures 2 and 3 show the estimated costs of service and prices of gas in various countries/regions (developed by Jensen Associates for submission to the CEC in early 1989). If crude oil prices were to rise, as projected by the U.S. Department of Energy, gas feedstock prices for methanol fuel production would rise also—but then so would the cost of gasoline, against which the methanol fuel would compete.

There is no question that methanol-fuel plants could be built more cheaply in some locations than in others. That natural gas feedstock could be supplied at the current price of $1.10/mmBtu in some locations is also undisputed. The important issue is whether each of the 25 advanced-technology plants postulated by Difiglio could be built for the $680-million-per plant cost he estimates, in regions where gas would also be available for $1.10/mmBtu.

Finally, it should be noted that if the LNG trade has taught us anything, it is that large, international, capital-intensive projects can be best handled through vertically integrated organizations in

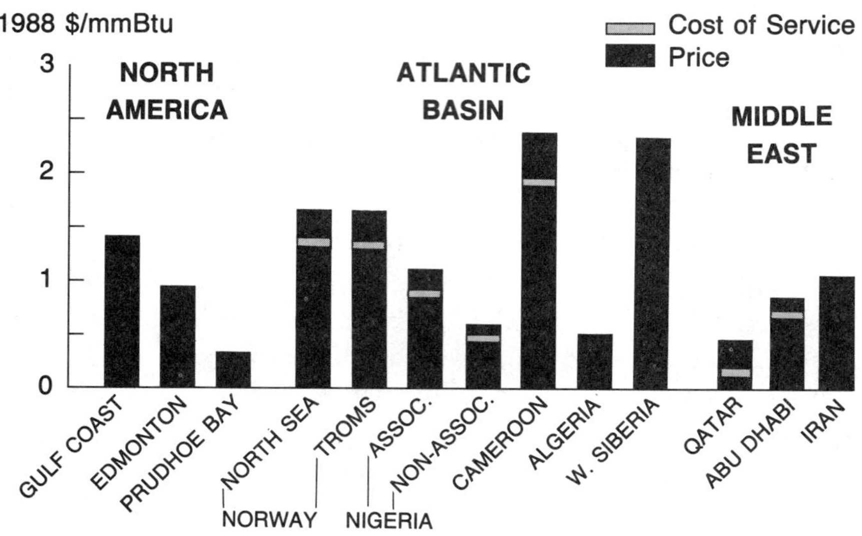

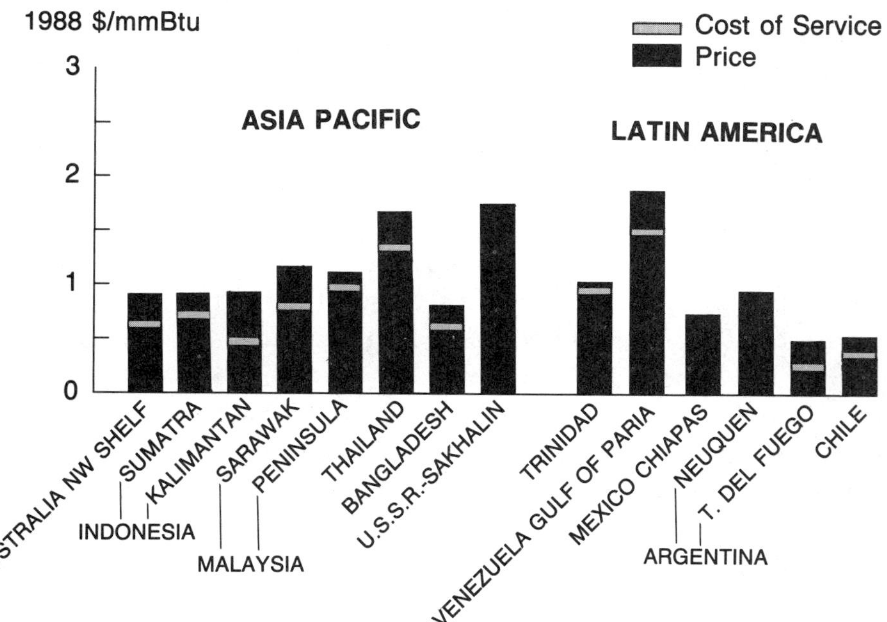

Source: Jensen Associates, Inc., 1988

which all of the participants, from the host country to the final supplier, share the risks and rewards in all parts of the chain. For a variety of reasons U.S. policy and regulatory practice have not been accustomed to this approach to obtaining energy supplies. Therefore, if imported methanol is to play a significant role in the U.S. transportation sector, new approaches may have to be adopted.

Commentary

MICHAEL D. JACKSON

Carmen Difiglio estimates in his paper that with current technology (2,500-ton/day plant) methanol can be produced at 42 cents/gal while the price drops to 28 cents/gal with advanced technology (10,000-ton/day plant using partial oxidation). These prices are consistent with studies performed by the California Advisory Board on Air Quality and Fuels (Acurex Corporation, 1989), California Energy Commission (CEC, 1989), and the U.S. Environmental Protection Agency (EPA, 1989). Spot prices of methanol during the last few years have ranged from 20 to 80 cents/gal. The California Methanol Fuel Reserve bought methanol in 1988 for 45 cents/gal, which included local storage of about 3 cents/gal per month. In 1989 it dropped to 40 cents/gal.

In addition to the major cost elements of a methanol plant such as financing, operations, and feedstock costs identified by Difiglio, overseas transportation and dealer markup must be considered. With current technology, he estimates the price of M85 (85 percent methanol, 15 percent gasoline) at the retail pump to be 80 cents/gal, or a gasoline equivalent price of about \$1.30/gal. With advanced technology, prices are projected to drop to 69 cents/gal, which is equivalent to a price of \$1.11/gal for gasoline. M85 currently sells in California at Chevron and Arco stations for between 69 and 80 cents/gal. Difiglio's estimate for the price of methanol produced with advanced technology is conservative. Partial-oxidation technology is now fairly well accepted and proved. Also a 10,000-ton/day plant would probably be constructed from four 2,500-ton/day trains.

Introducing methanol or any other alternative fuel will take many years, since building the feedstock supply and infrastructure will be a lengthy process. Replacing one million gallons of gasoline, for instance, would require the construction of 25 methanol plants producing 10,000 ton/day. This could take at least 8 to 12 years, and to obtain the necessary capital it will be necessary for investors to anticipate reasonable returns and minimal financial risks.

Difiglio believes there is an ample amount of remote natural gas to produce large quantities of methanol, even if one excludes such countries or regions as the USSR or the Middle East, which have substantial gas resources. Studies performed by the Acurex Corporation for CEC (Lowell et al., 1989) concur with Difiglio's conclusions. There appears to be enough gas in other areas of the world so that methanol production would not need to rely on gas reserves of OPEC countries or the Soviet Union.

Difiglio thinks that diversifying the sources of fuels used in the U.S. transportation sector will enhance U.S. energy security. Methanol could be produced by a number of countries and, in contrast to oil production, a single region of the world would not control most of the sources. Difiglio also argues that methanol producers would be less likely to curtail output because of the large investments in production facilities. This has been true with chemical-grade methanol—whether the market would continue to be competitive if a large methanol fuel market develops is unclear. Finally, Difiglio argues that reduced oil demand as a result of methanol's introduction will lower world oil prices.

Difiglio concludes that the introduction of methanol as a regional air quality strategy would lead to improved energy security for the United States. The state of California has also made this argument. However, I believe more data are needed before drawing this conclusion.

"Emissions of Greenhouse Gases From the Use of Gasoline, Methanol and Other Alternative Transportation Fuels" is a refinement of Mark DeLuchi's early work on estimating the effect on global warming of producing and using different fuels (DeLuchi, Sperling, and Johnston, 1987). Others have also reviewed the possible effect of using alternative fuels on the greenhouse effect (Unnasch et al., 1989; Gushee, 1989). All of these studies tried to account for greenhouse gas emissions from production, to transportation, to end-use of various transportation fuels.

The methodology used by DeLuchi and others is to estimate along each step of the fuel production process the amounts of carbon dioxide (CO_2), methane (CH_4), and nitrous oxide (N_2O) released to the atmosphere. This requires detailed energy and carbon balances for each step in the process, from extraction, to preparation, to conversion. Estimates are then made of the relative warming effects of these various gases in comparison with carbon dioxide and the results are reported in equivalent carbon dioxide emissions (g/mi). This method estimates the contribution of one vehicle and does not account for any changes that might occur in refining operations if alternative fuels were widely introduced. Furthermore, the results are very dependent on disaggregating the various energy flows accurately during fuel production and use. Although simple in concept, it is a very tedious task.

DeLuchi performed this analysis for light-duty vehicles operating on gasoline, methanol, natural gas, electricity, hydrogen, and ethanol. Similar calculations were performed for heavy-duty vehicles operating on diesel, methanol, natural gas, hydrogen, and ethanol. Vehicle fuel efficiency is the primary factor affecting total fuel cycle carbon dioxide emissions.

DeLuchi's assumptions differed significantly from analyses by the Acurex Corporation (Unnasch et al., 1989) in the allocation of flared natural gas to petroleum refining, the allocation of energy equally across the entire petroleum product slate, and process efficiencies for the conversion of natural gas to methanol. In the first area DeLuchi appears to have used more recent figures than were available to Acurex. The allocation of refinery energy equally over the product range appears to be an over-simplification, since, for example, gasoline production is much more energy-intensive than is that of fuel oil. Recent studies by Bechtel have indicated methanol process efficiencies of around 67 percent (Bechtel, 1988) compared with DeLuchi's estimate of 56 to 65 percent.

Accounting for all of the carbon dioxide and applying appropriate warming factors to methane and nitrous oxide, DeLuchi found that, in general, none of the alternative fuels, with the possible exception of hydrogen, provided any significant improvement from the current gasoline baseline. Electric vehicles were found to be worse than gasoline if the current U.S. electricity generation mix is used. However, this would improve if more solar or nuclear energy is used. Table 1 briefly summarizes DeLuchi's results for light-duty vehicles. For comparison, the results of recent studies by Acurex (Unnasch et al., 1989) are also shown for methanol and compressed natural gas compared with the baseline gasoline. Even though there are some substantial differences in assumptions, results of the analyses are fairly consistent in their trends.

Table 1
Light-Duty Vehicles Compared with Gasoline-Fueled Vehicles

	DeLuchi	Acurex
Methanol	0.0 % Better	2 to 23 % Better
CNG	13.1 % Better	14 to 29 % Better
EVs	8.8 % Worse	
Hydrogen	43.6 % Better	
Ethanol	28.4 % Worse	

Source: Unnasch et al., 1989

Other sources of greenhouse gases identified by DeLuchi are automobile chlorofluorocarbons, carbon dioxide associated with manufacturing vehicles, and ozone. Acurex estimated the effect of automotive chlorofluorocarbons use and found that their emissions roughly exceed a vehicle's combined lifetime emissions of fuel-related carbon dioxide, methane, and nitrous oxide. DeLuchi estimates the carbon dioxide associated with manufacturing a vehicle at about 20 percent of the vehicle's total lifetime carbon dioxide emissions. This may be more significant for natural gas vehicles if the manufacture of the compressor and the tank are included. Other gases such as ozone and water vapor have not been included in these analyses. DeLuchi points this out and concludes that their inclusion would be difficult, due to the problems in estimating their amounts and their effect on global warming.

In summary, DeLuchi's paper does a very good job at quantifying the greenhouse effects of alternative fuels, compared with baseline fuels. His results, which are consistent with previous studies, indicate that all near-term alternatives—such as methanol, ethanol, compressed natural gas, or electric vehicles—do not substantially reduce greenhouse gas emissions from the transportation sector. Conversely, fuels such as methanol and natural gas do not worsen the effect and do appear to provide a slight benefit.

The most effective way to control greenhouse gases from the transportation sector would be to improve vehicle fuel efficiency. Doubling fuel economy would reduce carbon dioxide by about 50 percent. Alternative fuels may play a role if engines using them can be designed more efficiently than gasoline engines. Finally, there are many uncertainties in these analyses. It is necessary to look at the range of possibilities and draw conclusions based on a comparison of the trends.

References

Acurex Corporation (1989). California Advisory Board on Air Quality Fuels, Economics Report, Vol. 4, Draft Final Report. Mt. View, Calif.

Bechtel, Inc. (1988). California Fuel Methanol Cost Study, Vol. 2, Final Report. San Francisco, Calif.

CEC (1989). AB 234 Report: Cost and Availability of Low-Emission Motor Vehicles and Fuels, Vol. 1 and 2. California Energy Commission, CEC Report P500-89-008A, Sacramento, Calif.

DeLuchi, M.A., D. Sperling, and R.A. Johnston (1987). A Comparative Analysis of Future Transportation Fuels. University of California, Institute of Transportation Studies, Research Report UCB-ITS-RR-87-13, Berkeley, Calif.

EPA (1989). Analysis of the Economic and Environmental Effects of Methanol as an Automotive Fuel. U.S. Environmental Protection Agency.

Gushee, D.E. (1989). Carbon Dioxide Emissions from Methanol as a Vehicle Fuel. Congressional Research Service Report 88-89-14S, Washington, DC.

Ho, S.P. (1989). Global Warming Impact of Ethanol Versus Gasoline. Prepared by Amoco Oil Company for presentation at National Conference of Clean Air Issues and America's Motor Fuel Business, Washington, DC.

Koyama, K., and C.B. Moyer (1989). The Transition to Alternative Fuels in California. In *Alternative Transportation Fuels: The Environmental and Energy Solution*, edited by D. Sperling. N.Y.: Quorum Books.

Lowell, D.D., et al. (1989). Methanol as a Motor Fuel: Review of Issues Related to Air Quality, Demand, Supply, Cost, Consumer Acceptance and Health and Safety. Prepared by Acurex Corporation for California Energy Commission, CEC Report P500-89-002, Sacramento, Calif.

Unnasch, S., et al. (1989). Comparing the Impacts of Different Transportation Fuels on the Greenhouse Effect. Prepared by Acurex Corporation for California Energy Commission, CEC Report P500-89-001, Sacramento, Calif.

Part IV

Health and Safety

CHAPTER 9

A Health and Safety Assessment of Methanol as an Alternative Fuel

PAUL A. MACHIELE

Abstract

For air quality and energy security reasons, a great deal of attention is currently being given to methanol as a possible means of meeting U.S. needs for transportation fuels. It is important that human health and safety issues be considered when analyzing the appropriateness of alternative fuels. This paper discusses the health and safety implications of methanol fuels (M100 and M85) and places them in context with the current risks of using gasoline.

Introduction

Methanol fuels have been gaining favor in recent years as alternatives to petroleum fuels for a number of reasons. These include its excellent combustion characteristics and its possible contributions to improved air quality and U.S. energy security. However, the desirability of methanol fuels has recently come into question for a variety of safety reasons. Concerns about explosive fuel tanks, invisible flames, formaldehyde emissions, and increased fuel toxicity have all been raised as potential barriers to the use of methanol fuels. Frequently, those who raise these concerns fail to consider the benefits that methanol fuels offer compared with the gasoline fuels in current use. Similarly, they often fail to recognize that relatively simple changes in vehicle design, fuel handling procedures, and fuel additives can be developed to reduce possible risks.

In the discussion that follows, the extent of current fuel safety risks from the use of gasoline should be borne in mind. These are compared with the risks that we can project for the use of methanol fuels. Assumptions are made regarding the effects of certain characteristics of methanol, the usage patterns that are anticipated with methanol, and the safety precautions that would need to be considered if methanol fuels come into widespread use. This paper will discuss concerns about flammability, acute toxicity, and, finally, air toxics.

Flammability

Current Fire Risks With Gasoline

There are an estimated 500,000 collision and non-collision vehicle fires annually in the United States for which arson is not suspected. These fires result in approximately 2,070 fatalities, 12,200 serious injuries, and $870 million in property damage (FEMA, 1986; EPA, 1988). In roughly 58 percent of the collision fires and 38 percent of the non-collision fires, gasoline is the material first ignited (FEMA, 1986). In addition, in 1.6 percent of non-vehicle fires (approximately 27,000 fires), gasoline is the material that is first ignited. An even greater proportion of fires ultimately involves gasoline.

Since injuries and fatalities in vehicle collisions involving fire often occur as a result of the collision itself, not all of the fatalities and injuries can be attributed to the ensuing fires. It is difficult to assess whether death or injury was caused by the collision or the fire, but it has been estimated that as many as 60 percent of the fatalities may be due to the fire, and this is assumed to apply to injuries as well (Data Link, Inc., 1988). Table 1 estimates the number of fires that are attributable to gasoline and the number of injuries and fatalities that are in turn attributed to those fires.

Table 1
U.S. Annual Vehicle Fires and Deaths and Injuries Due to Fire in Cases Where Fuel Is the Material First Ignited

	Gasoline		
	Collision	Non-collision	Total
Fires	10,200	180,900	191,100
Deaths	710[a]	148	858
Injuries	2,950[a]	2,110	5,060

	Projection for M100		
	Collision	Non-collision	Total
Fires	1,020	18,100	19,100
Deaths	36[a]	7	43
Injuries	143[a]	106	249

	Projection for M85		
	Collision	Non-collision	Total
Fires	7,650	135,700	143,300
Deaths	320[a]	67	387
Injuries	1,330[a]	950	2,280

[a] Assumes just 60 percent of deaths and injuries in collisions where fire occurs today are due to the fire.

Note: Excludes cases of suspected arson.

Source: FEMA, 1986; EPA, 1988

Despite the fact that diesel fuel represents approximately 20 percent of U.S. transportation fuel consumption, it accounts for just 3.0 percent and 1.2 percent, respectively, of the collision and non-collision related fires in which gasoline or diesel fuel is the first material ignited (FEMA, 1986). Although there are a number of differences in the usage patterns of diesel fuel and gasoline, the low fire rate for diesel fuel can be mostly attributed to its extremely low volatility.

The Projected Fire Risks From Methanol Fuels

A number of the properties of methanol cause it to be both less likely to ignite and less likely to cause injury if it does ignite. These properties include its volatility, lower flammable limit (LFL), vapor density, diffusivity in air, and a number of properties that affect the rate at which it gives off heat when it burns. The volatility of a fuel determines in large part the rate at which vapor is produced from exposed fuel, and thus the likelihood that flammable conditions may exist. This has a major effect on the frequency with which ignition occurs (Data Link, Inc., 1988; Machiele, 1989).

The volatility of pure methanol (M100) is 4.6 psi Reid vapor pressure, compared with 8 to 16 psi for gasoline, and 7 to 16 psi for M85 (Machiele, 1989). Gasoline is, and M85 will be, seasonally blended in order to improve vehicle operation in winter. This process results in a wide range in fuel volatility. It is anticipated that most gasoline marketed in the summer months will have a volatility of 9.0 psi, and if splash-blended, M85 would then have a volatility of approximately 7.5 psi (Machiele, 1989).

Based on collision-fire data with gasoline vehicles, Figure 1 shows that, compared with gasoline, the lower volatility of M100 may be expected to result in about a 70 percent reduction in collision-related vehicle fires; M85 may be expected to result in a 20 percent reduction. In addition, analysis of diesel and gasoline fire data suggests that the effect of fuel volatility may even be greater in non-collision fires. The ratio of diesel-to-gasoline fires in collision

Figure 1
Fuel Volatility vs. Car Fire Rate [3,5,6]

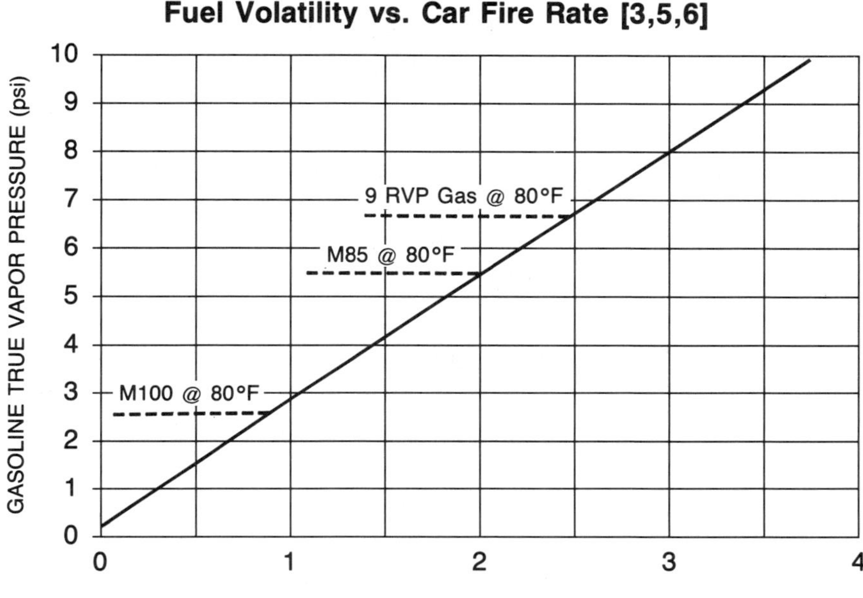

Source: NHTSA, 1988; MVMA, 1984; Doner, 1972

situations is roughly twice as large as in non-collision situations (FEMA, 1986).

The LFL determines the minimum concentration of fuel vapor in air required for ignition. The higher the LFL, the more unlikely that ignition will occur. The LFL for M100 is about 6.0 volume percent in air, compared with 1.4 percent for gasoline, and about 2 percent for M85 (Machiele, 1989). Thus, a concentration in air of more than four times that of gasoline is required with M100 to achieve ignition. This alone could have a significant effect on the frequency of fires; in combination with the low volatility of methanol, the effect should be dramatic.

The high vapor density of gasoline (2 to 5 times that of air) causes its vapor to travel along the ground to ignition sources, whereas the low vapor density of M100 (1.1 times that of air) causes it to disperse more evenly (Machiele, 1989). This also could cause ignition with M100 to be less likely. The vapor density of M85 ranges from 1.1 to 5, and most closely resembles that of gasoline, since roughly 60 percent (on a volume basis) of the vapor initially emitted from exposed M85 is gasoline hydrocarbon (Snow et al., 1989).

The diffusivity of a fuel vapor determines how rapidly a flammable concentration of vapor will disperse to non-flammable levels under conditions in which natural and artificial ventilation is limited. The diffusion coefficient is roughly 2.5 times greater for M100 than for gasoline (0.5 ft²hr versus 0.2 ft²hr), but since ventilation usually dominates natural diffusion, this is seldom a significant factor (Machiele, 1989).

Should a fire occur, methanol is less likely to cause injury, death, and property damage because of its low heat of combustion and high heat of vaporization, as well as its low volatility and modest boiling point. These properties combine to cause M100 to burn at a rate about 40 percent of gasoline's and release heat at a rate estimated to be 20 percent of that of gasoline. Similarly, M85 is estimated to burn at a rate roughly 50 percent of gasoline's and release heat at a rate 30 percent of that of gasoline (Machiele, 1989). Therefore, if a fire occurs, the rate of death, injury, and property damage resulting from the fire is estimated to be much lower with methanol fuels than with gasoline. See Figure 2.

Based on reduced estimates of fires and the possible reduction in hazard for fires that do occur, the use of methanol fuels could result in a dramatic reduction in vehicle fires and related fatalities and injuries each year. M100 fires might be 10 percent as likely to occur as gasoline fires, while M85 fires might be 75 percent as likely to occur. In addition, the death and injury rate, should a fire occur with M100 and M85, might be 50 percent and 60 percent, respectively, of that for gasoline. Based on these assumptions, Table 1 shows that the use of M100 could result in 172,000 fewer vehicle fires each year, 815 fewer fatalities, and 4,800 fewer injuries. The

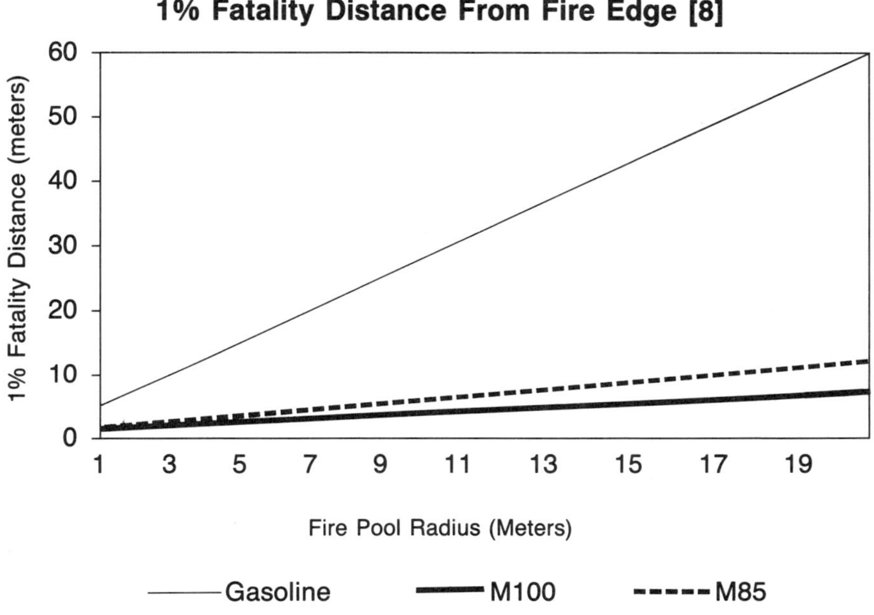

Figure 2
1% Fatality Distance From Fire Edge [8]

Source: Webb, 1988

use of M85 could result in 47,800 fewer vehicle fires, 471 fewer fatalities, and 2,780 fewer injuries.

Fuel Tank Flammability

There has been a great deal of concern expressed about the possibility of a fuel tank explosion with M100 and the serious hazard which that might represent. Methanol is flammable inside fuel tanks at temperatures ranging from 45 to 109 degrees F. Summergrade gasoline and M85 become flammable inside fuel tanks at temperatures below 10 and 14 degrees F, respectively (Machiele, 1989). For winter-grade fuels the temperatures are much lower.

Despite the possibility of a fuel tank explosion with M100, it is not expected to be a frequent occurrence. Fuel tanks are isolated environments, with only a limited number of possible ignition sources. In addition, limited testing of methanol fuel tanks has shown that, even if ignition occurs, the explosion is minor and often contained by the fuel tank. It is not the huge fireball that has often been suggested (Battista, 1989).

Nevertheless, precautions should be taken to prevent a fuel tank ignition. A number of simple vehicle design modifications can greatly reduce this possibility. These include the use of flame arresters on tank fill necks and vents; modification of in-tank fuel pumps and fuel level sending units to prevent electrical sparks; and the use of foam fillers in the fuel tank to prevent a flame from propagating through the tank. Foam fillers may be the most effective option since they prevent an explosion, regardless of the ignition source, and can be installed relatively cheaply and easily. These foams are currently used in some military, racing, and special security applications.

Flame Luminosity

The possibility that M100 fires would be invisible to the naked eye is another issue that has raised a great deal of concern with M100. Pure methanol burns with a flame that is not visible under well-lit or daylight conditions if nothing else is burning along with it. There is, therefore, the possibility that where methanol is burning on a surface such as concrete in daylight conditions, a person may unknowingly enter the fire. Situations such as this are not likely to be common, but they do represent a significant concern. Something should be added to produce a luminous flame. Gasoline added to methanol in M85 serves this purpose, but additives with fewer drawbacks would be preferable. Early efforts to find more acceptable additives have focused on various hydrocarbons, but these have met with little success (Fanick and Smith, 1984; Kirshenblatt and Bol, 1988). However, more recent efforts by one of the large oil companies has achieved better results with various organic compounds in concentrations in the hundreds-of-parts-per-million range. These results are still preliminary and proprietary, but provide a bias for believing that additives to provide acceptable flame luminosity will be available in the near future for use in M100.

Acute Toxicity

Due to the large numbers of situations where people may be exposed, fuel toxicity to humans is an issue of great importance when assessing the safety of transportation fuels. All transportation fuels can be toxic if ingested, inhaled, or if there is contact with the skin or eye. The question, therefore, concerns the level of exposure that might occur, and the level of exposure that is necessary to cause toxic symptoms. In 1987 there were an estimated 22,856 incidences of acute exposure to gasoline in the United States severe enough for people to seek medical advice. Four hundred twenty moderate or major (life-threatening) injuries and four deaths were associated with these exposures (Litovitz, 1988). Although not as great a problem as vehicle fires, acute fuel toxicity is a significant concern.

Ingestion

If gasoline is ingested, a minor exposure (such as 5 to 10 ml) can result in such mild effects as burning of the gastrointestinal tract and abdominal pain. Larger exposures (27 to 40 ml) can cause more severe effects such as fever, convulsions, and unconsciousness (Occupational Health Services, Inc. [OHS], 1985; Machiele, 1989). The normally fatal dose for an average adult ranges from 115 to 470 ml, but much smaller exposures can be fatal if gasoline is aspired into the lungs (Machiele, 1989; Battelle, 1989).

In comparison with gasoline, methanol is considered to be somewhat more toxic. Ingestion of as little as 18 ml can result in such toxic effects as headache and abdominal pain, while slightly larger amounts may cause blindness and unconsciousness (Machiele, 1989). The normally fatal dose ranges from 60 to 240 ml, but death can occur with as little as 26 to 70 ml (Health Effects Institute, 1987). The relative toxicity of M85 is currently unknown, but it probably resembles that of M100. In addition to concerns over the increased toxicity of methanol, M100 does not have a taste, color, or odor that might deter ingestion or cause a person to seek medical treatment. However, since the toxic effects of methanol occur 10 to 48 hours after ingestion, there is time to seek treatment (Machiele, 1989). But if the person is unaware of the ingestion or does not perceive the need for treatment, this time delay is of little benefit.

Two-thirds, or 15,360, of the gasoline exposures in 1987 involved ingestion, either alone or with another form of exposure, and resulted in about 223 moderate or major injuries (Litovitz, 1988; Battelle, 1989). Of these exposures, 41 percent were pediatric (age 0 to 5) and 59 percent were non-pediatric (Litovitz, 1988). Of the pediatric exposures more than 90 percent occurred as a result of non-automotive applications of gasoline (gasoline stored in and around the home for lawn/garden and grease/oil solvent purposes) (Litovitz, 1988). Of the non-pediatric exposures, 92 percent occurred as a result of siphoning (frequently to provide fuel for non-automotive applications) (Battelle, 1989). Despite concerns about methanol's toxicity compared with that of gasoline, a number of factors will limit the ingestion exposures to methanol fuels. Furthermore, precautionary measures can be taken to limit the number of exposures.

Since methanol is not currently intended for use in lawn/garden applications (and should not be until precautions are taken to protect against accidental ingestion), the exposures that occur as a result of these applications—which account for most of the pediatric exposures—would continue to occur from the use of gasoline, not methanol. Second, methanol, which is a polar molecule, is a relatively poor solvent for oils and greases, which are typically nonpolar. Thus, many of the cases of exposure to gasoline from its use as a solvent in and around the home will continue to occur with gasoline. Third, since methanol fuel will seldom be used around the home, there should be much less need to siphon fuel from vehicles, causing siphoning-related incidences to occur less frequently. Fourth, vehicle fuel tank modifications to limit fuel-tank flammability (flame arresters, foam fillers) should hinder or prevent siphoning from vehicles. If not, other methods to prevent siphoning from vehicle fuel tanks should be simple to implement. Fifth, fuel additives, such as corrosion inhibitors, detergents, anti-foaming agents, and many others used in gasoline today may also be used in methanol to give it a taste, color, and odor. If not, additives specifically for methanol can be developed toward these ends. These should provide the precautionary characteristics for M100 that currently exist with other fuels. Finally, educational efforts explaining the hazards of ingesting methanol should be implemented to reduce the chances of ingestion and the chances that an ingestion incident may go untreated.

Based on the above discussion, it seems reasonable to conclude that the hazard due to ingestion of methanol fuels should not be significantly greater than for gasoline today, and, in fact, may be less. As many as 90 and 95 percent of the pediatric and adult exposures, respectively, which occur today with gasoline may not occur with methanol. However, if ingestion of methanol does occur, the rate of injury and death may be twice that of gasoline. Based on these assumptions, Table 2 shows that the number of exposures to methanol fuels may be 14,000 fewer than with gasoline today, and, while the number of injuries may be 190 fewer, a few additional fatalities may result. This should not be misinterpreted as a solution to the situation today. An additional 5,700 exposures and 80 injuries should occur even with full implementation of the methanol fuels in automotive applications as a result of non-automotive exposures to gasoline.

Skin/Eye Contact

Skin or eye contact with all the transportation fuels have similar effects. Though typically not a serious concern, eye contact can cause irritation, redness and swelling, and in some cases minor tissue

Table 2

**U.S. Annual Incidences of, and Deaths and Injuries
Due to, Motor Vehicle Fuel Related Exposure**

	Gasoline	Methanol[a]		M85[a]	
Ingestion:					
Incidences[b]	15,360	1,190	(5,680)	1,190	(5,680)
Pediatric	6,300	630	(5,100)	630	(5,100)
Non-pediatric					
Siphoning	8,340	420	(0)	420	(0)
Other	720	140	(580)	140	(580)
Serious Injury	223	35	(80)	35	(80)
Death	0	2?	(0)	2?	(0)
Skin Contact					
Incidences[b]	2,900	1,450	(1,450)	1,450	(1,450)
Serious Injury	43	43	(22)	43	(22)
Death	0	0	(0)	0	(0)
Eye Contact					
Incidences[b]	5,100	2,550	(2,550)	2,550	(2,550)
Serious Injury	50	25	(25)	25	(25)
Death	0	0	(0)	0	(0)
Inhalation					
Incidences[b]	2,100	210	(1,050)	840	(1,050)
Serious Injury	105	21	(53)	85	(53)
Death	4	0	(2)	3	(2)

[a] Projected incidences.
[b] Includes exposures by multiple routes.

Note: Numbers in parentheses represent non-automotive exposures to gasoline, which are assumed to occur despite the use of methanol fuels in automotive applications.

Source: Litovitz, 1988; Battelle, 1989

damage (Machiele, 1989). With skin contact, concern arises only with prolonged exposure or repeated exposure with little time between exposures. Infrequent splashes on the skin are of little concern, but should be cleaned off. Gasoline (and probably M85 to some extent) tends to dry and defat the skin, resulting in chemical burns. If exposure is of long duration and a sufficient amount is absorbed into the body, damage to the heart, lung, kidney, and liver may result with possible life-threatening consequences (OHS, 1985; Hansbrough et al., 1985). With methanol the concern is that the skin may be exposed for a sufficient length of time to cause its absorption by the body to produce the toxic effects that result from ingestion. Typically, a body surface area the size of a hand must be immersed in methanol for three hours before toxic symptoms will begin to occur, and there is a four-hour interval before death is possible (Machiele, 1989). No unique toxic symptoms are known to occur with M85.

One-third, or roughly 8,000, of the gasoline exposures in the United States in 1987 involved skin or eye contact. Associated with these exposures were roughly 93 moderate or major injuries (Litovitz, 1988). Many of these were not vehicle related, but rather due to lawn/garden and solvent uses of gasoline. The majority of the severe injuries probably fell into two main categories: those resulting from misuse of gasoline as a grease and oil solvent, and those which resulted from gasoline spilled on clothing, for which no corrective action was taken. As already discussed, it does not appear likely that methanol will be used in and around the home as a grease and oil solvent, and as a result, methanol use will not result in accidents from its misuse for these purposes. Furthermore, since methanol evaporates completely much more quickly than gasoline, minor spills on clothing should not result in any severe injury, even if no corrective action is taken. Though methanol will usually evaporate before a toxic dose can be absorbed into the skin, a large, untreated spill onto clothing may cause a serious condition to develop due to the substance's high rate of absorption. A comparison to gasoline is difficult, since no information was found on the rate of absorption of gasoline into the skin, or at what degree of skin contact gasoline becomes toxic. If skin contact resembles ingestion, then methanol may be considered more toxic if there is severe exposure. There is expected to be little difference between M85 and M100, with the exception that M85 may be absorbed into the skin more quickly than M100, and thus toxic effects may be more likely to occur (Ferry et al., 1982).

Based on the above discussion, Table 2 indicates that with widespread use of methanol there would be 4,000 fewer incidences of skin and eye contact and 25 fewer injuries due to methanol than with the present use of gasoline. However, there would also be 4,000 exposures and 47 serious injuries associated with the non-automotive uses of gasoline. The net difference is an increase of 22 injuries associated with skin contact.

Inhalation

The acute toxicity of gasoline inhalation is such that exposure to low concentrations (0 to 500 ppm) may cause symptoms such as eye irritation, headache, and sore throat. More severe exposure (900 to 3,000 ppm) may result in dizziness, breathing difficulty, and bronchial pneumonia. Higher concentrations may cause intoxication, unconsciousness, and heart damage (Machiele, 1989; OHS, 1985; Rosenberg and Gasper, 1980). Death may result with exposure to 5,000 to 16,000 ppm (Rosenberg and Gasper, 1980).

In comparison, the toxicity of methanol is relatively mild until the exposure exceeds the body's ability to clear the metabolites of methanol. Exposure to low concentrations (0 to 3000 ppm) may result in irritation of mucous membranes, headache, and mild eye irritation. Exposure to higher concentrations, however, may begin to exceed the body's capacity to metabolize methanol, and may result in much more severe effects, such as blindness and death, depending on the duration of the exposure (Machiele, 1989). Table 3 provides a comparison between the minimum fatal exposure and the exposures that might occur in common situations. Ordinary exposures are likely to result in physical burdens similar in magnitude to those that the body experiences through normal dietary intake, and which do not even approach lethal doses.

With respect to inhalation, the toxicity of M85 is expected to most closely resemble that of gasoline. Although symptoms of both

Table 3
Methanol Body Burden for Selected Situations

Exposure Scenario	Assumed Condition	Added Body Burden of Methanol
Minimum lethal dose		300 mg/kg
Splash of fuel covering one hand	Dry in 2 min.	2.5 mg/kg
Personal garage	600 mg/m³, 15 min. Twice resting breathing rate	1.8 mg/kg
Self-service refueling	50 mg/m³, 4 min., Twice resting breathing rate	0.04 mg/kg
12-ounce diet beverage	555 mg aspartame	0.3 mg/kg
Dietary intake of aspartame (w/o diet beverages)	Normal Diet	0.3–1.1 mg/kg/day
Background body burden		0.5 mg/kg

Source: Health Effects Institute, 1987

methanol and gasoline exposure are possible (as well as unique symptoms that have not been identified), since the vapor given off by exposed M85 is composed of roughly 60 percent gasoline hydrocarbon, gasoline's effects should predominate (Machiele, 1989).

Nearly 10 percent, or 2,100, of the gasoline exposures reported in 1987 involved inhalation (Litovitz, 1988). Roughly half of the exposures reported to the National Capitol Poison Center were related to automotive uses of gasoline, while the other half were related to lawn/garden uses and deliberate abuse (often referred to as "gasoline sniffing") (Litovitz, 1988). A much greater number of exposures actually occur, but such minor exposures as those that occur during vehicle refueling are typically not reported. Associated with these exposures were roughly 105 moderate or major injuries and four fatalities. The available information did not allow for detailed analysis of the causes of the injuries and fatalities, but a significant number probably involved intentional abuse; the narcotic effect of exposure to gasoline has historically caused it to be abused.

Due to the low volatility of M100, the concentrations of typical exposures should be just 12 to 35 percent of those associated with gasoline (depending on whether the gasoline is a summer or winter blend) (Machiele, 1989). In addition, the diffusion coefficient for methanol is 2.5 times that for gasoline, causing vapor to disperse more quickly in stagnant air situations. Also, due to methanol's low vapor density (1.1 times that of air), methanol vapor does not tend to accumulate along the ground where already unconscious people may be exposed. For these reasons, accidental exposure to methanol vapor is not expected to be as severe as it is with gasoline. In addition, the intentional exposures observed with gasoline are not expected to occur with methanol.

For M85, the benefits are much smaller. The volatility of M85 is expected to be just slightly (20 percent) less than that of gasoline. Since the major part of the vapor given off by M85 is gasoline hydrocarbon, its diffusion coefficient, vapor density, and narcotic effect are expected to closely resemble those of gasoline.

A unique concern with M100 is that it does not have an odor by which it can be identified should a toxic concentration exist. Though experience with gasoline suggests that an odor is not sufficient to prevent a toxic exposure from occurring, it nevertheless seems likely that one or more of the additives mentioned earlier can simply and inexpensively enhance methanol-fuel safety by providing it with a recognizable odor (as is done with natural gas).

The hazard from the inhalation of M100 is probably less than for gasoline today, while that for M85 may in fact be higher. It can be assumed that just 20 and 80 percent of the vehicle-related inhalation exposures that occur with gasoline today will occur with M100 and M85, respectively, due to their lower volatility. However, in cases of severe exposure, the rate of injury may be as much as twice as great. The fatality rate for M100, however, is projected to be half that of gasoline, due to the lack of intentional abuse. Table 2 shows that there would be 1,900 and 1,300 fewer inhalation exposures, 84 and 20 fewer injuries, and four and one fewer fatalities with M100 and M85, respectively, compared with gasoline. However, there would be an additional 1,050 exposures, 53 injuries, and two deaths associated with the non-automotive uses of gasoline.

Air Toxics

Chronic exposure to transportation fuels and their exhaust emissions is probably a greater concern than acute exposure to the fuels themselves. The five emissions of greatest concern with these fuels are fuel vapor, particulate matter (including products of incomplete combustion [PICs] and polycyclic organic matter [POM]), benzene, formaldehyde, and 1,3-butadiene. A summary of these emissions is shown in Table 4.

Table 4
Air Toxics Emissions

| | -----1986 U.S. Cancer Risk Estimates----- | | |
	Gasoline	Optimized M100[a]	FFV on M85[a]
Fuel Vapor	17–68	0	9–34
Particulate[b]	1–176	0–2	0–50
Benzene	97–150	2–4	29–45
Formaldehyde[c]	43–81	30–58	90–171
1,3–Butadiene	221–252	2	80–91

[a] Estimate assuming complete replacement of gasoline.
[b] Including PICs and POM.
[c] Includes both direct formaldehyde emissions and indirect, photochemically generated formaldehyde.

Source: EPA, 1989; Adler and Carey, 1989

Fuel Vapor

There are two main concerns with fuel vapor: carcinogenicity, and possible mutagenic, teratogenic, reproductive, and developmental effects. Based on rat and mouse studies, gasoline vapor is currently classified by the EPA to be a group B2, probable human carcinogen. Information concerning possible mutagenic, teratogenic, or reproductive effects is inconclusive (Machiele, 1989).

Chronic exposure to methanol vapor is not considered to be carcinogenic, although it may have mutagenic, teratogenic, reproductive, and developmental effects. This is based on methanol's similarity to ethanol, for which these effects have been documented (based on acute exposures), and on limited animal studies as well. The studies that have shown mutagenic, teratogenic, and developmental effects with methanol were performed on animals at levels that would be acutely toxic to humans. Minor reproductive effects were seen in rats at the threshold limit value for methanol of 200 ppm, but not at concentrations of 50 times that level. The natural background body level of methanol and the amount of methanol typically seen in the human diet suggest that chronic effects associated with methanol are not likely to be significant given typical exposures (Table 3). This is an issue, however, that requires further study. To the extent that M85 vapor consists of both gasoline (two-thirds) and methanol vapor (one-third), the effects of both of these fuels can also be expected with M85.

Particulate

Though they have not yet been classified by the EPA as carcinogenic, particulate emissions are known to contain carcinogens. Diesel fuel represented only about 20 percent of the transportation fuels consumed in the United States in 1986. However, it is believed that as many as 860 cancer cases may have resulted from its use. With gasoline, there is much less particulate produced, but the PICs and POM produced are still a significant concern, and are estimated to have caused as many as 176 incidences of cancer in 1986 (Adler and Carey, 1989). Since methanol is a simple molecule with no carbon-carbon bonds, it produces no particulate when it burns. However, some lubricating oil burns when fuels are combusted in the engine, creating the possibility of a very slight contribution to particulate emissions even with M100. Since M85 is mostly methanol, roughly 72 percent of the particulate emissions associated with gasoline are eliminated with M85 (EPA, 1989).

Benzene

Benzene emissions are classified by the EPA as a group A human carcinogen. On average, approximately 1.5 percent of gasoline is benzene. As such, all forms of evaporative emissions of gasoline contain a significant amount of benzene. In addition, incomplete combustion of both benzene and other aromatics in gasoline causes a significant fraction of exhaust emissions to consist of benzene. Based on these emissions, there were an estimated 97 to 150 incidences of benzene-related cancer associated with gasoline's use in 1986 (Adler and Carey, 1989).

There is no benzene in M100, and no benzene emissions occur. There may be a very small amount of benzene in the exhaust of M100 vehicles, however, due to lubrication oil or other effects. Because of the gasoline in M85, from 23 to 86 percent of the benzene that occurs with gasoline can be expected to occur with M85, depending on the source of the emissions (EPA, 1989). Weighted together, it can be assumed that approximately 30 percent of the benzene emissions from gasoline will continue to occur with M85.

Formaldehyde

At very low concentrations (less than 1 ppm), formaldehyde can irritate the eyes, nose, and respiratory tract (Moulis, 1989). In addition, the EPA has classified formaldehyde as a group B1, probable human carcinogen. Formaldehyde in the atmosphere occurs both directly as a result of exhaust emissions (10 to 50 percent for gasoline today), and indirectly as a result of photo-oxidation of other hydrocarbons in the atmosphere (50 to 90 percent for gasoline today) (EPA, 1989). As a result of both sources, formaldehyde generated from gasoline vehicles was estimated to result in approximately 43 to 81 incidences of cancer in 1986 (Adler and Carey, 1989). Because of the potential for increased incidences of cancer, the higher formaldehyde emissions from methanol-fueled vehicles have raised a great deal of concern in recent years. However, as a result of possible improvements in methanol-engine and emission-control technology, and reduced, indirect formaldehyde generation (43 and 80 percent reductions in indirect formaldehyde for M85 and M100, respectively, due to reduced levels of reactive hydrocarbon emissions), use of M100 is projected to result in approximately a 30 percent reduction in the rate of formaldehyde-related cancer, compared with that of gasoline, while use of M85 may result in a 109 percent increase (EPA, 1989).

1,3-Butadiene

Based on animal studies, 1,3-butadiene is classified by EPA to be a group B2, probable human carcinogen. 1,3-butadiene is estimated to comprise just 0.35 percent of the exhaust hydrocarbon from gasoline vehicles, but, based on estimates for 1986, could have been responsible for more than 200 incidences of cancer (Adler and Carey, 1989). No data are currently available concerning the concentration of 1,3-butadiene in the exhaust of methanol vehicles, but current estimates suggest that it is only present in minute quantities in the exhaust of M100 vehicles. Due to the gasoline component, roughly 36 percent of the cancer incidences associated with 1,3-butadiene from gasoline vehicles is expected to occur with M85 vehicles (EPA, 1989).

Summary and Conclusions

No fuel can be made to be perfectly safe, but significant safety improvements over the use of gasoline can be expected with the use of methanol fuels. The greatest benefits occur with the use of

M100, as a result of significant reductions in fuel-related fires and toxic emissions. By implementing simple and inexpensive vehicle design modifications, and by utilizing fuel additives, most concerns about the unique safety hazards of M100 can be eliminated or greatly reduced. Blending in gasoline to form M85 is currently one simple means of eliminating some of the unique problems with M100, but at the same time many of the benefits of methanol are thereby reduced or eliminated, and overall fuel safety is degraded compared to M100. Regardless of whether or not methanol fuels are adopted for use on a widespread scale, consideration should be given to imposing a variety of precautionary restrictions on gasoline in order to reduce the number of hazards that result from its use.

References

Adler, J.M., and P.M. Carey (1989). Air Toxics Emissions and Health Risks from Motor Vehicles. U.S. Environmental Protection Agency, AWMA Paper No. 89-34A.6, July 1989.

Battelle Institute (1989). Incidence of Siphoning-Related Gasoline Ingestion. U.S. Environmental Protection Agency, Draft Report under EPA Contract No. 68-02-4294, February 1989.

Battista, V. (1989). Further Investigation of the Safety of Methanol Fuels. Transport Canada, Paper presented at the 1989 Summer National Meeting of the American Institute of Chemical Engineers, Philadelphia, PA, August 22, 1989.

Data Link, Inc. (1988). Study of Motor Vehicle Fires. Report prepared for the National Highway Traffic Safety Administration, February 1988.

Doner, J.P. (1972). A Predictive Study for Defining Limiting Temperatures and Their Application in Petroleum Product Specifications. Coating and Chemical Laboratory, NTIS AD-756420, November 1972.

Environmental Protection Agency (1989). Analysis of the Economic and Environmental Effects of Methanol as an Automotive Fuel. Office of Mobile Sources, U.S Environmental Protection Agency, Special Report, September 1989.

Environmental Protection Agency (1988). Summary and Analysis of Comments Regarding the Potential Safety Implications of Onboard Vapor Recovery Systems. Office of Mobile Sources, U.S. Environmental Protection Agency, August 1988.

Environmental Protection Agency (1987). Methanol Fuel Safety: A Comparative Study of M100, M85, Gasoline, and Diesel Fuel as Motor Vehicle Fuels. U.S. Environmental Protection Agency, Draft Technical Report, November 1987.

Fanick, R.E., and L.R. Smith (1984). Survey of Safety Related Additives for Methanol Fuel. U.S. Environmental Protection Agency, Report No. EPA 460/3-84-016, November 1984.

Federal Emergency Management Administration (FEMA) (1986). National Fire Incident Reporting System.

Ferry, D.G., et al. (1982). The Percutaneous Absorption of Methanol after Dermal Exposure to Mixtures of Methanol and Petroleum. Proceedings, Fifth International Alcohol Fuel Technology Symposium, Vol. 3.

Hansbrough, J.F., et al. (1985). Hydrocarbon Contact Injuries. *Journal of Trauma*. Vol. 25, No. 3, March 1985.

Health Effects Institute (1987). Automotive Methanol Vapors and Human Health: An Evaluation of Existing Scientific Information and Issues for Future Research. Health Effects Institute, May 1987.

Kirshenblatt, M., and M.A. Bol (1988). Investigation into Methanol Fuel Formulations. Sypher Mueller International, Inc., Society of Automotive Engineers Paper 881599.

Litovitz, T. (1988). Acute Exposure to Methanol in Fuels: A Prediction of Ingestion Incidence and Toxicity. National Capitol Poison Center, for API, October 31, 1988.

Machiele, P. (1989). A Perspective on the Flammability, Toxicity, and Environmental Safety Distinctions Between Methanol and Conventional Fuels. U.S. Environmental Protection Agency, Paper No. 43b, presented at the 1989 Summer National Meeting of the American Institute of Chemical Engineers, Philadelphia, PA, August 22, 1989.

MVMA (1984). Gasoline Fuel Survey.

Moulis, C.E. (1989). Formaldehyde Emissions from Mobile Sources and Potential Human Exposures. U.S. Environmental Protection Agency, AWMA Paper No. 89-34A.1, July 1989.

Occupational Health Services, Inc. (OHS) (1985). Material Safety Data Sheet: Gasoline/Automotive. Occupational Health Services, Inc., September 1985.

Rosenberg, S.E., and J.R. Gasper (1980). Potential Health and Safety Impacts from Distribution and Storage of Alcohol Fuels. Argonne National Laboratory for U.S. Department of Energy, June 1980.

Snow, Richard, et al. (1985). Characterization of Emissions from a Methanol-Fueled Motor Vehicle and Methanol-Gasoline Blended Fuels. *JAPCA*. 35:1168–1175.

Webb, R.F. (1988). Assessment of the Safety of Transportation, Distribution and Storage of Methanol Fuels. Transport Canada, September 1988.

CHAPTER 10

Health Effects of Methanol: An Overview

LAWRENCE FISHBEIN
CAROL J. HENRY

Abstract

The proposed use of methanol ("wood alcohol") by itself or in gasoline mixtures as an alternative fuel has made it necessary to review the current state of knowledge of its health effects. This paper focuses on the acute effects of methanol exposure, principally as a result of accidental poisonings due to ingestion (e.g, siphoning or mislabeled beverages). A sequence of effects is discussed that takes into account differences in human sensitivity to methanol's toxicity. These effects include intoxication, permanent blindness, neurologic impairment, coma, and death. Much of this review includes important issues and features of methanol toxicity discussed and some recommendations made at a recent workshop on "Methanol Vapors and Health Effects: What We Know and What We Need to Know," held in Washington, D.C., June 14–15, 1989.

Methanol is metabolized first to formaldehyde, then to formate (formic acid), and finally to carbon dioxide and water. Although the basic manner in which methanol is metabolized is similar in all species studied, differences in the rates of metabolism lead to certain individuals being more susceptible to methanol toxicity. The metabolism of methanol in humans suggests that a dietary deficiency of folate (a nutritional requirement), or decreased levels of folate due to pregnancy or a variety of other causes, may lead to an accumulation of formate, which is believed to be largely responsible for the acute effects noted in human exposure to methanol.

In contrast to the relatively large amount of information describing the acute or short-term toxic effects of methanol, much less is known about the long-term or chronic effects of low-level exposure to methanol. Therefore participants at the workshop suggested further research on methanol's long-term effects. The lack of *definitive* data on chronic, reproductive, developmental, neurobiological, potential immune and cardiovascular effects, was pointed out and recommendations were made for further study in these areas. Since methanol would be used in combination with unleaded gasoline, research to examine the health effects of methanol-gasoline mixtures, including exposure by inhalation and skin absorption, was also suggested.

Introduction

It is important to find out if the proposed introduction of methanol-fueled vehicles may have an adverse health impact on exposed populations. Therefore, participants at the recent workshop discussed metabolism, mechanism of toxicity, pharmacokinetics, and the comparative toxicology of methanol. Regulatory concerns associated with increased methanol use were also considered. Finally, the importance of assessing the levels of methanol (and formaldehyde) exposure currently experienced by the general population was stressed.

Acute Toxicity

Nearly all available information about toxicity in humans concerns the consequences of acute exposure. Although acute toxicity can occur via ingestion, inhalation, and skin exposures, ingestion has been the most frequent and serious route, with many case histories of poisonings recorded (Health Effects Institute, 1987; American Petroleum Institute, 1988). Acute methanol toxicity in humans follows a well-defined pattern. Very high exposure proceeds via narcosis (intoxication), followed by sleep, and, at a sufficiently high and rapid ingestion of methanol, death. Lower toxic exposure is manifested by a sequence of events resulting, initially, in a transient mild depression of the central nervous system. A latent period follows, which generally lasts 12 to 24 hours. Subsequently, a syndrome develops that consists of metabolic acidosis, accompanied by ocular effects of varying severity, depending on exposure and a variety of physical symptoms. These may include dizziness, nausea, vomiting, severe abdominal pain, and difficulty in breathing. This may progress to coma, and death, usually from respiratory failure.

The time of onset of symptoms in victims is quite variable, ranging from several hours to several days. Human susceptibility to the acute toxic effects of methanol by ingestion is variable. For example, ingestion of one to two ounces of methanol may cause blindness; yet one-half ounce has caused death in some individuals, while 16 ounces has caused blindness in some and has not proved fatal to others. (Four ounces, or 95 grams, of methanol has been lethal in 40 percent of the cases. For a 70 kg person, this dose is equivalent to about 1.4 grams methanol/kg body weight). The currently accepted value for the lethality of methanol is 0.3 to 1.3 grams/kg, in untreated cases. Although rapid absorption from inhalation and skin exposures to methanol can also lead to blindness and death, very few cases of serious poisoning appear to occur by these routes.

Extrapolation of 1987 data, reported by 63 U.S. poison centers to the American Association of Poison Control Centers, estimates current annual incidence of methanol exposures at 6,400. A large percentage of ingestion cases are due to siphoning by male teenagers and young adults (American Petroleum Institute, 1988).

A limited number of case reports, as well as epidemiologic studies, suggest that extended human exposures to methanol may cause effects that are qualitatively similar to those from relatively high levels of acute exposure (Health Effects Institute, 1987). There are dietary sources of methanol, including consumption of fresh fruits and vegetables (fruit juices average 140 mg methanol/liter with a range of 12 to 640 mg/liter), and consumption of foods sweetened with aspartame, which hydrolyzes in the digestive system, releasing about 10 percent of the molecule that is available for absorption as methanol. An aspartame-sweetened beverage, with an aspartame content of 555 mg/liter, contains the equivalent of 60 mg methanol/liter. This is considerably less than the average methanol content of fruit juice, which is 140 mg methanol/liter (RSI, 1989). It must be stressed that dietary sources of methanol, as noted above, are not considered to lead to any adverse effects.

Mechanism of Methanol Toxicity

Methanol is distributed uniformly to all organs and tissues in direct relation to their water content, regardless of the exposure route. While the basic metabolic pathway for methanol is similar in all species, there are interspecies differences, with respect to the rates of reaction. Methanol is metabolized to formaldehyde, which is then oxidized to formate (formic acid) and finally to carbon dioxide (Health Effects Institute, 1987):

$$CH_3OH \xrightarrow{1} HCHO \xrightarrow{2} HCOOH \xrightarrow{3} CO_2 + H_2O$$

Formate metabolism occurs via a *folate*-dependent pathway in the liver. Non-primate species, including rodents, generally metabolize formate very efficiently, at any dose, and thus escape toxicity attributable to formate build-up. Humans and other primates are much less efficient at metabolizing formate, and are thus more vulnerable to methanol toxicity at higher exposures (e.g., above 5,000 ppm). At normal levels, formaldehyde, *per se*, is not believed to exert toxic effects in this pathway.

At the recent workshop, Dr. J. Clary (RSI, 1989) further summarized data concerning formaldehyde toxicity. About 2.5 ppm of formaldehyde is *normally* found in the blood of humans. If one assumes that the half-life of formaldehyde in blood is 1.5 minutes, and that formaldehyde is equally distributed in the body blood pool, then the adult human *normally* produces and uses (or detoxifies) over 50,000 mg formaldehyde per day, or about 2,500 mg/hour. Available data suggest that this level of endogenous formaldehyde is metabolized normally with no adverse effects. It has also been suggested that humans could handle 100 to 150 mg/kg/hr (up to 5,000 ppm) of methanol, if it were completely converted to formate. However, 1,000 mg/kg/hr (two to three ounces/hr) of methanol may produce adverse effects, due to accumulation of formaldehyde and the overloading of the normal detoxification system. It should be noted that nasal tumors in rats have been reported following inhalation of formaldehyde at very high local concentrations.

As noted earlier, the metabolism of methanol in all species is folate-dependent. The metabolism of methanol in humans suggests that folate deficiency in humans, and the resultant formate accumulation, may play a role in *acute* methanol toxicity. However, it is not known whether folate deficiency also plays a role in developing chronic effects from long-term exposure to low levels of methanol. It has been suggested that methanol may exert two different syndromes, formate-associated acute toxicity and methanol-associated chronic toxicity.

The implication of folate dependency for the metabolism of formate in the metabolic pathway for methanol in humans raises two questions. Will folate-deficient populations be more vulnerable to developing adverse effects from methanol exposure? Could increased folic-acid utilization, resulting from exposure to methanol, lead to folic-acid deficiency and its resultant adverse health effects, such as megaloblastic anemia?

Approximately 10 percent of the U.S. population is considered to be marginally folate-deficient. Groups of individuals prone to folate deficiency include pregnant women; elderly people whose diet is inadequate; alcoholics; some users of anti-convulsant and contraceptive drugs; and some individuals with celiac disease or Crohn's disease, gluten-enteropathy, and rheumatoid arthritis.

Chronic Toxicity

In contrast to the acute toxic effects of methanol, much less is known about the possible effects of low-level chronic exposure to methanol by any route. Although there have been a number of chronic and carcinogenicity studies recently reported by the New Energy Development Organization (NEDO, 1986), these studies have not been fully reported and the complete report is not currently available. NEDO (1986) studied rats and mice exposed to 10, 100, or 1,000 ppm methanol for 20 hours/day by inhalation, for 18 months in mice, or 24 months in rats. NEDO concluded that there was no evidence of carcinogenicity in animals exposed to 1,000 ppm methanol. The NEDO (1986) chronic inhalation studies in monkeys involved exposure of eight animals, per dose level (0, 10, 100, and 1,000 ppm), to methanol for 22 hours/day for up to 2.5 years. Although some slight changes were noted in the liver and kidney, as well as some pathological changes in the nervous system in animals exposed at the 1,000 ppm level, these changes were judged to be transient and probably reversible. As with the earlier NEDO studies, lack of a detailed report precludes an unambiguous interpretation.

Reproductive and Development Effects

Data on the reproductive and developmental effects of exposure to methanol in rodents and monkeys are limited and ambiguous. A NEDO (1986) two-generation reproductive study showed no reproductive effects in rats exposed to methanol at 1,000 ppm, 20 hours/day for 12 months, nor were reproductive effects noted in monkeys exposed by inhalation to 130 mg/m³ (100 ppm) for 22 hours/day, for 30 months.

The effects of methanol exposure on circulating testosterone, luteinizing hormone (LH), and follicle-stimulating hormone (FSH) levels were studied in rats exposed to 200, 2,000, or 10,000 ppm methanol for eight hours/day, five days/week, for one, two, four, or six weeks (Cameron et al., 1984). Although significant reductions were seen in serum testosterone levels after six weeks at 2,000 ppm methanol, the high-dose group (10,000 ppm) showed no changes in serum testosterone levels. LH levels were significantly increased, after exposure to 10,000 ppm methanol for six weeks. There were no significant changes in FSH. The significance of these studies has been questioned, along with a recommendation for confirmation or reevaluation (RSI, 1989). Subsequent studies by Cameron et al. (1985) indicated that testosterone levels recovered 18 hours following cessation of exposure. By the end of the first week of daily exposure, there was no longer a significant reduction in testosterone levels, suggesting that there may have been some adaptation.

Relatively few studies have been reported concerning the potential developmental effects of methanol, by any route of exposure. Nelson et al. (1985) noted significant reduction in the fetal weight of male and female rat pups exposed *in utero* to 10,000 or 20,000 ppm of methanol, by inhalation for seven hours/day, during days 7 to 15 of gestation. While treatment-related malformations were reported at 20,000 ppm, as well as a suggestion of an effect at 10,000 ppm, no adverse effects were noted at 5,000 ppm. Slight maternal toxicity was observed, at the 20,000 ppm inhalation dose.

Treatment-related malformations in rat pups exposed *in utero* to 5,000 ppm methanol by inhalation, for 22 hours/day, during days 7 to 17 of gestation were noted in one of the NEDO (1986) studies. Other toxic effects, including fetal death, were also reported in this group of exposed animals. However, this dose was also toxic to the mother, which may have influenced the incidence of malformation. An additional NEDO (1986) study reported no teratologic effects in monkeys exposed to 1,300 mg/m³ (1,000 ppm) methanol vapors for 22 hours/day, for up to 30 months.

Neurobiological and Other Effects

Methanol has long been known as a neurotoxin. There have been several Russian studies, reported in 1959 and 1967, claiming neurobehavioral effects in humans exposed to very low levels of methanol vapors that resulted in stimulation of visual and peripheral olfactory receptors (Health Effects Institute, 1987), but these have not been substantiated. There is a dearth of data regarding the effects of chronic exposure to methanol on the central nervous system. Studies in humans have been initiated to determine effects of methanol on learning, memory, performance, cognition, and neurophysiology (Health Effects Institute, 1987; RSI, 1989).

Recommended Research Needs

The workshop participants decided that the most important areas for further study (not in order of priority) were research to elucidate the mechanism of ocular toxicity to methanol; the development of physiologically based pharmacokinetic model(s) for methanol toxicity in order to reconcile experimental dose rate and route of exposure in animals with actual or predicted human exposures; the investigation of the potential reproductive, developmental, and neurotoxicological effects of long-term methanol exposure; investigation of the effect of methanol exposure on folic-acid deficiency in sensitive populations such as pregnant women; and finally, research on the potential immunological, cardiovascular, and hepatic effects of methanol and their dose-response characteristics.

Another suggestion for further research was an examination of the health effects of methanol-gasoline mixtures, since some data suggest that combining methanol and gasoline accelerates dermal absorption. Examining the possible role of methanol in the absorption and uptake of important gasoline components such as benzene, toluene, and xylene, by both dermal and inhalation routes of exposure, was suggested. Studies of the effects of emissions from actual methanol fuel-burning engines, including the potential inhalation toxicity of formaldehyde, were also recommended. In order to prevent accidents due to accidental siphoning of methanol fuels, it was pointed out that innovative closures and packaging techniques must be developed (American Petroleum Institute, 1988).

References

American Petroleum Institute (1988). Acute Exposure to Methanol in Fuels: A Prediction of Ingestion Incidence and Toxicity. API Publ. No. 4477, Washington, DC.

Cameron, A.M., K. Zahlsen, E. Haug, O.G. Nilsen, and K.B. Eik-Nes (1985). Circulating Steroids in Male Rats Following Inhalation of n-alcohols. *Archives of Toxicology.* 8:422–424.

_____ (1984). Circulating Concentrations of Testosterone, Luteinizing Hormone and Follicle Stimulating Hormone in Male Rats after Inhalation of Methanol. *Archives of Toxicology, Supplement.* 7:441–443.

Health Effects Institute (1987). Automotive Methanol Vapors and Human Health: An Evaluation of Existing Scientific Information and Issues for Future Research. Health Effects Institute, Cambridge, MA.

NEDO (New Energy Development Organization) (1986). Toxicological Research of Methanol as a Fuel for Power Station.

Nelson, B.K., W.S. Brightwell, D.R. Mackenzie, A. Kahn, J.R. Burg, W.W. Weigel, and P.T. Goad (1985). Teratological Assessment of Methanol and Ethanol at High Inhalation Levels in Rats. *Fundamentals of Applied Toxicology.* 5:727–736.

RSI (1989). Risk Science Institute: Summary Report of the Workshop on Methanol Vapors and Health Effects: What We Know and What We Need to Know. June 14–15, 1989. Washington, DC.

Commentary

MICHAEL J. MURPHY

The papers by Paul Machiele and Lawrence Fishbein give valuable and comprehensive reviews of the possible safety and health hazards of methanol fuel. We need to decide whether all new hazards should be avoided while old hazards benefit from a kind of grandfather clause. A better approach would seem to be a comparison of the relative hazards of traditional and alternative fuels. Such a comparison should consider that the overall safety hazard is the combination of the risk, or the probability that an effect will occur, and its severity, or the degree of harm caused by the effect if it does occur.

For example, the probability of a fire occurring is apt to be lower for methanol than for gasoline, because of the lower vapor pressure, the higher minimum flammability limit, and the higher ignition temperature of methanol as compared to gasoline. Yet the probability of personal contact with and injury from a fire is apt to be greater for methanol because of the non-luminous nature of the methanol flame. However, a methanol fire radiates much less heat, so it causes less harm. Both the risk that a fire will occur and the severity factors must be considered in making an overall judgment of the hazard involved.

An attempt should be made to investigate the hazards of both new and old fuels with equal vigor. For example, if the health hazard of skin contact is a concern, then similar amounts of data on both existing and proposed fuels should be obtained. Any such comparison must consider a multitude of complex technical issues.

The issue of methanol's corrosiveness has been exaggerated in many popular accounts. Battelle research has shown that common aluminum alloys differ greatly in their rate of corrosion from methanol and some aluminum alloys are quite resistant to methanol corrosion. Regarding heat from fire, even if the volume of fuel in a methanol-fueled vehicle is about twice as large as in a gasoline-fueled vehicle with comparable range, the amount of energy contained is about the same—therefore, the total amount of heat released by a potential fire is also about the same.

No methanol-fuel-related health problems have been observed at any of the Urban Mass Transportation Administration methanol bus demonstration sites as of December 1989. Battelle's measurements of methanol vapor levels during the fueling operations at these sites show exposures far below Occupational Safety and Health Administration limits. Inasmuch as these measurements were for operators whose full-time jobs entailed fueling large vehicles, it is unlikely that owners of private automobiles would experience a significant exposure to methanol.

Some of the reported methanol health effects in animal models have occurred at very high dosages, as much as 25,000 times the expected human exposure from ordinary use of methanol as an alternative fuel. Although there are valid experimental reasons for using high dosages, it is necessary to extrapolate back to a more typical dose before making a comparison of the relative hazards involved.

In conclusion, in considering these health and safety issues it is important to perform health and safety analyses vigorously and in similar depth for old fuels, such as gasoline and diesel fuel, and for new fuels, such as methanol, compressed natural gas, and reformulated gasoline. Such analyses must not, however, become merely a continuing call for ever more research.

Commentary

HUGH L. SPITZER

To provide perspective on the Machiele and Fishbein papers, I would like to summarize some results from the workshop "Methanol Vapors and Health Effects, What We Know and What We Need to Know," which was held last June and co-sponsored by the American Petroleum Institute, the U.S. Environmental Protection Agency, the Health Effects Institute, and the Risk Science Institute. Scientists from government, industry, and universities discussed the health effects of methanol and identified further research needs. It was generally agreed that there is a dearth of data on methanol exposure and that this data gap is critical.

The workshop participants identified a number of health-related issues that should be explored as the use of methanol as an alternative fuel is being considered. Concern was expressed that using toxicity data obtained from rodent studies to assess potential human health effects might be misleading. While rodents suddenly exposed to high concentrations of methanol suffer no adverse effects, human exposure can result in blindness or death. There is, therefore, a pressing need to understand the extent to which we can use available experimental animal data to predict methanol's effects on human beings. There is also a need to develop an animal model for methanol, which would be an adequate predictor of human health effects.

Another concern expressed at the workshop was the adverse effects of methanol on sensitive populations. Folic acid, a vitamin, is required for a number of critical processes in the human body, such as the synthesis of hemoglobin, the oxygen-carrying component in red blood cells. Without adequate folic acid a person becomes anemic. Folic acid is also essential in the metabolism of methanol to CO_2 and water. Four sensitive populations were identified: pregnant women not taking vitamin supplements; elderly people on poor diets; young women just entering puberty; and alcoholics.

Potential effects on these sensitive populations may be serious, given a recent report that women not taking vitamin supplements (including folic acid) early in pregnancy had a greater number of babies with defects than women taking vitamins. There is a critical need to understand if exposures to methanol could adversely affect those with marginal folic-acid deficiency in these sensitive populations, and if so, at what concentrations of methanol.

Paul Machiele's presentation suggested that ocular toxicity, resulting from accidental exposure to methanol, probably would not be a problem. There are no data, however, to support this conclusion. Since the eye is the target organ for methanol toxicity, it is, therefore, critical that we understand if direct exposure of the eye to methanol could be harmful.

Concern about dermal exposure also was dismissed by EPA, based on two assumptions. First, methanol is a poor solvent; and second, adverse effects would require hours of exposure. While pure methanol is a poor solvent, M85 (15 percent gasoline and 85 percent methanol) may be a good solvent. Current American Petroleum Institute studies strongly suggest that dermal exposure of hands and arms might lead to adverse effects in less than 10 minutes.

Formaldehyde, a metabolite of methanol in the human body, is a product found in tailpipe emissions of cars that burn methanol. It has been shown that rodents exposed to high concentrations of formaldehyde vapor develop nasal tumors. By inference, humans exposed to formaldehyde vapor may also be at risk of getting nasal cancer. It is generally acknowledged that the resulting tumors in rodents are associated with the irritating effect of formaldehyde in the nose. Recent studies at the Chemical Institute of Industrial Toxicology (CIIT) suggest that it may be necessary to reevaluate some assumptions regarding the potential risk to humans from methanol. Preliminary studies at CIIT have found that monkeys exposed to high concentrations of formaldehyde have effects (cell proliferation in the trachea) that might be associated with some early stages in the carcinogenic process. The amount of tissue at risk in a human being may be much greater than that predicted from experiments with rodents.

Part V

Costs of Methanol Systems

CHAPTER 11

Economics of Fuel Methanol: Asking the Right Question

MICHAEL F. LAWRENCE

Introduction

Jack Faucett Associates has researched, written, and reviewed several studies[1] that estimate the future market price of alternative fuels. Before presenting Jack Faucett Associates' latest estimates of the future price of methanol fuel, this paper will discuss the research procedures used in making these estimates and will explain why they differ.

Interest in the capability of alternative fuels to replace petroleum fuels grew rapidly after the sharp oil price rises of the 1970s. Also, due to concerns about air quality, policy analysts began to consider the need for petroleum substitutes as early as 1973. The U.S. Environmental Protection Agency (EPA) funded two parallel studies, one conducted by Exxon Research and Engineering[2] and the other by the Institute of Gas Technology,[3] to identify and rank alternative transportation fuels that could replace gasoline and diesel fuel. The criteria used in these studies focused on resource availability, extraction and processing concerns, effects on fuel distribution, storage and dispensing systems, and emissions. Some new studies, funded by state and federal agencies and by energy companies, have also been performed.

Although analysis of the use of alternative transportation fuels originally focused on crude oil supply or price issues, it has expanded to include environmental, economic, and social cost effects as well. Technical/engineering issues have been explored to the extent that there is a consensus: vehicles can be built to use alternative transportation fuels safely and efficiently, and raw materials and conversion technology are available to produce sufficient volumes of alternative fuels, particularly compressed natural gas (CNG) and methanol.

The EPA first asked for estimates of the future price of methanol fuel in the mid-1980s. The EPA was by then concerned that future ozone standards might be met only by reducing vehicle miles traveled or by switching to an alternative fuel with more desirable emissions characteristics. Compressed natural gas, electricity, and alcohol fuels, such as methanol and ethanol, were found to be the most desirable gasoline substitutes. Methanol was cheaper than ethanol and offered the most compatibility with the current fuel system.

To determine whether the use of alternative fuels would be economically feasible, policymakers tried to determine its future price. In order to do this, answers were required to such questions as: What would the feedstock be? Where would it be available? How much would it cost? What would the conversion cost be? What would the delivered cost be? How much would it sell for?

Researchers have addressed each of these questions during the past three years and their results reflect the uncertainties inherent in making forecasts about an industry that does not yet exist. Although the chemical methanol industry can provide products for demonstration projects, a new industry—requiring an investment of perhaps $50 billion—would be required to develop methanol fuels for the transportation sector. Studies suggest that delivered methanol cost, assuming reasonable returns to investors, could range from levels close to current gasoline prices to about 50 percent higher.

Methanol Fuel and Public Policy

Private sector and public sector decision making are quite different. Private sector decisions are internal to the firm and based on maximizing profit, identifying the best opportunity available, and accepting a level of risk. In the public sector both costs and potential benefits must be considered; therefore, the approach to decision making must be different.

Recent economic studies of methanol fuels have focused on analysis for investors. With assumptions about raw material and capital and operating costs, investors can determine what market price would allow a sufficient return, compared with the next best potential investment. In public sector analysis such elements as cost-benefit analysis, risk analysis, alternative analysis (cost-effectiveness), and market-failure analysis must be included. Cost-benefit studies require analysts to consider all possible effects of a proposed policy. Some costs and benefits can be measured in dollars. Often, however, intangible costs and benefits that do not lend themselves to quantification must be incorporated into decision making.

The studies undertaken to date have focused on the costs of producing methanol and have failed to consider fully the other side of the ledger. In particular, all potential benefits of methanol must be considered. These include environmental quality and a quantification of the health effects of methanol, compared with those of gasoline; market efficiency or the quantification of the possible effect that the availability of methanol could have on the price of gasoline; an estimate of oil's cost to the environment, such as the cost of developing reformulated gasoline, more vehicle controls, and fuel volatility; the energy security benefits offered by having an alternative fuel, as well as a diversified supply source of fuel;

and, finally, the more efficient use of resources offered by increased use of natural gas.

During the past two decades the United States has been constantly threatened by possible oil shortages and oil price increases. The oil market has been more stable recently than in the past. The United States' recent commitment to protect Persian Gulf shipping channels speaks clearly about the importance that U.S. policymakers put on national energy security. The existence of an alternative source of supply would provide benefits by reducing these risks. Natural gas, for use as compressed natural gas or methanol, is available in large quantities throughout the world. Many countries that are not major oil suppliers could supply gas and this would provide some decentralization of energy supply.

Carrying out a public policy always entails some risk, but failing to act is also risky. The administration's alternative-fuels program provides an innovative, low-risk approach by which the United States could introduce methanol fuel in the late 1990s. Consumers would use methanol if the price was competitive with that of gasoline, but DiFiglio and Lawrence (1987) have shown that methanol consumption would decline rapidly if methanol's price exceeded that of gasoline on a gasoline-equivalent basis. Under the administration's program and some regional programs, automobile manufacturers would produce vehicles capable of using methanol or gasoline, and service stations would supply methanol as their capital stocks were replaced. Additional infrastructure investments would need to be made to accommodate the product if, and only if, methanol proved to be cost-competitive with gasoline or if the use of methanol was mandated. However, if these small investments—on the order of $100 to $300 per methanol-compatible vehicle and $5,000 in incremental service station investments (Singh, 1989)—were not made, consumers would not have the option of using methanol.

Public policy should also evaluate the usefulness of other fuels. The administration's alternative-fuels program provides flexibility. It would be up to manufacturers to determine what kind of compatible vehicles to produce, but it appears that with current technology, each manufacturer would select a gasoline-methanol flexible-fuel vehicle as the most cost-effective and marketable choice. The administration's program addresses the so-called chicken-and-egg dilemma, whereby neither vehicle manufacturers nor fuel producers would make necessary investments until the other industry had done so. The U.S. Department of Energy (DOE) has begun to study many of these potential benefits, but much more research is required. The focus of this effort must shift from private sector investment studies to public sector cost-benefit-risk analysis.

Feedstock

The feedstock debate is over. If methanol is to be produced for the transportation sector, it will be produced from natural gas. The questions that remain are which gas resources will be used and what their price range will be. The key issues in the natural gas feedstock analysis are proven reserves, potential reserves, location, opportunity cost, local consumption, development benefits, wellhead price (user cost), and gathering cost.

In general, countries that are rich in natural gas and considered to be potential producers of methanol include members of the Organization of Petroleum Exporting Countries (OPEC), except Iran; arctic regions in Alaska, Canada, and Norway, which are included in the Organization for Economic Cooperation and Development (OECD); Central and South American countries, such as Chile, Argentina, and Mexico; South Pacific nations, such as Malaysia, Australia, and New Zealand; and, finally, Trinidad and Tobago, Thailand, and China. (Table 1 shows the potential supply-and-demand regions.) The Soviet Union and Iran have been excluded from this analysis, although they are, of course, the two largest sources of natural gas reserves and their resources may someday be available for methanol production. The proven reserves of all these potential suppliers are in excess of 1,375 trillion cubic feet (tcf) and account for only about one-third of the world's proven reserves. The proven gas fields in countries considered to be potential suppliers could produce as much as 500 billion gallons of methanol per year if all were producing at maximum feasible levels with depletion rates generally about four times current rates.

Therefore, the natural gas supply appears to be more than sufficient to support a U.S. methanol market. However, it is not clear from gas reserves data alone which countries would supply feedstock for methanol plants. In each case other opportunities for this gas, local costs of conversion of the gas to methanol, political considerations, and distance from the final market would determine which countries would supply methanol. Studies that focus only on selected countries might fail to analyze the best option. Countries are building and expanding methanol plants and considering other ways to extract development benefits and hard currency from these resources.

Estimating prices for the gas supplied to methanol plants is difficult. The possible demand for methanol production far exceeds other potential markets for this gas. DOE commissioned Jack Faucett Associates to estimate the marginal cost of producing gas in each of the possible supply regions. For each supply region gas development and operating costs were estimated for region-specific drilling efforts, possible success rates, and average well productivity in the 1980s and using 1985 actual U.S. costs per well. The marginal supply price for natural gas was derived as a function of the average price, the estimated rate of depletion of reserves, and the assumed rate of return. (Lawrence and Kapler, 1989)

Estimated yearly capital expenditures on hydrocarbon development in each supply region were based on the average annual number of wells drilled in 1983–85, average well depth, and the location of the gas onshore, offshore, or in arctic areas. Capital expenditures included drilling and equipping costs, non-drilling costs (lease equipment, overhead, and so on), and gathering costs. Since country-specific expenditures outside the United States are largely unavailable, U.S. average expenditures per well for each well-depth category were used, with an adjustment for arctic regions.

**Table 1
Natural Gas
Regional Analysis**

Supply Regions	Demand Regions
OECD	US
Alaska	Canada
Canadian Arctic	Western Europe
Norwegian Arctic (Troll)	Japan
Australia/New Zealand	OPEC
PERSIAN GULF OPEC	Rest of World
Iran	
Other Persian Gulf	
OTHER OPEC	
Algeria/Libya	
Nigeria/Gabon	
Venezuela	
Ecuador	
Indonesia	
OTHER	
Mexico	
Malaysia	
Thailand	
Trinidad & Tobago	
Tierra del Fuego	
China (PRC)	

Note: Although Iran and the Soviet Union are usually excluded as a matter of policy, they could become major methanol suppliers.

Source: Lawrence and Kapler, 1989

Total annual hydrocarbon investment was allocated between gas and oil, according to the percentage of the total number of successful wells accounted for by gas wells in 1983–85. This method means that all development expenditures on oil wells containing associated gas were allocated to oil. No capital costs were assigned to the associated gas. If an international gas market develops as a result of the use of gas in the transportation sector, some of this investment should be allocated to gas production. Procedures that use the inverse of demand elasticities for joint products (Ramsey pricing) could be applied.

Average gas prices were converted to marginal gas prices at the wellhead and are shown in Table 2. These results indicate that there is an abundance of gas, which can be produced at wellhead costs that compete very favorably with the price of oil. Approximately 87 percent, or 26 tcf per year, can be supplied for less than $1/mcf at the selected supply regions. Table 2 also indicates a wide disparity in supply price among regions.

The highest-cost regions identified are Argentina and Chile (Tierra del Fuego) and the arctic regions, followed by Australia, Thailand, and Mexico. Argentina and Chile's high costs are due to excessive unit operating costs—more than four times those shown for Alaska—resulting from a very low average well productivity. Operating costs are also unusually high in the arctic regions because of harsh conditions. The Canadian Arctic and Norwegian Arctic have large per-unit investment costs and significant concentrations of nonassociated gas, according to estimates. The Alaskan price, a function of high operating costs and relatively high investment expenditures, is increased by estimates of a sizeable expansion in capacity. High supply prices for Australia, Thailand, and Mexico are attributable to deep wells and the estimated low well productivity.

Most OPEC countries and Malaysia have very low costs because of very high estimates of well productivity. High productivity significantly reduces per-unit investment and operating costs. Venezuela and Ecuador, also low-cost countries, are estimated to have lower productivity than the rest of OPEC, with correspondingly higher per-unit investment and operating costs. However, this is offset by the negligible investment and user costs in these countries because of an estimated small percentage of nonassociated gas. This pattern also applies to Trinidad and Tobago.

Table 2
Natural Gas Supply Price, Wellhead Regional Analysis
Price Per mcf
(1987 Dollars)

OECD	
Alaska	1.76
Canadian Arctic	1.65
Norwegian Arctic	1.65
Australia	1.01
New Zealand	0.45
PERSIAN GULF OPEC	
Iran	0.73
Other Persian Gulf	0.16
OTHER OPEC	
Algeria	0.30
Libya	0.07
Nigeria/Gabon	0.13
Venezuela	0.13
Ecuador	0.13
Indonesia	0.25
OTHER	
Mexico	0.51
Malaysia	0.03
Thailand	0.77
Trinidad & Tobago	0.05
Tierra del Fuego	1.44
China (PRC)	0.96

Source: Lawrence and Kapler, 1989

Even if it can be assumed that the price estimates are accurate, it could be misleading to conclude that cheap gas is available. Although gas that is processed in its country of origin may be exchanged at a rate close to the supply price, the development of a significant international market in gas would tend to establish a world market price that balances supply and demand. The world price would lie somewhere on the aggregate world supply curve, thus yielding a producers' surplus, or a return above supply price, to the participating suppliers who enjoy lower costs. The stronger the demand for unprocessed natural gas, the higher the market-clearing world price would be. The DOE gas supply analysis discussed earlier focused on wellhead costs, not plant gate costs for gas. However, estimates of past gas production costs include the cost of some capital equipment for gathering the gas. The DOE capital cost study (DOE, 1989) includes some gathering cost and the cost of moving the gas to the plant gate. Further analysis of this gas-gathering cost is necessary. Some published gas price estimates include all gathering costs. Future analysis should be more explicit as to the estimates of these costs and as to their assignment to gas cost, or plant cost. Alternatively, gathering cost could be maintained as a separate cost item.

Processing

Two recent studies of methanol plant capital and operating costs have recently been made. Although these studies are similar, there were some distinct differences in both procedures and results. The studies by Bechtel, Inc., and Chem Systems, Inc., were funded by an oil industry consortium and DOE, respectively. The key issues involved in estimating gas processing costs are gathering costs, technology, capital requirements, capital availability, capital charges, operating cost, and useful life.

Chem Systems included some gas-gathering costs in its capital estimates, whereas Bechtel did not. These are important in some countries, and care should be taken to ensure that they are correctly accounted for. In some regions, such as Libya, high gathering costs may limit the desirability of methanol production. Although both studies assumed larger plants than current world scale, the technology and thermal efficiencies assumed differed. Bechtel assumed gas requirements of .1 mmBtu/gal of methanol produced, whereas Chem Systems assumed that technological advances would improve efficiency so that gas requirements would be only .09 mmBtu/gal. Chem Systems employed a 10 percent real return on investment, whereas Bechtel assumed a 17 to 20 percent return. Chem Systems assumed that operating plant life was only 15 years, whereas Bechtel assumed 23 years. Finally, the major difference was that Bechtel analyzed six specific sites, whereas Chem Systems developed four categories of cost, based on availability of infrastructure and local construction costs.

Chem Systems divided the possible natural gas supply regions into four categories, based on the investment required to establish a large, advanced-technology methanol plant in each. Category one required the smallest investment, as these sites are already developed and are situated within an established industry environment, such as the U.S. Gulf Coast. The remaining categories required increasing amounts of investment, because of high labor rates or low productivity, unusual terrain, or lack of infrastructure. Category four required the largest investment, because of the offshore location of natural gas and high gathering costs or because of harsh climatic conditions.

The unit cost of producing fuel-grade methanol in a new advanced-technology 10,000-mt/d plant, as estimated by Chem Systems, is expressed as a function of site category and the cost of natural gas. These data are shown in Table 3.

Bechtel estimated the operating cost of producing a gallon of methanol to range from 6 cents/gal to 9 cents/gal, depending on conversion technology and plant location. Capital recovery charges were estimated to range from 14 cents/gal to 31 cents/gal. The total major operating and capital costs estimated by Bechtel are shown in Table 4. Thus, Bechtel's processing costs are 3 cents/gal to 12 cents/gal higher than Chem Systems costs for a developed site, and 2 cents/gal to 7 cents/gal higher for remote harsh condition sites. This, coupled with an assumption of 10 percent lower productivity (.09 mmBtu/gal versus .1 mmBtu/gal) for all Bechtel sites, leads to significant differences in the cost of producing methanol.

Table 3
**Chem Systems Estimates of the Cost of Producing Methanol
Using Advanced Technology in Four Different Locations**
(cents per gallon)

Site Category	Methanol Production Cost	
	20 % Capital Recovery	30 % Capital Recovery
1	0.09 G[a] + 0.17	0.09 + 0.23
2	0.09 G + 0.19	0.09 + 0.24
3	0.09 G + 0.24	0.09 + 0.33
4	0.09 G + 0.33	0.09 + 0.45

[a] The cost of natural gas in $/mmBtu.

Source: U.S. Department of Energy, 1989

Several 10,000-mt/d plants would be required to supply a major methanol market. Research to understand how capital markets would respond to this level of demand has been limited. Constraints in construction capacity have been identified (Difiglio, 1989) that would limit the potential market in the year 2000.

Transportation

The cost of shipping methanol to the United States, as estimated by Chem Systems, depends on the tanker capacity, the distance involved, and the tanker's fuel consumption. New vessels would be needed, since any significant U.S. automotive use of methanol would require shipments that exceed current tanker capacity. The vessels would have to be built especially for methanol cargoes. Therefore, estimates of shipping costs must allow for empty tanks on the return trip.

Table 5 estimates the costs of shipping methanol from a number of possible supply regions to the United States (destination, Los

Table 4
Bechtel's Estimates of Producing Methanol in Six Different Locations Using Several Conversion Technologies
(cents per gallon)

Bechtel Location	Chem Systems Site Category	Non-Gas Operating Cost	Capital Recovery	Total
Texas	1	.06	.14	.20
Canada				
(Non-arctic)	1	.05	.16	.21
Trinidad	1	.06	.23	.29
Saudi Arabia	2/3	.07	.29	.36
Australia	3/4	.09	.31	.40
Alaska	4	.09	.26	.35

Source: U.S. Department of Energy, 1989; Bechtel Corporation, 1988

Angeles). Although methanol is now generally transported in tankers that do not exceed 40,000 deadweight tons (dwt), the estimates were based on a 250,000-dwt tanker, in view of the possible sharp increase in volume that would accompany the significant use of methanol as an automotive fuel. Costs range from 1 cent/gal to 3 cents/gal and are comparable to rates currently charged for large-scale oil transportation.

In order for U.S. ports to accommodate large vessels, the current supertanker receiving capacity would have to be expanded. Some analysts have assumed that no such facilities would be available and, therefore, that methanol would have to be carried to U.S. ports in smaller vessels. If this was the case, the shipping cost could increase by 6 to 9 cents/gal. Bechtel estimates transportation costs to be 5 to 9 cents/gal, except for Alaska. It assumes that a new pipeline will be built in Alaska and that the transportation cost would be 52 cents/gal.

Table 5
Methanol Shipping Costs
(1987 $/gal)

OECD	
Alaska	0.008
Canadian Arctic	0.008
Norwegian Arctic	0.027
Australia/New Zealand	0.027
PERSIAN GULF OPEC	
Iran	0.022
Other Persian Gulf	0.022
OTHER OPEC	
Algeria	0.025
Libya	0.025
Nigeria/Gabon	0.027
Venezuela/Ecuador	0.013
Indonesia	0.023
OTHER	
Malaysia	0.025
Thailand	0.026
Mexico	0.015
Trinidad & Tobago	0.014
Tierra del Fuego	0.019
China (PRC)	0.021

Note: Assumptions: Product shipped to Los Angeles
Heavy fuel oil @ $13.50 per barrel
Methanol transported in 250,000-dwt tankers

Source: Lawrence and Kapler, 1989

Delivered Costs of Methanol

Combining feedstock, conversion, and shipping cost estimates, it is possible to construct aggregate supply curves for delivered methanol. A typical supply curve is presented in Figure 1. Points on the supply curves can then be selected to demonstrate marginal supply

Figure 1
Methanol Supply Curve

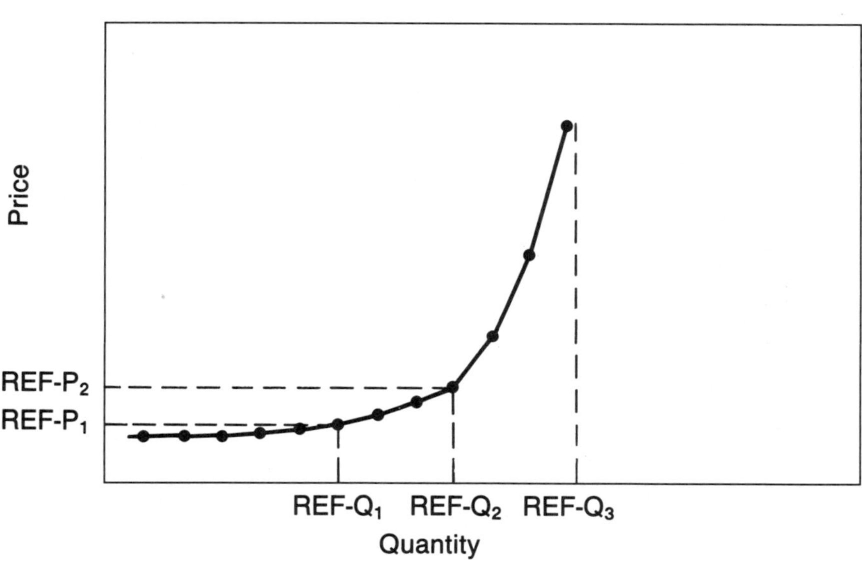

Note: The methanol supply curve includes natural gas feedstock costs, process-ing cost, and transportation cost to Los Angeles. A different curve is estimated for each supply region.

Source: Lawrence and Kapler, 1989

prices of methanol corresponding to various alternative-fuel de-mand scenarios for the United States in the year 2000. One such scenario is shown in Table 6.

The scenario in Table 6 assumed an annual U.S. demand for 61 billion gallons of methanol. (Sixty-one billion gallons of methanol would replace 2.2 million barrels of gasoline per day.) The natural gas that would be required as feedstock is 5.5 tcf. Methanol could be delivered to the United States at a supply price of 28 cents/gal—the equivalent of a gasoline supply price of 50 cents/gal. A distri-bution markup must be added to this wholesale price to arrive at a retail price. Distribution costs might range from 10 cents/gal to

Table 6
Results: Answers to the Wrong Questions
Market Price 28 cents/gal Methanol to Los Angeles Port

	SUPPLY REGION				
	Other Persian Gulf[a]	Nigeria/ Gabon	Venezuela/ Ecuador	Trinidad/ Tobago	Total
Methanol Supplied (billion gallons/year)	30	9	19	3	61
Gas Used (tcf)	2.7	0.8	1.7	0.3	5.5
Shipping Cost ($ Gal Methanol)	.022	.027	0.13	0.14	
Processing Cost ($ Gal Methanol)	0.24	0.24	0.24	0.17	
Gas Netback					
($ Gal Methanol)	0.02	0.01	0.03	0.10	
($ mcf)	0.20	0.14	0.30	1.07	

[a] Other Persian Gulf countries are Kuwait, Qatar, Saudi Arabia, and the United Arab Emirates.

Source: Lawrence and Kapler, 1989

20 cents/gal. (EEA, 1989; EPA, 1989). Using the marginal supply curves for each country, it was estimated that the Persian Gulf supplied 50 percent and Venezuela/Ecuador supplied 31 percent of the methanol delivered to the United States. It was also estimated that West Africa provided about 15 percent and that Trinidad provided less than 10 percent.

Conclusions

This study has shown that there is an abundance of low-cost natural gas feedstock available for conversion into methanol for the U.S. transportation sector. Reasonable conversion and shipping costs, as a result of using advanced technologies, would make these alternative fuels obtainable at supply prices competitive with those of gasoline. It would be possible, therefore, for the United States to displace as much as 2.5 million barrels of gasoline consumption per day by the year 2000. The introduction of methanol, however, would require additional investment in vehicle conversion and in fuel-delivery infrastructure.

The estimates in this paper are for supply prices. They will probably be increased by rents and transfers, the size of which will be determined by market conditions. Market prices for alternative fuels will depend on gasoline prices, because the oil market will set the standard for rents paid to resource suppliers. High gasoline prices, resulting from robust demand, would enable natural gas suppliers to extract higher rents than would be possible in a depressed oil market.

Notes

1. Projects have been undertaken for the U.S. Department of Energy, the U.S. Environmental Protection Agency, and the California Energy Commission.

2. F.H. Kant, et al., Feasibility Study of Alternative Fuels for Automotive Transportation, Exxon Research and Engineering Co., U.S. Environmental Protection Agency, Report EPA-460/3-74-009, June 1974 (three volumes).

3. J.B. Pangborn and J.C. Gillis, Alternative Fuels for Automobile Transportation—A Feasibility Study, Institute of Gas Technology, U.S. Environmental Protection Agency, Report EPA-460-3-74-012, July 1974 (three volumes).

References

Bechtel, Inc. (1988). California Fuel Methanol Cost Study (consultant report).

Difiglio, C., and M. F. Lawrence (1987). Economic and Security Issues of Methanol Supply. In *Methanol: Promise and Problems*. Society of Automotive Engineers.

DOE (1989). Assessment of Costs and Benefits of Flexible and Alternative Fuel Use in the U.S. Transportation Sector, Technical Report Three: Methanol Production and Transportation Costs. U.S. Department of Energy.

EEA (1989). Fuel Methanol Infrastructure Analysis for the South Coast Air Basin. Energy and Environmental Analysis, Inc.

EPA (1989). Analysis of the Economic and Environmental Effects of Methanol as an Automotive Fuel. U.S. Environmental Protection Agency.

Lawrence, M.F., and J.K. Kapler (1989). Natural Gas, Methanol, and CNG: Projected Supplies and Costs. In *Alternative Transportation Fuels*, edited by D. Sperling. Greenwood Press.

Mawdsley, A.D., W.E. Stevenson, and R.G.J. Zimmerman (1989). Capital Servicing Costs of Large-Scale Fuel Methanol Plants. Bechtel Financial Services.

Singh, M.K. (1989). A Comparative Analysis of Alternative-Fuel Infrastructure Requirements. Society of Automotive Engineers Paper 892065.

CHAPTER 12

The Cost-Effectiveness of Substituting Methanol for Gasoline: What Federal Action Is Needed?

ALBERT McGARTLAND

Introduction

In recent years increasing attention has been given to the potential role of alternative fuels (defined, for the purposes of this paper, as any fuel other than gasoline or diesel fuel) as a replacement for petroleum-based fuels in the transportation sector. The impetus originally came from concern about U.S. dependence on oil imports, arising from the 1973 Arab oil boycott and the sharp increase in world oil prices. More recently, attention has shifted to the potential of alternative fuels as a means of reducing the pollution emitted by cars and trucks. This interest culminated in the

The author would like to thank, without implicating, Arthur Fraas, Jeff Alson, Mike Gold, Robert Hahn, Tom Lareau, Michael Shelby, and Margaret Walls. Portions of this paper were presented at the 1988 annual meeting of the Western Economic Association, at a session organized by Tom Lareau.

Bush administration's proposal for an alternative-fuel program in those cities with the most serious air quality problems.

When it becomes cheaper to operate a car or truck with one of the alternative fuels, market forces would be expected to respond by offering these alternatives to consumers. Supporters of market intervention, however, point out that the market may not respond, since consumers will require both an adequate fuel distribution network and alternative-fuel vehicles. Both fuel producers and auto manufacturers, they argue, will be reluctant to undertake the substantial capital investment—on the order of $10 to $20 billion to realize economies of scale—without some assurance that there will be an adequate market for their vehicles and fuel. These advocates of government intervention, therefore, argue that there is a "chicken-and-egg" problem that will slow the introduction of alternative fuels.

This paper takes a somewhat different approach by considering whether it would be effective to introduce alternative fuels before a well-operating market would do so, in order to deal with the air quality problems that now exist with gasoline fuels. Since the cost-effectiveness of methanol is very uncertain, the federal government role in promoting methanol must be well defined. The market must be permitted the flexibility to opt out of any technology that proves to be too expensive.

Background

The Clean Air Act required that all areas meet the National Ambient Air Quality Standard (NAAQS) for ozone by December 31, 1987. Although substantial areas of the country already meet these NAAQSs, roughly 80 metropolitan areas with a population of 90 million do not. While continuing improvement in air quality beyond 1987 is projected as new cars continue to displace older ones, at least 30 "hard-core" areas will not attain the NAAQS for ozone. As a result, additional control programs in these areas will require far more comprehensive and costly measures than those already adopted—or it will have to be conceded that the NAAQS for ozone cannot be attained with available technology in these 30 areas.

The strategy of the U.S. Environmental Protection Agency (EPA) in its current proposal for revision of the Clean Air Act is to enforce existing control requirements, promulgate new emissions limits, and require states to adopt additional measures to control emissions in these non-attainment areas. In implementing this strategy, EPA proposes stringent national standards to reduce emissions from some sources—such as limiting gasoline volatility and banning the use of solvents in some consumer products.

Nationwide measures impose substantial control costs on both attainment and non-attainment areas, however, and may yield only modest air quality improvements in the non-attainment areas. Therefore, EPA will also require states to adopt local control measures, in non-attainment areas with serious air quality problems. These measures may include regulation of small industrial sources, such as wineries, auto body shops, and bakeries. As indicated in Table 1, the cost of controlling these sources varies widely, with control costs for some sources substantially exceeding the costs imposed to date on industrial plants and motor vehicles.

In addition, EPA's proposed Clean Air Act amendments would require a "neat fuels" (methanol or compressed natural gas [CNG]) program and an "oxygen-blend" (gasoline blended with alcohols) program in the non-attainment areas having the worst air quality problems. Under the neat fuels program the administration bill would require the sale of one million neat fuel vehicles per year in the nine worst areas. EPA believes that current-technology alternative-fuel vehicles using M85 (85 percent methanol and 15 percent gasoline) or compressed natural gas could yield a 20 to 50 percent reduction in vehicle emissions of ozone-forming hydrocarbons (EPA, 1988) and thereby reduce ozone levels in these cities by 2 to 3 percent by 2005. Advanced-technology vehicles using these fuels could achieve tailpipe emissions reductions ranging from 50 to as much as 80 percent and could also achieve substantial reductions (30 to 80 percent) in ozone (EPA, 1987). A long-term program

Table 1
Available Volatile Organic Compounds Reductions[a]

Control Option	Potential Reduction in Emissions	Cost Per Ton Reduced	Annual Costs
	(% of 1982 Inventory)	($/ton)	(millions $)
Mobile Sources			
Gasoline Marketing			
Option 1: Stage II Control	1.1–1.6	1,050	50–80
Option 2: Onboard Control	1.9	2,730[b]	180
Other Evaporative Controls			
(11.5 psi)	1.7	n/a[c]	n/a
Gasoline Volatility	0.4–5.5	n/a	n/a
Additional Inspection			
and Maintenance	0.3	2,000–4,000	132
Industrial Sources			
Consumer Solvents	2–4	n/a	n/a
Pesticides (Reduction in			
Volatiles Content)	1–2	n/a	n/a
New Industry Controls			
—Auto Body Shops	0.2	0–4,000	96
—Bakeries	0.2	2,000–5,900	204
—Metal Rolling Scrubber	0.9	8,160	1,600
—Chemical Manufacturing			
(distillation)	0.5	1,950	234
—Coke Oven By-product	0.8	100	17
—Municipal Sewage Products	0.1	62,000	781
—Wood Furniture Coating	0.4	200	12

based on advanced-technology M100 (pure methanol) or compressed natural gas vehicles, it is estimated, could yield a steady-state reduction in ambient ozone levels of 6 to 11 percent. (Calculations of total volatile organic compounds reduction assume that alternative-fuel vehicles account for 30 percent of total miles traveled.)

Table 1
(Continued)

Control Option	Potential Reduction in Emissions	Cost Per Ton Reduced	Annual Costs
	(% of 1982 Inventory)	($/ton)	(millions $)
Controls in Non-attainment Areas Only	1–2		
—External Floating Roof Tanks	n/a	0–3,316	n/a
—Solvent Metal Cleaning	n/a	– 360 to 260	n/a
Enhanced New Source Review	0–1	n/a	n/a
Tighter Existing Regulations	0–2	n/a	n/a
Transfer, Storage, and Disposal of Hazardous Waste	1.7–4.5	220–1,400	741
Barge and Tanker (Only Ports)	1–2	n/a	n/a
TOTAL[b]	17–45		3,826

[a] Volatile Organic Compounds are precursors to ozone and, in general, their control is the most effective way to reduce ozone.
[b] National costs divided by tons reduced in non-attainment areas.
[c] Estimate not available.
[d] Total reflects overlap of related strategies.

Source: Office of Information and Regulatory Affairs, U.S. Office of Management and Budget, 1989

Cost-Effectiveness of Alternative-Fuels Programs for Improving Air Quality

In order that the overall merits of alternative-fuels programs may be judged, this paper compares the costs of an alternative-fuels strategy for controlling emissions with that of other control strategies. Assuming that the question is how to achieve the EPA-specified NAAQS at least cost, this approach is appropriate (Baumol and

Oates, 1971) and it is not necessary to evaluate directly the air quality gains from the use of alternative fuels.

The cost per ton of reducing reactive hydrocarbons is calculated using the following formula:

$$\frac{(\text{Change in Cost of Fuel}) + (\text{Annualized Increased Capital Cost})}{(\text{Emissions Reduced per Year})}$$

A more detailed formula is presented in the Appendix. This equation was used to calculate control cost (per ton) for two alternative-fuels programs: adoption of a fleet of M85 vehicles in the year 1997 and adoption of advanced-technology (M100) vehicles in the year 1997. These results are presented in Table 2. Estimates from other sources for some of the key variables in the cost equation were used to make this calculation. Continuing research will provide better data to make these calculations in the near future.

Cost of Neat Fuel Vehicles

It is assumed that under a program similar to that proposed by the administration, manufacturers would be able to realize economies of scale. For these large production runs, a methanol-fueled car using M85 would cost an additional $300 (CEC, 1989). EPA believes that, given a sufficient volume of orders, the price for an advanced-technology vehicle would be about the same as for a gasoline-fueled vehicle.

For a baseline gasoline-fueled vehicle, an emissions rate of 1.72 g/mi is assumed. This is the current emissions standard. The future standard is likely to be close to 0.94 g/mi, which is mentioned in most Clean Air Act proposals. Presumably, a 0.94 g/mi car would be more expensive to manufacture and, therefore, the relative cost of a methanol-fueled car would decrease.

The Price of Methanol (M85 and M100)

The price of chemical-grade methanol has fluctuated dramatically during the last two years, from 28 cents/gal in mid-1987 to 60

Table 2
Cost-Effectiveness of Methanol
at a Range of Energy Prices
(1988 dollars)

Price of Methanol[a]	Price of Gasoline[a]	Cost (per ton) of HC Reduced[b]
M85		
0.53	0.95	3,300– 7,800
0.58	0.95	7,600–12,000
0.63	0.95	12,000–16,000
0.53	0.66	16,000–20,000
0.63	0.66	24,000–29,000
M100		
0.53	0.95	– 3,900
0.58	0.95	– 1,800
0.63	0.95	260
0.53	0.85	– 50
0.58	0.85	2,000
0.63	0.85	4,000
0.53	0.66	3,900
0.58	0.66	5,700
0.63	0.66	7,800

[a] Prices exclude taxes. Taxes are not a social cost of production and therefore should not be included in a cost-effectiveness analysis. Taxes on methanol and gasoline are roughly equivalent on a Btu basis. For gasoline, average taxes are estimated to be 24 cents/gal; for methanol, taxes are 12 cents/gal.
[b] All costs[a] are based on gasoline cars' meeting current emissions standards of 1.72 g/mi.

Source: Calculations performed by ABT Associates, 1989, using equation in Appendix.

cents/gal in early 1989 to a mid-1989 price of 47 cents/gal. This price volatility is due to a combination of factors, including the increased use of methanol as a feedstock for MTBE (a gasoline additive) and an increase in European demand. The best price to use in estimates is the long-run equilibrium price that balances long-term supply and demand (the price that will provide a sufficient return to producers to justify investment in plant and equipment).

During 1989 several studies were completed on the cost of producing methanol, including a study by the U.S. Department of Energy (DOE) on alternative-fuel use and studies carried out in connection with California's proposal to require the sale of flexible-fuel vehicles by 1995. Estimates from these studies generally suggest a cost for methanol, delivered to Los Angeles, ranging from 39 cents to 46 cents/gal. SRI International (1989), however, projects costs ranging from 44 cents/gal to more than 60 cents/gal. The wide range in the projected cost reflects production costs from a range of sites with important differences both in the existing infrastructure and in the difficulties of carrying out the project at the site. In fact, substantial volumes of gas are available at the lower-cost sites, so that costs of production would likely fall at the lower end of the range.

The differences in these estimates reflect the effect of two key assumptions: the pricing of the natural gas used as a feedstock and the appropriate return on investment. In all of these studies the lowest-cost sources of methanol use remote natural gas—gas that does not have access to pipelines or suitable alternative uses—as a feedstock. In some areas (for example, Nigeria and the Middle East) substantial quantities of natural gas are currently being flared or vented. DOE (1989) and California Energy Commission (CEC, 1989) studies assume that this gas would be available as a feedstock at costs ranging from 50 cents to $1/mcf. However, SRI International (1989) assumes that there will be alternative uses for this gas—as a feedstock for chemical plants, as liquefied natural gas, or in the generation of electric power. As a result, SRI assumes a higher cost for natural gas used as a feedstock—ranging from a low of roughly $1/mcf in the late 1990s to more than $2/mcf by 2010.

In addition, DOE and CEC estimate fairly small increases in the cost of producing methanol during the next two decades, because of the substantial supplies of natural gas available and because of their assumptions about the relatively low-opportunity cost of the available natural gas. SRI assumes, though, that an increase in world oil prices and the resulting attractiveness of alternative uses of natural gas will drive up the price of methanol. Note that the formula to calculate methanol's cost-effectiveness (presented earlier) treats a decrease in the price of gasoline the same, mathematically, as an increase in the price of methanol. It is the price spread between the two fuels that determines the operating costs.

Finally, regarding return on investment, DOE and CEC typically require a before-tax return of 15 to 20 percent. SRI assumes that a project would have to yield a before-tax return of 25 to 30 percent because of the risks entailed in the substantial investment required. These differences account for a difference of 5 to 10 cents/gal in the estimated production cost for methanol.

An additional adjustment must be made to reflect the lower energy content of a gallon of methanol compared with a gallon of gasoline. This difference in energy content is partially offset because most experts expect methanol-fueled cars to get more miles per unit of energy than gasoline-fueled cars. A variety of sources have suggested that flexible-fuel cars will achieve 7.5 percent more miles per Btu than similar cars running on gasoline (CEC, 1989). A dedicated methanol-fueled car running on M85 in 1995 will achieve roughly 15 percent more miles per Btu. Finally, an advanced-technology M100 car will achieve 20 to 30 percent more miles per Btu than a gasoline-fueled car (CEC, 1989). To convert the cost of methanol to the energy-equivalent cost of gasoline, a factor of 1.8 has been used here for M85 vehicles, and a factor of 1.6 has been used for dedicated vehicles.

The Price of Gasoline in the Future

The range of future gasoline prices (wholesale) was taken from SRI International (1989) and includes 1997 estimates of a low price of 90 cents/gal and a high price of $1.19/gal. (All prices are quoted in 1988 dollar equivalents.) Other uncertainties, such as the ability of methanol-fueled cars to maintain their emissions reduction performance for the life of the vehicle, are not reflected. Other issues not considered in this discussion include the relationship between the cold-start problem and emission performance; the higher Reid vapor pressure that results when mixing M85 and gasoline; and the region-specific effectiveness of hydrocarbon control.

The Cost of Reducing Hydrocarbons With a Methanol Program

Given all these assumptions, a calculation of cost per ton was produced, which indicates that the cost is very sensitive to the price of the alternative fuel, relative to the price of gasoline. (See Table 2.) Small changes in this relative price can dramatically affect the estimated control cost (per ton). For example, the cost of using M100 varies from roughly $4,000/ton to $8,000/ton. This range reflects only a small variation in the prices of gasoline and methanol. It does not reflect any uncertainty about future vehicle costs, potential actions of OPEC, or other costs and benefits of methanol.

This framework assumes that attaining the NAAQS for ozone is the policy goal. Oates, Portney, and McGartland (working draft) have questioned the wisdom of establishing *uniform national* ambient air quality standards. They find that regional costs and benefits differ greatly and that constraining each region to a national air quality goal imposes unnecessary costs.

Even if a national standard is determined by EPA, the costs of attaining the current ozone standard may not be justified by the benefits. Recent analysis funded by EPA and performed by Resources for the Future suggests that the benefits per ton of hydrocarbon reduced are far less than the costs necessary to reach the NAAQS for ozone in non-attainment areas (Krupnick et al., 1988). It is worth considering whether many urban areas are willing to make the sacrifices needed to attain the ozone standard. The data suggest an unwillingness on the part of many urban areas to impose the numerous small, but expensive, controls needed to attain the standard. If greater decision making by the states is appropriate (and states should have the flexibility to select their own air quality), then the cost-effectiveness criteria presented here may not be appropriate.

It is worth considering other options available for controlling reactive hydrocarbons. Other strategies involve costs ranging from $100/ton to more than $8,000/ton (see Table 1). Apart from being expensive, measures such as imposing restrictions on the use of cars in hard-core non-attainment areas are also likely to be intrusive and inconvenient.

A Proposal for Government Action

The cost-effectiveness of methanol cannot be predicted. Methanol could easily prove to be either extremely costly or competitive with gasoline. The wide range of plausible cost-per-ton estimates provides little information for decision makers. With projected increases in the price of gasoline, methanol fleet use may become an increasingly attractive control option. If gasoline prices increase by 5 percent a year, large methanol programs will be cost-effective before 1997. The author's assumption that the real price of methanol in 1997 will be 53 to 63 cents/gal is critical to this analysis. Underlying this estimate is the assumption that remote gas would have no other economic use and would be priced accordingly (based on the Jack Faucett Associates report [1987] and DOE analysis [1988]).

In addition to the controls presented in Table 1, other control options (that may be much cheaper) are rapidly evolving. Although the emissions reductions per car using a reformulated gasoline may be less than reductions per vehicle using M100, emissions from every car will be reduced. (A methanol program will cover only a fraction of the total fleet.) Thus, this author finds it difficult to recommend the *mandating* of methanol. Locking into any technology with uncertain costs does not make sense. Other cheaper alternatives may be available (especially if methanol's costs turn out to be high).

Does this mean that the United States should forgo any progress on attaining the NAAQS for ozone? If the government does not pursue methanol as a control strategy, should it do nothing?

Once it becomes cheaper to operate a car or truck with an alternative fuel, market forces would be expected to respond by offering this alternative to consumers. However, the market may fail to capture many of the "benefits" associated with methanol use (in particular, the lower emissions of reactive hydrocarbons). Given the positive externalities associated with methanol, it is easy to show that the market does not weigh all the costs and benefits of determining when to introduce methanol. The positive externality associated with methanol use (compared with gasoline) generates a social benefit (negative cost) that the market should recognize but does not.

Unfortunately, the government cannot make the decision either. As argued above, it is not clear that even when these positive externalities are included in the decision making that the costs of a methanol program will not be greater than benefits. Regulators simply do not have enough information.

The good news is that a properly designed methanol program can include a scheme that will provide the market with the correct incentive to adopt methanol. By internalizing the externalities, this scheme will allow the private sector to weigh all the benefits of methanol against the corresponding costs.

The idea is not new. Much has been written about market incentives for the control of externalities. For a variety of reasons, however, the debate has centered on the costs and benefits of methanol—not on what the optimal methanol policy should be. The two are different issues. There are two basic approaches to internalizing the externality: allocating marketable permits (or emission reduction requirements) or setting prices on emissions. The remainder of this section outlines how these approaches might be applied to methanol.

Marketable Permits

Ideally, formal marketable permit systems, in each non-attainment area, could establish a market value for hydrocarbon reductions.

If every urban area had a well-operating system, EPA could permit a coalition of methanol fuel and vehicle suppliers to sell emission credits for every ton of hydrocarbon reduced by an urban fleet of methanol-fueled vehicles.

A more pragmatic approach would be for the federal government to assign nationwide emissions reductions to major industrial polluters and to let the industries determine how to achieve these reductions. (Buying reductions from other industries would be permitted.) If a coalition of methanol fuel and vehicle suppliers believe methanol promotion is the cheapest way to accomplish these reductions, the market, on its own, will supply methanol.

Emissions Fees

In an ideal world, each urban area would establish emissions taxes at just the right level to attain the ozone standard. These taxes would include a tax on the use of gasoline-fueled automobiles, which could be a combination of taxes on gasoline and on automobiles. Under a simple scheme, a tax on gasoline would provide incentives to reduce driving (and pollution). If the use of M100 resulted in an 80 percent reduction in reactive hydrocarbons, it would receive a tax that was 80 percent less than the tax imposed on gasoline. A national program would adjust the existing highway trust fund tax according to the pollution potential of each motor fuel.

All of these rough market-incentive mechanisms are at least approximations of a proper incentive to bring about the introduction of methanol. As future energy market conditions evolve, one of them could ensure that methanol would be supplied at the "right" time.

The "Chicken-and-Egg" Problem

As discussed earlier, those who support market intervention argue that the adoption of alternative fuels will be slowed because consumers will require both an adequate fuel distribution network and alternative-fuel vehicles, but both fuel producers and auto manufacturers will be reluctant to undertake the substantial capital investment—on the order of $10 to $20 billion to realize scale economies—without some assurance that there will be an adequate market for their vehicles and fuel. Thus, these advocates of government intervention argue that there is a chicken-and-egg problem that will slow the adoption of alternative fuels. While no one can deny the existence of this problem, the market has shown a tremendous ability to overcome such problems. Television and TV shows, radio, compact disc players, computers and software, even diesel cars and diesel fuel are all examples of the market overcoming a chicken-and-egg problem.

Conclusion

This paper has outlined the basic conditions under which an economic case can be made for promoting alternative fuels. Alternative-fuels programs offer potentially large air quality gains—gains ranging from a factor of two to an order of magnitude greater than most of the other control measures currently under consideration.

However, the control cost (per ton) varies dramatically with changes in the price spread between gasoline and the alternative fuel. Given the enormous uncertainty about future fuel prices, this author believes that mandating controls is probably unwise. If methanol prices prove to be higher than expected—and gasoline prices lower—the cost of an alternative-fuels program could be enormous. However, alternative-fuels prices may be much lower than expected, a situation that could mean the government should have intervened.

Establishing a level playing field (regarding regulatory and tax burdens) is generally the best way to promote efficient vehicle and fuel-control strategies. One obvious way to incorporate a credit is through an emissions tax-subsidy that reflects differences in the emissions of all fuels and vehicles. As an alternative to an emissions tax, the government could (1) assign to each industry pollution-reduction targets that recognize the reductions achievable under an alternative-fuels program and (2) allow trading. Once the emissions reduction targets were assigned, the market could determine the best mix of fuels and technologies to meet these emissions reductions.

The Bush administration has chosen the latter approach. It makes sense to recognize the potential of alternative fuels but yet preserve the market's ability to adopt technologies and fuels only when it is efficient to do so. It would be unfortunate if Congress instituted a less flexible approach.

Appendix

The formula detailed below was used in this paper to calculate the cost-effectiveness of each low-polluting fuel. The cost per ton of hydrocarbon reduced is equal to:

$$\frac{M(x)*G/M*((MG*P(meth))\text{-}P(gas)) \; + \; x*C*(i*(1+i)n)/((1+i)n\text{-}1))}{M(x) \, * \, g/m(\text{gasoline car}) \, * \, t/g \, * \, (\text{percent reduced})}$$

where: $M(x)$ = Number of miles driven, on average, by x number of cars. (Typically this depends on the alternative-fuel strategy. Fleet cars may be used more intensively than the average car.)

G/M = Gallons of gasoline used to drive 1 mile.

MG = The number of gallons of methanol needed to achieve the same amount of miles as one gallon of gasoline.

$P(meth)$ = Price of methanol per gallon. (For CNG calculations, substitute price of CNG on a gallon-equivalent basis.)

$P(gas)$ = Price of a gasoline per gallon.

C = Increased cost of methanol (CNG) vehicle.

i = Interest rate.

n = Years of useful life (7 years).

g/m = The grams of reactive hydrocarbons emitted per (gasoline car)   mile for a gasoline car.

t/g = Tons per gram.

Percent = Percent of emissions reduced by using the alter-
Reduced   native-fuel vehicle.

References

Alcohol Week. Various issues, 1985–1989.

Baumol, W.J. and W.E. Oates (1971). The Use of Standards and Price for Protection of the Environment. *Swedish Law and Economics Journal*. Stockholm. March, 1971, pp. 42–54.

CEC (1986). California's Methanol Program: Evaluation Report, California Energy Commission.

CEC (1989). AB 234 Report: Cost and Availability of Low Emission Motor Vehicles and Fuels (draft). California Energy Commission.

DOE (1988). Assessment of Costs and Benefits of Flexible and Alternative Fuel Use in the U.S. Transportation Sector. U.S. Department of Energy, Report No. DOE/PE-0080.

DOE (1987). *Annual Energy Outlook, 1986*. U.S. Department of Energy.

Energy and Environmental Analysis, Inc. (1986). Distribution of Methanol for Motor Vehicle Use in the California South Coast Air Basin (report to EPA).

EPA (1988). Air Quality Benefits of Alternative Fuels, prepared for the Vice President's Task Force on Alternative Fuels by the U.S. Environmental Protection Agency.

EPA (1987). Guidance on Estimating Motor Vehicle Emission Reductions for the Use of Alternative Fuels and Fuel Blends. U.S. Environmental Protection Agency, Report No. EPA-AA-TSS-PA-87-4.

Vice President's Task Force (1987). Report of the Alternative Fuels Working Group.

Jack Faucett Associates (1987). Methanol Prices During Transition, (Final report to the Office of Mobile Sources), U.S. Environmental Protection Agency.

Krupnick, A.J. (1988). An Analysis of Selected Health Benefits from Reductions in Photochemical Oxidants in the Northeastern United States (report prepared for the Ambient Standards Branch,

U.S. Environmental Protection Agency, by Resources for the Future).

Oates, W., P. Portney, and A. McGartland (working draft). National or Local Standards for Environmental Quality: A Tale of Two Cities.

SRI International (1989). The Economics of Alternative Fuels and Conventional Fuels.

CHAPTER 13

Production Costs for Alternative Liquid Transportation Fuels

JAMES L. SWEENEY

Introduction

With the current policy interest in fuels that release less volatile organic compounds and other pollutants than gasoline, it is useful to assess probable costs of these alternatives in comparison with costs of gasoline derived from crude oil. This paper uses work done for the National Research Council (NRC), which was called upon by the U.S. Department of Energy to outline a broad research and development program aimed at producing liquid transportation fuels from U.S. resources. The Committee on Production Technologies for Liquid Transportation Fuels, chaired by John P. Longwell and organized by James Zucchetto,

Partial funding for this research was provided by the Stanford University Center for Economic Policy Research (Energy, Natural Resources, and Environment Program).

completed the task in November 1989. This paper draws heavily—
and sometimes verbatim—on Chapter 3 and Appendix D of that
committee's report (1990).

As part of the NRC study, probable future costs of several fuel
production technologies were estimated. Using assumptions specified by the committee, literature-documented costs of the processes
were compiled by Bernard Schulman and Frank Biasca (1989) of
SFA Pacific, Inc., and were reviewed, refined, and updated by various committee members. The results were used to estimate production costs on a consistent basis across feedstocks and technologies,
based on current technical understanding.

While the NRC examined many liquid alternatives to gasoline, this paper presents estimates only for a set of "clean," or
low-polluting, fluids—methanol, ethanol, and compressed natural
gas (CNG). Methanol estimates are provided for methanol
manufactured from several different feedstocks: natural gas, coal,
and wood fiber. Ethanol is estimated for production using corn
as a feedstock. Electric vehicles or hybrid electric vehicles have
not been evaluated. Estimates for production from U.S. resources
are presented first, followed by estimates of methanol production
abroad, using natural gas as a feedstock. The emphasis is on comparison of costs among low-polluting fuels and between these fuels
and gasoline.

Structure of Analysis

The cost analysis relies on estimates of energy prices, technological assessments specific to the production processes, and
assumptions about the economic environment. These considerations are combined to construct estimates of the total cost of
producing and using the alternatives. Costs are expressed in terms
of crude-oil-equivalent prices: the crude oil price that would make
driving a gasoline-fueled spark-ignition vehicle just as expensive
as driving an automobile powered by the specific fuel. Figure 1
provides an overview of the cost calculations.

Figure 1
Flow Chart of Cost Calculations

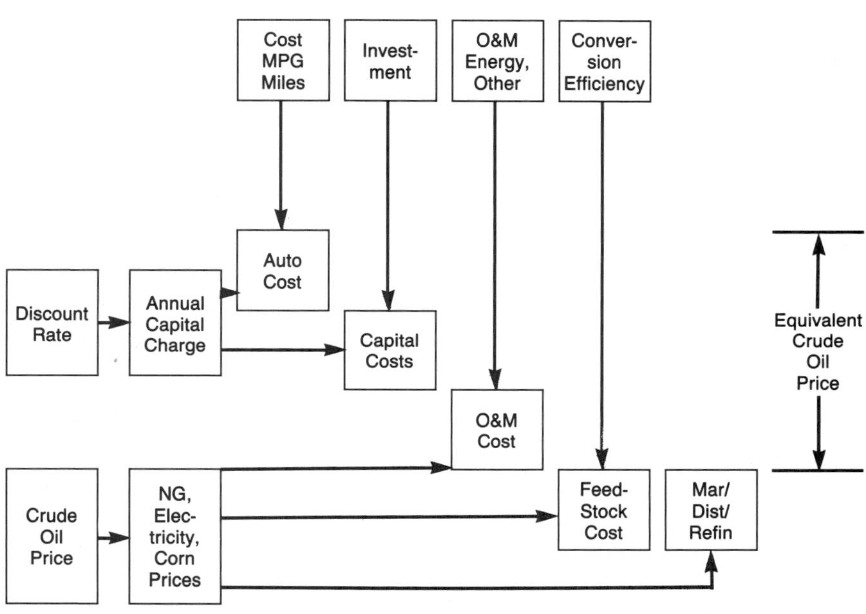

Source: J.L. Sweeney, 1989

Technological assumptions for each process include the types and conversion efficiencies of feedstocks used, the capacity and investment cost of a typical facility, the non-energy operating and maintenance (O & M) cost, the energy O & M inputs, expressed in terms of the types and quantities of energy used (other than as feedstocks), and the quantities of by-products produced. A gasoline equivalency factor is included that measures the quantity of the specific fuel required to produce as much energy as one gallon of gasoline. Evaluations are based on technical parameters that could be expected to prevail for a mature technology and industry, not those relevant to pioneering plants. Because several of the technologies have not been commercially implemented or have not been implemented on a large-volume basis, there is a high degree of uncertainty about the technical parameters. In addition, because methanol and compressed natural gas (CNG) would increase the cost of automobiles, the additional capital cost of a new dual-fuel automobile is included. Costs are those associated with a dual-fuel vehicle because consumers would avoid purchasing dedicated non-fleet vehicles unless and until a sufficient infrastructure of filling stations were in place.

Economic assumptions (indicated in Figure 1) include prices for the various energy and non-energy feedstocks, expressed, where appropriate, as a function of the prevailing crude oil price; prices of energy used in plant operation; the cost of capital and the corresponding annual capital charge factor; costs of crude oil refining, as a function of crude oil price; and costs of fuel distribution and marketing, as a function of cost of the specific fuel. Non-energy costs are estimated independently of oil prices, though real (inflation-adjusted) construction costs may vary with energy prices and with the development rates of alternative-energy facilities. The analysis assumes a gradual growth of the specific alternative-energy industry, not a crash program under which construction costs could rise rapidly.

For each technology, cost-per-equivalent oil barrel (in 1988 dollars) is calculated as the summary cost measure, based on the assumed oil price environment. The calculation begins with the crude oil price, defined as the average price of crude oil imported into the United States. The crude oil price implies prices of energy and non-energy inputs to the various production processes. In particular, prices of natural gas, electricity, and corn (as feedstock) are assumed to increase with crude oil prices. The specific functional relationships are discussed at a later point.

Product costs are calculated by adding feedstock costs, operating and maintenance (O & M) costs (energy plus non-energy), and annual capital cost, and by subtracting by-product credits. Per-gallon feedstock costs and energy O & M costs are calculated as a product of feedstock (or other input) quantity per gallon and the input price. Non-energy O & M costs are directly estimated. The annual capital charge is estimated as the investment cost multiplied by an annual capital charge factor. The net result is a refinery gate price. The addition of distribution and marketing costs leads to the pump price, net of local, state, and federal excise taxes.

Gasoline-equivalent cost is calculated by multiplying pump price by the gasoline equivalency factor and adding the per-gallon incremental vehicle cost for methanol or CNG-fueled vehicles (above the costs of a gasoline-fueled vehicle). Gasoline-equivalent cost is used to estimate cost-per-equivalent oil barrel by two approximate adjustments. Distribution and marketing costs of gasoline are subtracted from gasoline-equivalent cost. A refining credit, equal to the refining cost of gasoline, is also subtracted. This credit is based on the historical relationship between prices of gasoline and crude oil. It is not, strictly speaking, the cost of refining gasoline, since refineries produce many different products.

The procedure outlined above gives a crude-oil-equivalent cost that itself depends on the crude oil price. Crude oil price is generally assumed to be $20/b, slightly above the current level. In addition, an exogenously determined crude oil price of $40/b is tested, along with an endogenously determined crude oil price.[1]

Specific Factors

Feedstock Quantities Used

Table 1 lists the six combinations of feedstock and product, along with the identifying symbol and the quantity of each feedstock, per gallon of the final product produced.

Gasoline Equivalency Factor

The gasoline equivalency factor has a value of 1.8 for methanol, reflecting a 10 to 18 percent efficiency gain for methanol-fueled automobiles and a value of 1.5 for ethanol, assuming no efficiency gain over gasoline. For CNG, all analysis is conducted in terms of

Table 1
Identifying Symbols and Feedstock Quantities Used

Feedstock, Product	Feedstock	Quantity (per gallon)
NG > Methanol	Natural Gas	0.095 mcf/gal
Coal > Methanol	Coal	11.3 lbs/gal
UCG > Methanol	Underground Coal Gasification	N.A.
Wood > Methanol	Wood	13.40 lbs/gal
	Oxygen	3.35 lbs/gal
CNG	Natural Gas	0.12 mcf/gal equiv.
Corn > Ethanol	Corn	0.40 bu/gal

Source: J.L. Sweeney, 1989

output per equivalent gallon of gasoline, so that the gasoline equivalency factor is equal to 1.0.

Domestic Production Input Prices

Table 2 summarizes specific input price assumptions for domestic fuel production. Relationships between crude oil price and other energy prices are calibrated from the base case and high world oil price forecasts published by the Energy Information Administration (EIA) in its 1989 *Annual Energy Outlook (AEO)*. The *AEO* base case forecasts assumed for the year 2000 a world oil price of $28/b (1988 dollars), and the high world oil price forecasts assumed a world oil price of $35/b.

It is assumed that conversion facilities for coal and natural gas are located on the gulf coast and that facilities for underground coal gasification and ethanol production are located at the resource site. Natural gas compression would occur at the point of end use. Thus natural gas used as a methanol feedstock is priced at an estimated wellhead price, and natural gas compressed into CNG is priced at a delivered cost to the industrial sector.

Table 2
U.S. Domestic Input Price Assumptions

	General Function	Value @ Poil = $20
Natural Gas Price ($/mcf)	$3.91 + 0.05857 × (Poil − 28) to a maximum of $5.00	$3.44
Corn Price ($/bu)	$2.50 + 0.01786 × (Poil − 28)	$2.36
Electricity Price ($/kwhr)	$0.049 + 0.00020 × (Poil − 28)	$0.047
Coal Price ($/ton)	$38.00	
Wood Price ($/dry ton)	$32.40	
Oxygen ($/ton)	$57.05	
NG Delivery Cost ($/mcf)	$0.94	
where Poil = Price of crude oil ($/b)		

Source: J.L. Sweeney, 1989

EIA estimated wellhead natural gas prices at $3.91/mcf in the base case and $4.32/mcf in the high world oil price case. Costs delivered to industrial users were estimated to be 94 cents/mcf higher in both cases. Our analysis linearly extrapolated the relationship between oil prices and wellhead natural gas prices, but it did not allow wellhead natural gas prices to exceed $5/mcf (1988 dollars), because coal gasification becomes competitive above this price. Figure 2 plots the wellhead natural gas price function (natural gas as a function of crude oil price) used throughout most of the study as well as the price function for a sensitivity case with natural gas price function reduced by $2/mcf at each oil price.

EIA estimated the average electricity price delivered to industrial users at $0.0490/kwhr in the base case and $0.0504/kwhr in the high world oil price case. Our analysis linearly extrapolated this relationship between oil prices and electricity prices. Corn prices and corn by-product prices were also assumed to be linearly related

Figure 2
Natural Gas Price vs. Crude Oil Price

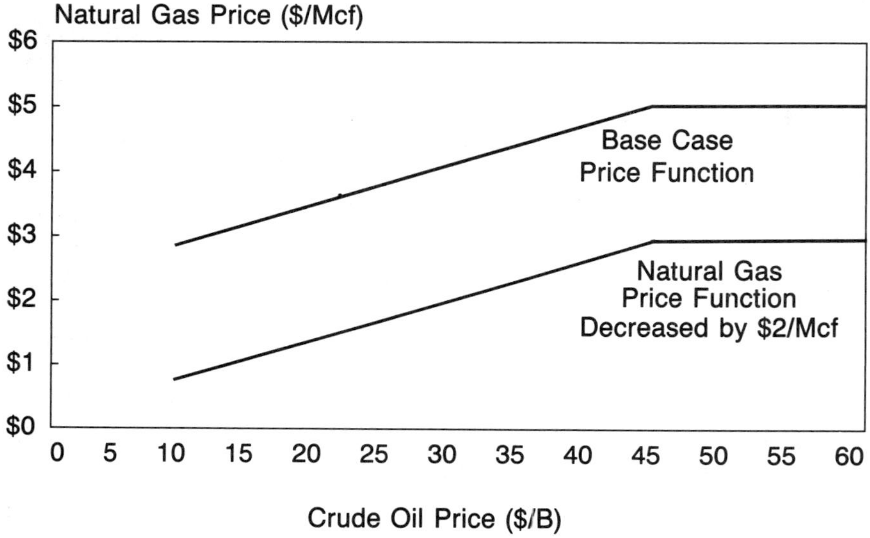

Natural Gas Price ($/Mcf)

Base Case
Price Function

Natural Gas
Price Function
Decreased by $2/Mcf

$6 $5 $4 $3 $2 $1 $0

0 5 10 15 20 25 30 35 40 45 50 55 60

Crude Oil Price ($/B)

Source: J.L. Sweeney, 1989

to oil prices. The relationship between corn prices and crude oil prices was based on an assumption that every bushel of corn requires the use of three-fourths of a gallon of oil in its production. Thus, if the crude oil price increased by $1/b, the per-bushel cost of growing corn would increase by the cost of the oil used in the growing and harvesting process: 0.75 • $1 / 42.

Refining and Distribution Costs

Refining costs of gasoline are assumed to increase with the price of crude oil. The relationship is calibrated so that the spread between gasoline and crude oil price would be $7/b if the crude oil price were $28/b, and would increase by $2/b for every $11/b increase in the crude oil price. The resulting equations are presented in Table 3.

Table 3
Refining, Distribution, and Marketing Costs

Gasoline Refining ($/b)	$7 + 0.18182 × (Poil − 28)
Gasoline Distribution/ Marketing ($/b)	$4 + 0.0200 × (Poil − 28)
Methanol/Ethanol Distribution/ Marketing ($/b)	$4 + 0.0200 × (Peq − 28)/EQ

Where: Poil = Price of crude oil ($/b)
　　　　Peq = Crude-oil-equivalent cost of methanol or ethanol
　　　　EQ = Gasoline equivalency factor

Source: J.L. Sweeney, 1989

Distribution and marketing cost (per actual barrel) is assumed to be $4/b of product if crude oil is $28/b. The same cost per barrel is applied to gasoline, methanol, and ethanol. Distribution costs are also related to the value of the product being distributed. Distribution costs are assumed to increase by $0.02/b for every $1/b increase in product costs. This increase is meant to reflect additional inventory costs of the higher-valued product. The same relationship is applied to all methanol, gasoline, and ethanol. The resulting equations are also presented in Table 3.

The distribution and marketing cost for compressed natural gas is separately analyzed, since the compression station and the distribution station would typically be one unit (Schulman and Biasca, 1989). Thus, capital cost for this technology is the cost of a single filling station that would deliver an amount of fuel equivalent to 1,200 gallons of gasoline per day and that would use 0.12 mmBtu of natural gas per equivalent gallon of gasoline. Annual operating and maintenance costs (non-energy) are estimated as 7 percent of the original investment cost. No separate additional distribution and marketing costs are assessed.

Incremental Automobile Costs

For methanol and CNG, the capital cost for new cars is anticipated to be greater than for gasoline. For these fuels, this additional capital cost is translated to an equivalent per-mile additional operating cost, so as to give the same discounted present value of costs to the consumer as would the one-time additional capital cost.[2] For this calculation, it is assumed that a typical car is driven 15,000 miles per year during the course of seven years. The per-mile equivalent cost is then translated to a per-gallon cost by multiplying by the average fuel efficiency of new cars, assumed to be 27 miles per gallon (mpg).

Discount Rates and Annual Capital Charge Factors

Two different real (inflation-adjusted) after-tax annual discount rates within a discounted cash flow (DCF) framework are considered—10 percent and 15 percent—with corresponding annual capital charge factors of 16 percent and 24 percent. The 10 percent rate is based on average historical returns required for equity capital in financial markets and on average historical returns earned by physical capital in U.S. industry. The 15 percent discount rate is based on estimation of typical hurdle rates (minimum estimated rates of return) required by corporations for investments or on costs of capital for risky projects.

Annual capital charge factors are calculated so that the net present value of the stream of capital charges, after taxes, is just equal to the initial investment cost, using the cost of capital as the discount rate. A project life is chosen as 20 years, plus the construction time. It is assumed that the investment costs are incurred two years before the middle of the first year of plant operation, or equivalently, 1.5 years before the project begins operating.[3]

In developing these factors it is assumed that the various processes face tax rules consistent with the current U.S. tax laws. In particular, it is assumed that the corporation pays a 35 percent tax rate on profits and that the investment is depreciated over time using a 10-year double-declining-balance depreciation schedule. The tax-deductibility of interest payments is incorporated implicitly in this analysis through use of the after-tax cost of capital.[4]

Plant Investment Costs

Plant investment costs are based on estimates of costs that might apply in a developed industry and are not based on costs of the first demonstration or pioneering plant, since some of these technologies have not been fully demonstrated and have not been commercialized. Table 4 displays estimates of investment costs for each technology.

Plant capacities have been chosen based on what is economically efficient for the various technologies. Capacities have not been

Table 4
Domestic Production Investment Costs and Capacities

	Methanol				CNG	Ethanol
Process Feedstock	**NG**	**Coal**	**UCG**	**Wood**	**NG**	**Corn**
Capacity (b/d, actual)	79,033	90,000	17,946	554	29	14,490
Average production (b/d)	75,081	81,370	16,225	501	29	13,101
Investment ($mm)	883	3,071	690	24	0.15	400
Investment/ Capacity	11,173	34,122	38,449	43,056	5,250	27,605
Investment/ Capacity (oil equiv.)	20,111	61,420	69,208	77,500	5,250	41,408

Source: J.L. Sweeney, 1989

standardized across the various technologies, since the appropriate scale varies radically across plants. Capacities in Table 4 are expressed both in terms of actual barrels per day and in terms of equivalent oil barrels per day.

As shown in Table 4, estimates of the investment cost per-equivalent oil barrel vary widely and generally exceed the $10,000 to $20,000 range typical of investments for crude oil extraction in the United States. They range from a low of $20,000 for methanol production (natural gas feedstock) to a high of $77,500 for methanol derived from wood. Although the CNG investment cost is shown as lower than the range cited, these figures only include the investment cost for a CNG delivery station and not the additional costs of the vehicles themselves or the investment cost of natural gas wells.

Cost Estimates for Domestically Produced Fuels

Based on the procedure outlined above, cost estimates have been developed for each fuel. Data are presented in Table 5 and are displayed in a sequence of graphs.[5]

The top part of Table 5 shows cost estimates expressed in units of dollars-per-gallon of product. For each fuel-feedstock combination, estimates are given for costs per gallon of feedstocks, non-energy operating and maintenance (O & M), energy O & M, and capital charges. The sum of these, minus any by-product credits, provides an estimate of the refinery gate price (the price charged at the point of bulk distribution). Adding distribution and marketing costs gives an estimate of pump price (the price to the motorist) net of taxes. Note that federal and state excise taxes on the product are not included in these estimates, and thus these figures should be compared with gasoline prices net of gasoline taxes.

Pump prices (net of taxes) for domestically produced methanol range from 60 cents/gal for natural gas as a feedstock, to $1.14/gal for wood as a feedstock. Coal-based methanol pump prices are estimated at 73 cents/gal for underground gasification, and 87 cents/gal

Table 5
Domestic Fuel Cost Estimates
($20 crude oil price, 10% DCF)

	Methanol				CNG	Ethanol
Process Feedstock	**NG**	**Coal**	**UCG**	**Wood**	**NG**	**Corn**
Product Cost ($/gal)						
Feed	0.33	0.21		0.31	0.41	0.94
O&M, Non-energy	0.06	0.16	0.18	0.20	0.15	0.16
O&M, Energy				0.02	0.03	0.11
Capital Charge	0.12	0.40	0.45	0.50	0.06	0.32
By-products						(0.47)
Refinery Gate Price	0.51	0.77	0.63	1.04		1.06
Distribution Marketing	0.10	0.10	0.10	0.11		0.11
Pump Price (net of taxes)	0.60	0.87	0.73	1.14	0.65	1.17
Vehicle Incr. Cost ($/gal)	0.07	0.07	0.07	0.07	0.34	0.00
Gasoline-Equivalent Cost (no tax)	**1.16**	**1.64**	**1.38**	**2.13**	**0.99**	**1.75**
Product Cost ($/equiv. barrel)						
Feed	24.77	16.23		23.64	17.34	59.40
O&M, Non-energy	4.20	11.78	13.61	14.86	6.30	9.83
O&M, Energy				1.83	1.33	6.93
Capital Charge	9.38	30.10	33.91	37.98	2.33	20.29
By-products						(29.70)
Refinery Gate Price	38.36	58.11	47.52	78.30		66.75
Distribution Marketing	7.37	7.77	7.55	8.18		6.72
Pump Price (net of taxes)	45.72	65.88	55.07	86.49	27.30	73.48
Netback to Crude Price						
Refining Credit	(5.55)	(5.55)	(5.55)	(5.55)	(5.55)	(5.55)
Gasoline Distribution/ Marketing	(3.84)	(3.84)	(3.84)	(3.84)	(3.84)	(3.84)
Product Cost (crude equivalent)	36.34	56.49	45.69	77.10	17.92	64.09
Vehicle Incr. Cost ($/b)	2.82	2.82	2.82	2.82	14.12	
Crude Oil Equivalent Cost ($/b)	**39.16**	**59.31**	**48.51**	**79.92**	**32.03**	**64.09**

Note: Refinery gate price is not a meaningful concept for CNG; therefore, no estimates are provided.

Source: J.L. Sweeney, 1989

for above-ground gasification. The pump price (net of taxes) for ethanol produced from corn is estimated to be $1.17/gal.

To make these pump prices comparable with those of gasoline, prices are multiplied by the gasoline equivalency factor, and the incremental automotive costs are added. The resultant gasoline-equivalent costs (net of taxes) for methanol range from $1.16/gal for methanol derived from natural gas to $2.13/gal for methanol derived from wood fiber. For ethanol derived from corn, the gasoline-equivalent cost would be $1.75/gal. The lowest cost alternative would be CNG, with a gasoline-equivalent cost of 99 cents/gal. Note that the CNG gasoline-equivalent cost includes the 34 cents/gal incremental cost of the dual-fuel automobile.

These gasoline-equivalent costs can be netted back to the crude-oil-equivalent costs by subtracting the gasoline-refining credit and the distribution/marketing cost for gasoline ($5.55 and $3.84, respectively) to give the crude-oil-equivalent costs, displayed on the bottom line of Table 5. The lowest crude-oil-equivalent cost for methanol is $39.16/b, roughly $20/b above the assumed $20 crude oil price. Both biomass-derived alcohols have crude-oil-equivalent costs estimated to exceed $60/b. Compressed natural gas has the lowest crude-oil-equivalent cost at $32/b.

The crude-oil-equivalent costs are graphed on Figure 3. Figure 3 is coded to indicate the progress of each technology toward commercial application. Technologies that have already been commercialized include natural gas used as feedstock to produce methanol, CNG used directly in automobiles, and corn distilled into ethanol. Technologies that have been successfully demonstrated on a commercial scale include coal and wood used as methanol feedstocks. There has been no successful demonstration on a commercial scale of methanol production through underground coal gasification.

Table 5 also shows components of the crude-oil-equivalent cost. For most technologies, feedstock cost and capital cost are the two largest cost categories. By-product credits are small, except for the production of ethanol from corn, a process that produces large

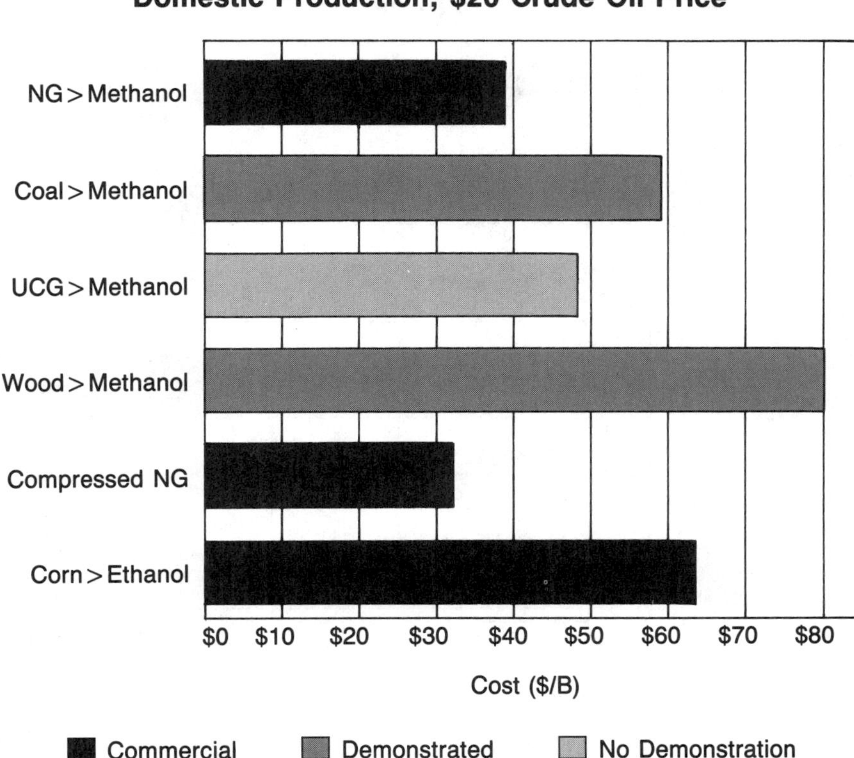

Source: J.L. Sweeney, 1989

organic by-products. The four processes for producing methanol suffer from either high annual capital costs (between $24/b and $30/b of equivalent oil cost, for methanol from coal or wood) or high feedstock cost ($28/b of equivalent oil cost, for methanol from natural gas).

Figure 4 has been provided to illustrate the importance of the discount rate. It provides cost data for both the 10 and 15 percent real discount rates and shows that the costs of these capital-intensive technologies are sensitive to the discount rate. The greatest sensitivity occurs for technologies that require the largest per-barrel investment.

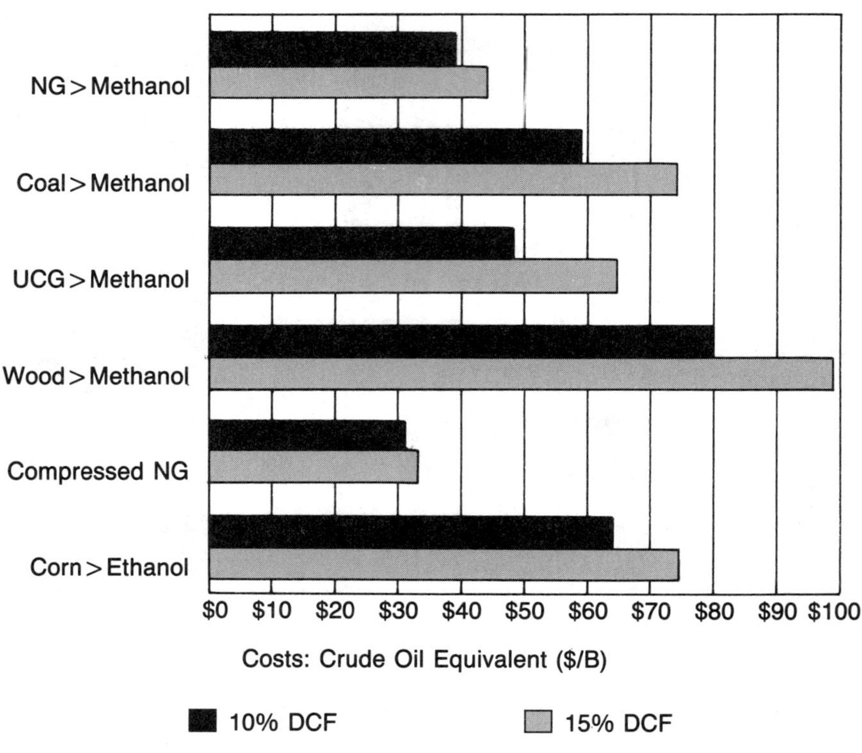

Source: J.L. Sweeney, 1989

Figure 5 shows the sensitivity of cost estimates to the price of crude oil. In this graph, two crude oil prices are assumed—$20/b and $40/b—as well as the endogenously determined oil price. This figure shows that the crude-oil-equivalent costs of the various technologies do depend on the crude oil price, but that the sensitivity is relatively small. Two factors explain the effect of the crude oil price: the cost of energy inputs and the marketing and refining credit. Higher oil prices result in a greater credit but also result in higher feedstock costs. The net effect of these two factors will determine

Figure 5
Impact of Crude Oil Price on Cost of Domestic Production

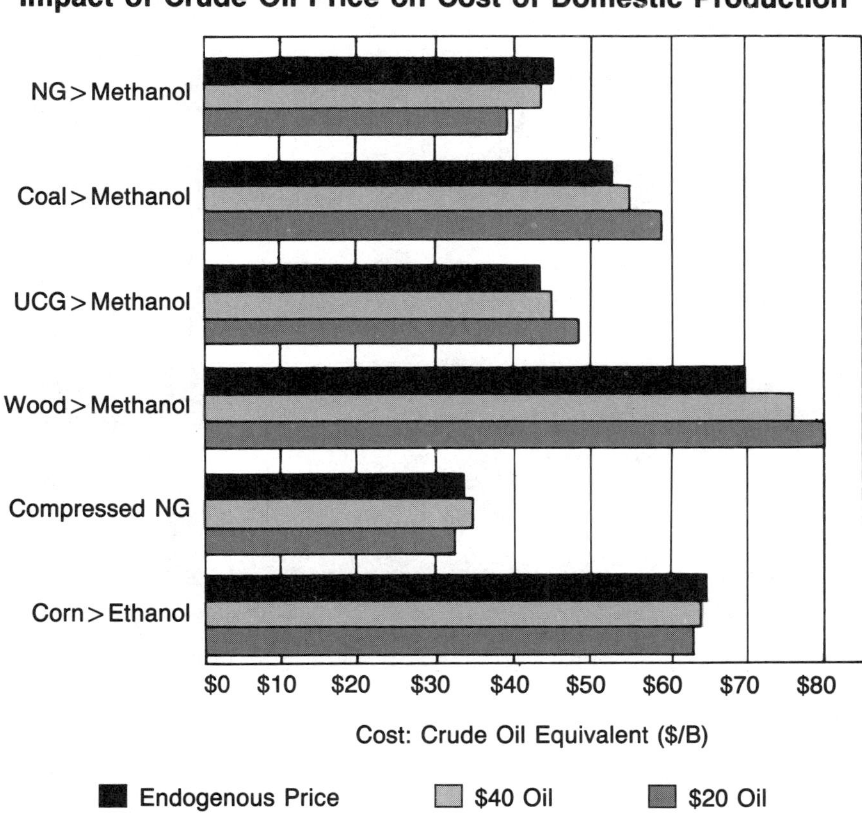

Source: J.L. Sweeney, 1989

whether higher oil prices lead to larger or smaller crude-oil-equivalent cost. For methanol produced from natural gas, higher crude oil prices lead to higher costs of methanol, because the influence of the crude oil price on the natural gas price is the dominant factor.

Figure 6 indicates the sensitivity of costs to the natural gas price function. The natural gas price function is reduced by $2/mcf at each oil price. This variation reduces the estimated cost significantly for the two fuels using natural gas as their primary feedstock. The

Figure 6
Impact of Natural Gas Price Function on Cost of Domestic Production; $20 Crude Oil Price

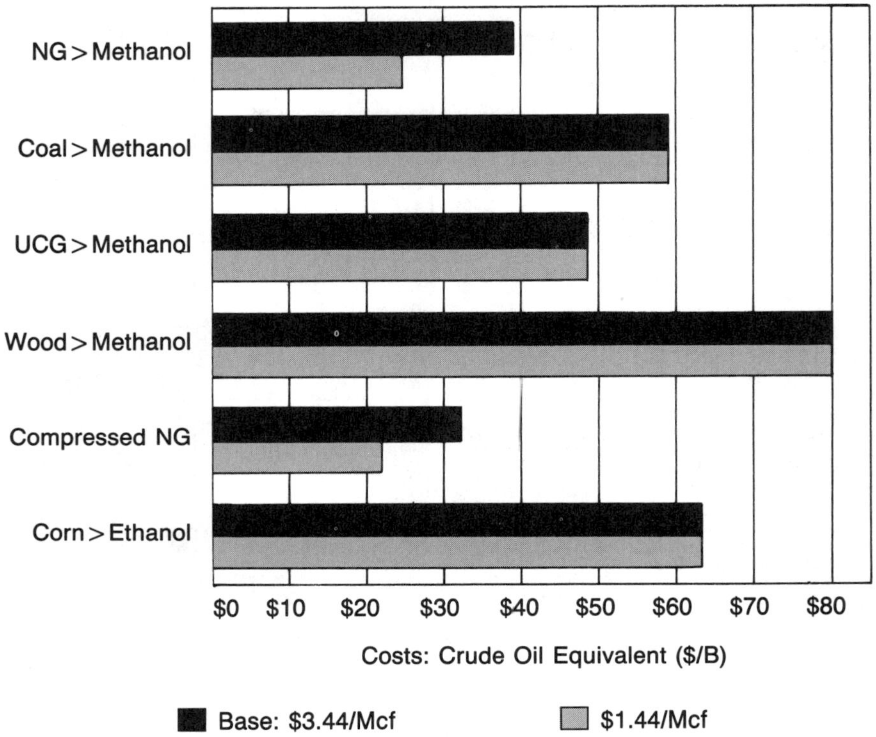

Source: J.L. Sweeney, 1989

drop in the natural gas price function brings the cost of methanol closer to that of CNG, but the CNG cost remains lower than that of methanol.

Sensitivity tests for domestically produced methanol (using natural gas as a feedstock) illustrate major uncertainties about methanol costs. Ranges of crude-oil-equivalent cost estimates are based on sensitivity tests for five parameters, varied one at a time. Results are presented in Figure 7.

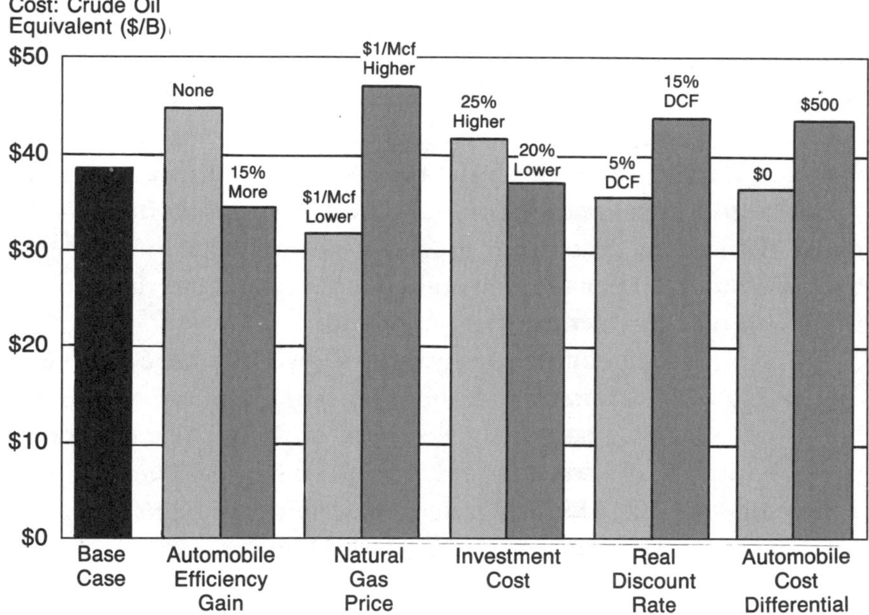

Source: J.L. Sweeney, 1989

The base case assumes that automobiles fueled with methanol will enjoy a 10 to 18 percent efficiency gain over equivalent vehicles fueled with gasoline. This assumption results in a gasoline equivalency factor of 1.8. Methanol vehicle efficiencies that are 15 percent greater than in the base case (bar labeled "15% More") and 15 percent smaller (bar labeled "None") are illustrated by the pair of bars in Figure 7 denoted as "Automobile Efficiency Gain." The former is based on a gasoline equivalency factor of 1.57, the latter on a gasoline equivalency factor of 2.06. This range of automobile efficiency variations leads to a $10/b variation in the crude-oil-equivalent cost of methanol.

The natural gas price of $3.44/mcf, in the base case, is decreased by $1/mcf and increased by $1/mcf for the set of bars labeled

"Natural Gas Price." This range of natural gas prices leads to a $15/b variation in the crude-oil-equivalent cost of methanol. The investment cost of a methanol production facility is increased by 25 percent and decreased by 20 percent, leading to a $3.50/b variation in the crude-oil-equivalent cost of methanol. The real discount rate is varied from a low of 5 percent to a high of 15 percent, for the fourth set of sensitivity tests, leading to a $9/b variation in methanol cost. Finally, the differential between the cost of a methanol-fueled automobile and that of an equivalent gasoline-fueled vehicle is reduced to zero from the base case of $200 per car and increased to $500 per car. This range leads to an $8/b variation in the crude-oil-equivalent cost of methanol.

The large range of price estimates shown in Figure 7 is based on varying one parameter at a time. The complete range of possible prices would be larger, since several of these parameters may vary simultaneously from the median estimates used here. Figure 7 thus suggests that a high degree of uncertainty remains about the costs of methanol produced in the United States, even if the future technological parameters are known.

Foreign Costs

Most analyses of methanol derived from natural gas conclude that methanol will not be produced in the United States but in countries with abundant low-cost natural gas. The following estimates of the cost of foreign methanol imported into the United States assume that the technological parameters that apply to U.S. production of natural gas will remain valid for other nations. However, production costs can be expected to vary in several ways: natural gas prices will be lower, investment costs will be higher, operation and maintenance costs will be higher, and discount rates may be higher. In remote or less industrialized areas endowed with abundant natural gas supplies, the market-clearing price of natural gas can be expected to be lower than in the United States. How much lower, however, is difficult to estimate.

A possible cost-analysis approach would attempt to predict the natural gas price from production costs by estimating production cost functions for natural gas in the various remote locations, estimating the natural gas demand for methanol production, and estimating the natural gas price based on the supply-demand equilibrium. This approach, however, would be appropriate only if the estimated demand functions included other possible future uses of the natural gas. For example, natural gas could be liquefied and exported to the United States, Japan, or other nations. Energy-intensive production processes (for example, fertilizer production) could be established in foreign locations to use natural gas near the point of production. Failure to include these alternatives would lead to underestimates of the price of natural gas in remote locations.[6]

Consequently, foreign natural gas cannot be expected to sell for long at prices below $1/mcf to $2/mcf, unless costs of all energy-conversion processes at the location are extremely high (as may be the case in some remote locations). For this study, no attempt has been made to estimate a single natural gas price. Rather a range of possible prices—$1/mcf, $2/mcf, and $3/mcf—has been considered, with a base case of $2/mcf.

Investment costs will be higher in remote or undeveloped areas, relative to U.S. costs. Several studies have estimated such variations. For this discussion, estimates are broadly consistent with those developed by Bechtel as part of the "California Fuel Methanol Cost Study" (1989). In some locations investment costs might exceed U.S. levels by 25 percent. Production in the Middle East may fall into this category. For remote locations, investment costs could exceed U.S. levels by 75 percent. These costs may apply to methanol production off the coast of Australia. A 25 percent differential was used as a base case, and a 75 percent differential was used as a possible case.

Operating and maintenance costs in remote or undeveloped areas could also be expected to exceed U.S. levels, although not by as much as investment costs would. For this study, it is assumed that the percentage differential for O & M costs, above U.S. levels, is one-fourth as great as the corresponding percentage differential for investment costs. Discount rates for investment in remote and/or undeveloped locations will be larger than U.S. levels if investors perceive significant political risk in those locations. If international corporations are the investors, then such a perception may be rational for many investments in developing countries. However, some countries could subsidize the costs of capital to promote industrial development. This study has continued to use a 10 percent discount rate as the base case and has done sensitivity tests using a 15 percent real discount rate (reflecting some, but not extreme, political risk) and using a 5 percent discount rate (reflecting subsidized capital costs). In addition, methanol produced abroad would incur costs of tanker transportation to the United States. Here it is assumed that these costs amount to $2/b of methanol.

Table 6 shows the crude-oil-equivalent cost for foreign methanol of about $35/b—still higher than costs of CNG, but lower than the costs of domestically produced methanol. The cost reduction from lower-priced natural gas more than makes up for the cost increase due to higher investment and O & M costs, plus the transportation cost to the United States.

Sensitivity tests around this base case, parallel to those underlying Figure 7, were conducted for foreign methanol production. Results are shown in Figure 8. In one pair of tests, automobile efficiency gains for methanol are increased by 15 percent (to roughly 30 percent) and reduced by 15 percent (to roughly no gains). Natural gas price is increased to $3/mcf or decreased to $1/mcf. In addition, free natural gas ($0/mcf) is also presented to illustrate total non-feedstock-related costs. Investment costs are increased to 75 percent above U.S. levels and decreased to U.S. levels. Real discount rates of 5 percent and 15 percent are tested. Automobile costs differentials of $0 and $500 are presented.

The lowest crude-oil-equivalent cost occurs in the unrealistic situation of a price of zero for natural gas.[7] This case does show, however, that the crude-oil-equivalent cost of methanol, even without

Table 6
Foreign Methanol Costs

Product Cost	$/gal	$/Equivalent Barrel
Feed	0.19	14.40
O&M	0.06	4.47
Capital Charge	0.16	11.72
Transportation to U.S.	0.05	3.60
Refinery Gate Price	0.45	34.19
Distribution/Marketing	0.10	7.28
Pump Price (net of taxes)	0.55	41.47
Vehicle Incr. Cost ($/Gal)	0.07	2.82
Gasoline Equiv. Cost (no tax)	**1.05**	**44.29**
Netback to Crude Price		
Refining Credit		(5.55)
Gasoline Distribution/Marketing		(3.84)
Product Cost (crude equivalent)		32.08
Crude Oil Equiv. Cost ($/b)		**34.91**

Note: The estimates assume that investment costs exceed U.S. costs by 25 percent, O & M costs exceed U.S. costs by 4.25 percent, natural gas sells for $2/mcf, and the real discount rates remain at 10 percent.

Source: J.L. Sweeney, 1989

considering the cost of the natural gas feedstock, is near the current crude oil price. Thus, it is unrealistic to expect foreign methanol to be competitive with gasoline derived from crude oil, unless the crude oil price increased greatly or methanol was highly subsidized, relative to gasoline (or gasoline was highly taxed, relative to methanol). In other cases, methanol costs vary from a low of $28/b to a high of $42/b. In this graph, factors are varied one at a time. For combinations of variables changing together, the range would be larger.

Figure 8
Foreign Methanol Cost Sensitivities

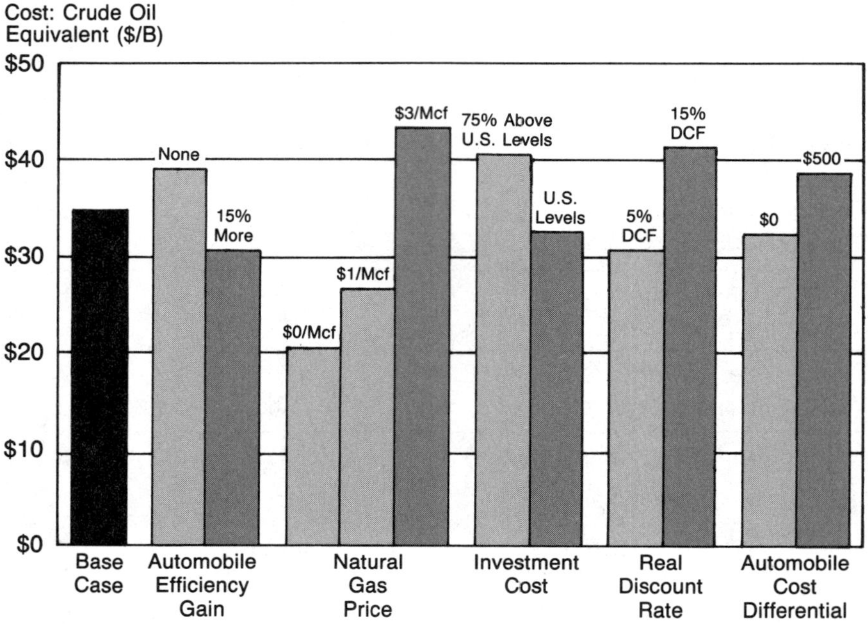

Cost: Crude Oil Equivalent ($/B)

$50
$40
$30
$20
$10
$0

None
15% More
$0/Mcf
$1/Mcf
$3/Mcf
75% Above U.S. Levels
U.S. Levels
5% DCF
15% DCF
$0
$500

Base Case
Automobile Efficiency Gain
Natural Gas Price
Investment Cost
Real Discount Rate
Automobile Cost Differential

Source: J.L. Sweeney, 1989

An additional set of sensitivity tests is based on variation of the three most critical parameters: natural gas price, investment cost, and discount rate. These tests allow several parameters to be varied simultaneously. Figures 9 and 10 show the results. For each case, the oil-equivalent per-barrel cost of methanol from natural gas is illustrated (Figure 9) with the various components of that cost (Figure 10). The first cost estimate is based on U.S. conditions, including the U.S. natural gas price of $3.44/mcf. No transportation cost is included for methanol. The next two estimates are also based upon U.S. investment and O & M costs, but the natural gas price is allowed to range from $1/mcf to $3/mcf, and a $2/b cost is added for transportation of methanol to the United States. These figures show the sensitivity of the total cost to natural gas prices.

The remaining estimates are based on conditions that might

Figure 9
Foreign Methanol Cost Sensitivities

Cost: Crude Oil Equivalent ($/B)

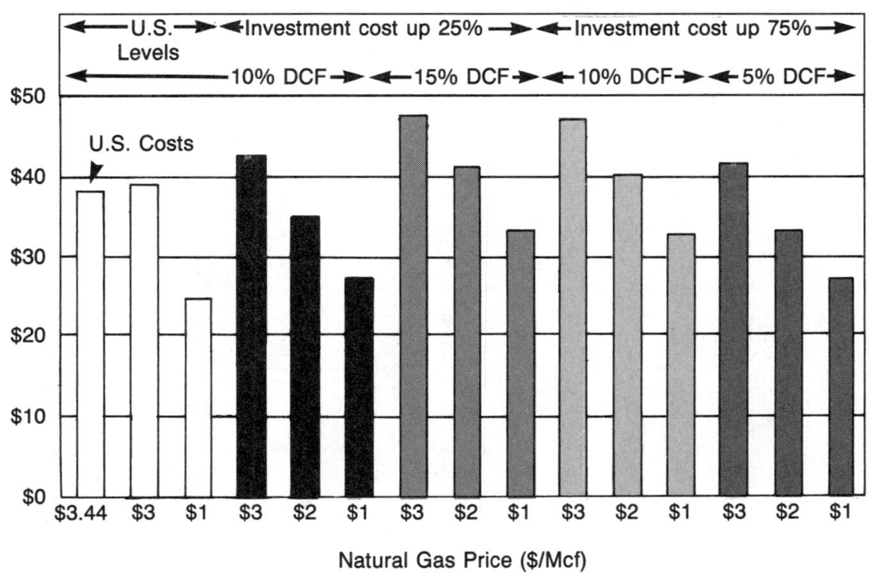

Source: J.L. Sweeney, 1989

exist in remote locations. Estimates are separated into four groups of three cases. Two groups include investment costs 25 percent higher than U.S. levels and corresponding increases in O & M costs. Methanol production in the Middle East may fall into this category. Discount rates of 10 percent and 15 percent are used to account for possibly greater investment risk outside the United States. For these two groups, crude-oil-equivalent costs range from $26/b (if natural gas was priced as low as $1/mcf and real discount rates were 10 percent) to $43/b (for $3/mcf natural gas price and a discount rate of 15 percent).

The final two groups include investment costs 75 percent above U.S. levels and 19 percent increases in O & M costs. Such a situation may apply to methanol production off the coast of Australia. A discount rate of 5 percent is shown, in addition to the 10 percent value, to illustrate costs that might exist if the host government promoted

Figure 10
Components of Total Cost, $/B
Foreign Methanol Production

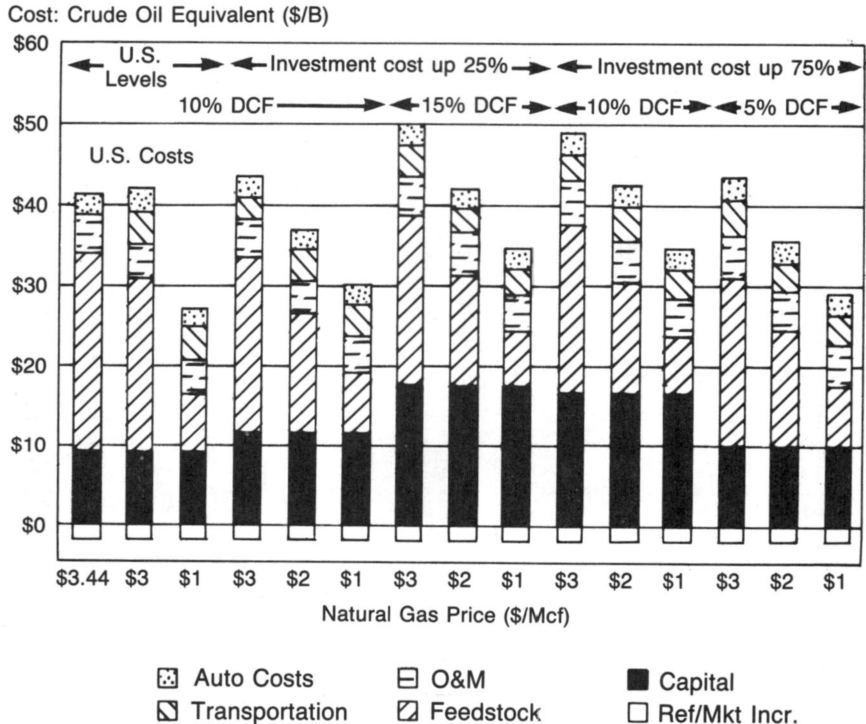

Source: J.L. Sweeney, 1989

development through low-interest loans or loan guarantees. For these two groups, crude-oil-equivalent costs range from $27/b (if natural gas was priced as low as $1/mcf and the cost of capital was subsidized by 5 percent) to $44/b (for $3/mcf natural gas price and a 10 percent discount rate).

Figures 9 and 10 show that the cost of methanol from remote locations can be expected to vary greatly from lower-48 U.S. estimates. Variations in investment cost, natural gas feedstock costs, and discount rates can greatly influence cost estimates. Tanker transportation costs are a smaller source of variation. Since different values for these three major cost determinants can be expected, methanol costs from foreign sources may vary greatly by the location of production. Market prices of methanol manufactured in different locations, however, can be expected to be comparable.

Figure 9 shows that U.S. production costs of methanol would be significantly higher than methanol production costs in those foreign locations where natural gas prices are low and where capital costs are only slightly above U.S. levels. This result suggests strongly that methanol produced in the United States is unlikely to be economical in comparison to methanol from remote foreign locations. To the extent that it is used in the United States, methanol is apt to be produced in foreign locations. Even for remote gas, the costs of producing methanol remain highly uncertain but will probably be considerably higher than the cost of producing gasoline.

Closing Observations

Costs of large-scale methanol production remain highly uncertain. Domestic production of methanol appears to be costly. Therefore, it is reasonable to anticipate that broad adoption of methanol-fueled vehicles would lead to reliance on imports from remote foreign locations that have abundant low-cost supplies of natural gas. At this time, however, the prices of natural gas in those locations and the investment costs for new methanol plants cannot be predicted with any certainty. Therefore, one cannot confidently predict whether methanol will have crude-oil-equivalent costs as low as $30/b or as high as $45/b.

The cost of three low-polluting fuels was estimated as part of the NRC study—methanol, ethanol, and CNG. Other possible technologies, such as hybrid electric vehicles, although promising, have not been evaluated by the NRC. The two alcohols have been strongly promoted by U.S. policy, while CNG has been given less attention. Estimates suggest that (without subsidy or mandate) the two alcohols would become competitive with gasoline only if the crude oil price greatly increased from its current level. CNG, however, would become cost competitive at a far lower crude oil price. The shorter range between fillings or the reduced trunk space of CNG automobiles would, perhaps, lead consumers to favor alcohol fuels over CNG. But the cost disadvantages of alcohol fuels may mean that they would not be the low-polluting fuel of popular choice except for government mandates or the use of financial incentives to induce consumers to use alcohol fuels rather than other alternative fuels.

Current proposals by the Bush administration and those by the state of California suggest a strategy based upon forcing significant numbers of alternative-fuel vehicles into the fleet. Most would be dual-fuel or flexible-fuel vehicles. But once these automobiles were in the fleet, consumers could still choose which fuels to use. Without large subsidies for methanol and ethanol or taxes on gasoline, conventional gasoline would still enjoy an operating cost advantage over these alcohol fuels. On purely economic grounds, gasoline could be expected to remain the fuel of choice even for those individuals who owned flexible-fuel vehicles adapted for alcohol fuels. On the other hand, owners of dual-fuel vehicles adapted for alcohol or CNG would find CNG to be less costly than gasoline and could be expected to use that fuel, at least for short trips.

With the clear possibility that methanol may not turn out to be economically competitive in large-scale production, even compared with other low-polluting fuels, it seems inappropriate at this time to create a policy presumption in favor of this fuel. Instead, policy flexibility should be maintained. It is appropriate to develop incentives and standards designed to protect the environment. But it is not appropriate to develop mandates favoring a fuel such as methanol despite the evidence that methanol may be more costly than CNG, and the economic incentives that exist for consumers to use gasoline over methanol even in those automobiles equipped to run on either fuel.

Notes

1. For endogenous price determination, crude oil price is not an input to the analysis but rather is calculated as that price that would make the specific process/fuel just competitive with crude oil. Crude oil price is chosen so that calculated cost per equivalent oil barrel is exactly equal to the crude oil price.

2. The specific equation to calculate the per-mile equivalent cost is as follows:

$$K = 15{,}000 \; Ceq \; \frac{\{1 - 1/(1 + r)^7\} \; \{1 + r\}}{r}$$

where K is the additional capital cost of a new automobile, Ceq is the per-mile equivalent cost, and r is the real consumer discount rate, assumed to be 10 percent annually. The per-gallon incremental automobile cost is thus Ceq • mpg, where mpg is the miles per gallon of fuel efficiency of new automobiles.

3. More precisely, it is assumed that investment costs are spread over time so that the present value of these expenditures, discounted to a point 1.5 years before the project begins operating, is equal to the total investment cost.

4. When present value calculations are conducted using the after-tax cost of capital, no additional deductions for interest payments should be included explicitly.

5. Note that costs quoted here are based on an exogenously determined $20 crude oil price, while the NRC study usually quotes costs based on endogenously determined crude oil prices. The NRC study asked at what crude oil price various technologies would be competitive with crude oil. This paper asks what would be the cost of alternative technologies if oil price remained at levels expected in the near-term future. The different goals explain the different estimating procedures.

6. Estimating natural gas prices as equal to production costs would be analogous to estimating the crude oil price from Saudi

Arabia at \$2/b, the approximate production cost of crude oil. The fallacy of that approach is obvious.

7. Natural gas prices below \$1/mcf could be expected to occur only in remote locations for which investment cost was far above the base case value (125 percent of U.S. levels). If gas prices and investment costs were that low, liquefying the natural gas and exporting LNG would be more profitable than producing methanol. That opportunity would lead to rises in the natural gas price to at least \$1/mcf.

References

Committee on Production Technologies for Liquid Transportation Fuels, Energy Engineering Board, National Research Council (forthcoming, 1990). *Fuels to Drive Our Future*. Washington, DC: National Academy Press.

DOE (1989). Annual Energy Outlook 1989. U.S Department of Energy, Energy Information Administration, Report No. DOE/IEA-0383(89), Washington, DC.

Schulman, B., and F. Biasca (1989). *Liquid Transportation Fuels from Natural Gas, Heavy Oil, Coal, Oil Shale and Tar Sands: Economics and Technology*. Mountain View, CA.

Commentary

THOMAS J. LAREAU

The papers by Lawrence, McGartland, and Sweeney provide a broad overview of the cost of introducing methanol as an alternative to gasoline. Two of the authors, Michael Lawrence and Albert McGartland, strongly favor an analysis that accounts for both market and nonmarket effects (pollution and energy security) but they do not attempt a cost-benefit analysis. James Sweeney extends the analysis to other alternative fuels and feedstocks, while McGartland includes an analysis of the cost-effectiveness of switching to methanol.

Past failures to anticipate energy price paths should indicate that future fuel prices are very uncertain. The uncertainty about the price differential hinges not only on uncertainties about the investment cost and the technology used in methanol plants, as the Sweeney paper brings out, but also on economic factors, particularly feedstock costs, capital costs, and energy prices. Ultimately, as the McGartland analysis makes clear, the cost of methanol substitution would depend primarily on the difference between the price of methanol and that of gasoline. When relative fuel prices are an important input, a sensitivity analysis is necessary. Sweeney and McGartland provide such analyses, while Lawrence does not.

The following comments are based on the slide presentation at the conference and on a previously published paper: Michael F. Lawrence and Janis K. Kapler, "Natural Gas, Methanol, and CNG: Projected Supplies and Costs," in Daniel Sperling, ed., *Alternative Transportation Fuels*, (N.Y.: Quorum Books, 1989).

Costs depend on the volume of production. Although many studies provide evidence of scale economies in methanol production and distribution, methanol's proponents in the policy debate usually assume a high-volume scenario. However, the low costs that are associated with economies of scale will take a long time to realize, perhaps as long as 10 to 15 years. Thus, the unit costs of the administration's alternative-fuel proposal, or other low-volume scenarios, will be higher than the costs predicted here.

The Lawrence analysis includes an estimate of natural gas supply and a calculation of methanol production costs.[1] However, this commentator finds his analysis to be incomplete and his conclusions to be misleading. In Lawrence's view methanol could be sold at a price that is competitive with today's gasoline price, and the potential benefits of substituting methanol for gasoline are substantial. It appears that Lawrence's cost analysis systematically applies assumptions to the Chem Systems cost algorithms that lead to low-cost estimates. Lawrence recognizes that the marginal extraction costs of natural gas feedstocks can and probably will differ from the actual selling price. Yet he does not take into account important market factors, such as netbacks where commodity markets exist, government pricing rules, taxes, and the probable existence of competing markets in developing countries for gas in the future. Mexico's refusal to sell natural gas to the United States is a good example of a country's desire to postpone resource sales until the price is optimal. Similarly, Lawrence assumes low capital costs, low shipping and distribution costs, and advanced-technology plants. At the same time he ignores investment risks.

Although some of Lawrence's lower-cost assumptions may turn out to be correct, it is not likely that all of his lower-cost assumptions will be realized. Furthermore, Lawrence does not explore the effects of input parameters other than his own, although the Chem Systems framework encourages the use of different input assumptions. Other studies, such as Sweeney's, assume different input prices and reach different conclusions.

The Sweeney cost analysis (based on a recent National Research Council [NRC] study) includes methanol produced from natural gas, coal, or biomass as well as ethanol and compressed natural gas. This study considers two oil price regimes—$20/b and $40/b. In addition, Sweeney computes the "endogenous" price of petroleum required for alternative-fuel price parity. It ranges from $33/b for compressed natural gas to $70/b for methanol made from wood. Sweeney's research also accounts for the interrelationships among energy prices, so that at higher petroleum prices other input costs (electricity, natural gas, refining and distribution, and corn) are adjusted upward. Finally, Sweeney provides a sensitivity analysis of methanol's cost with respect to oil prices, investment cost, the discount rate, the natural gas price function, automobile efficiency, and vehicle cost.

Sweeney concludes that it is impossible to predict whether methanol will have crude-oil-equivalent costs as low as $35/b or as high as $50/b. This range is substantially higher than Lawrence's cost estimate (or that presented by Carmen DiFiglio). According to Sweeney, it appears likely that methanol may not be competitive with conventional vehicle fuels unless oil prices rise substantially. Sweeney observed that while the costs of compressed natural gas (CNG) seem more attractive, CNG receives less consideration than methanol. Sweeney's conclusion from the NRC cost analysis, that although "it is appropriate to develop incentives and/or standards designed to protect the environment, it is not appropriate to mandate a specific fuel," appears to this commentator to be sound.

McGartland provides a cost-effectiveness framework to compare strategies—such as an inspection and maintenance program for limiting "super emitting" vehicles—with methanol substitution. Although McGartland's analysis suggests that methanol substitution may not be cost-effective, he concludes that the uncertainty is too large, at this time, to draw a final conclusion.

For those who believe energy security benefits are sizeable, methanol's dollar-per-ton cost-effectiveness that McGartland calculates would have to be adjusted downward. In other areas, however,

McGartland's estimates may be optimistic, and especially with regard to the effect of the assumed baseline emissions from comparable gasoline-fueled vehicles. McGartland uses 1.73 g/mi, as predicted by EPA's MOBILE 4 emissions model. With the administration's Clean Air Act amendment proposals, this baseline would be 0.94 g/mi. Analysis by regulatory organizations in California suggest baseline emissions of about 0.4 g/mi to 0.6 g/mi, depending on the stringency of exhaust standards. All else being equal, the latter baseline would increase McGartland's dollar-per-ton cost-effectiveness estimates by at least a factor of three.

McGartland ridicules the argument that methanol will not get its chance to compete, even if its price is attractive compared to that of conventional fuels, because there will be no vehicles available to use it. To accept such reasoning, he argues, rejects the evidence of rapid market penetration of complementary products (such as electricity and lighting), based on new technology.

Many experts now agree that there is insufficient data on which to base the choice of an alternative fuel and the best vehicular technology to go with it. Therefore, it is important to establish standards and then to allow the market to determine the best fuel (or fuels) and vehicle technology in conjunction with other strategies to control emissions. The cost of regulatory mistakes continues to rise. As McGartland and members of the EPA staff at this conference pointed out, large stationary sources and vehicles have already been subjected to stringent controls. Yet widespread violations of the ozone standard continue. It appears that improving air quality will be very costly, whatever path is chosen. Therefore, we must take advantage of market forces to hold down these costs as much as possible.

Note

1. Based on a recent study by Chem Systems, *Assessment of Costs and Benefits of Flexible and Alternative Fuel Use in the U.S. Transportation Sector: Technical Report Three: Methanol Production and Transportation Costs*, DOE/PE-0093 (Washington, D.C.: U.S. Department of Energy, 1989).

Commentary

MARGARET A. WALLS

A careful reading of the papers by James Sweeney, Albert McGartland, and Michael Lawrence emphasizes the uncertainty in estimating the costs of methanol. Sweeney shows how oil-equivalent methanol prices vary with changes in natural gas prices, discount rates, and other variables. His methanol prices ranged from $27/b to $44/b. McGartland's cost-effectiveness figures indicate that the expenditure required to produce hydrocarbon emissions under a methanol program could range from a negative number to as much as $29,000 per ton of hydrocarbons produced. These numbers incorporate only the Jack Faucett Associates and Department of Energy assumptions for natural gas feedstock costs. If McGartland had allowed the range to be a little wider, the upper end on his cost-effectiveness numbers might have gone even higher. Lawrence did not give ranges for his estimates of the cost of methanol, but that probably indicates even more uncertainty in his figures.

In fact, uncertainty is inherent here. Forecasting the price of any commodity is a risky business—remember the $60/b oil price forecasts made only 10 years ago? Two factors make forecasting methanol prices even more difficult than other energy prices: First, most analysts feel that using data from the past to forecast prices will not work because the size of a future methanol-fuel market is expected to be much larger than the current methanol-chemical market. In other words, the methanol supply curve will probably look different at much larger quantities. Second, trying to forecast natural gas prices for methanol feedstock purposes in areas where no natural gas market currently exists is especially difficult. No methanol study can predict what the opportunity cost of natural gas reserves may be in 10 or 20 years. The opportunity cost of remote natural gas will depend on possible alternative uses for it, domestically or for export. It could be used, for example, for electricity generation or heating; as a petrochemical plant feedstock; for reinjection into oil reservoirs; or for liquefaction and export to established gas markets.

The amount of gas consumed domestically depends on economic growth in a country or region, and the relative prices of other fuels. It is worth noting that many oil-producing countries have increased their domestic gas use recently, because of increases in electricity use and growth in their petrochemical industries. The value of gas for reinjection purposes depends on future oil prices and the rate of production decline in oil fields during the next 10 to 20 years. The future value of liquefied natural gas (LNG) in established markets depends on economic growth and relative energy prices in those markets, and also on the degree of concern for the environment. Electric utilities and industrial boilers may switch from oil and coal to natural gas, in order to meet stricter environmental standards.

The Gas Research Institute forecasts a LNG price in the United States of $3.87 per mmBtu in the year 2000. (GRI's oil and gas price forecasts tend to be low—its crude oil price forecast for 2000 is only $20/b.) The Department of Energy's estimate of LNG production costs net of the gas feedstock component, as well as shipping cost estimates, yields a natural gas feedstock cost range of $1 to $2 per mmBtu. This is a rough estimate of the opportunity cost of gas for methanol use. If methanol producers wanted access to natural gas in these remote locations, they would probably have to pay at least its opportunity cost for use as LNG. This range of $1 to $2 is roughly double the range used in the Environmental Protection Agency's cost estimates, and it is more than double the figure presented by Lawrence.

Why do we really care about what the cost of methanol will be? Although it is useful to try to estimate the cost, just as it is useful to forecast gasoline prices, the result should not affect policy. Regardless of methanol's predicted cost, policymakers should not mandate or subsidize its use. A more efficient way to achieve air quality goals would be to tighten standards and to allow emissions trading. Whether methanol emerges as the cheapest way of meeting the standards or whether the automobile manufacturers and oil companies can find another way, achieving the goal of improved air quality should be all that matters.

Another alternative—and this commentator realizes that it is only because she doesn't face political constraints that she can suggest it—would be to tax gasoline. If introducing a new fuel is really the best way to achieve air quality goals, let all the candidates compete equally by raising the price of gasoline. Again, if methanol comes out the winner, so be it. But let's not decide the outcome without running the race.

Commentary

GLYN D. SHORT

The papers dealing with methanol's costs reflect the current debate about the appropriate economic methodology for assessing the costs of alternative fuels. However, without agreement on some key points there can be no generally accepted conclusions.

There is now a consensus that large volumes of methanol fuels will not be produced in the contiguous United States in the near to medium future. Small quantities may become available where special circumstances exist, for example, from remote small gas fields. In general, however, the value of alternative uses of natural gas is too high in the United States for large-scale methanol production.

Using conventional methods of determining cost, the most competitive fuels available today are gasoline and diesel. Consequently, if alternative fuels of any type are to be introduced, the externalities of using conventional fuels should be factored in. Thus, the approach advocated by Lawrence has much to commend it. However, cost-benefit analyses of environmental issues are fraught with uncertainty and governed by subjectivity. While it may be possible for engineers, economists, and politicians to reach agreement on large-scale methanol production and transportation costs, placing a value on environmental benefits leads to endless debates.

The key question with regard to alternative fuels, therefore, is what premium society is prepared to pay to gain environmental benefits. After the desired environmental goals have been agreed to, the minimum production cost to attain those goals can be calculated. If this cost is excessive, then the use of methanol can be forgotten. The main determinants of methanol's production costs are generally agreed to be the natural gas feedstock price and the cost of capital. Most cost estimates range from 28 cents/gal to 35 cents/gal, delivered to Los Angeles. At this price level, methanol would compete effectively with petroleum-based fuels.

Given the huge surplus supply of gas, it is reasonable to assume that any producing country would price its gas so that it could profit from the production of methanol. If methanol production is viewed simply as a way of liquefying natural gas, the analogy of Australia's supplying liquefied natural gas to Japan becomes relevant. This arrangement indicates that natural gas prices can be unrelated to crude oil prices under certain circumstances.

It is in respect to the cost of capital that the debate heats up. The appropriate return on capital is a function of perceived risk and is complicated by political and environmental considerations. Thus, Lawrence, McGartland, and Sweeney suggest rates of return of anywhere from 8 to 30 percent. The relationship between gas prices and the rate of return on capital is clearly demonstrated in Sweeney's Figure 8. However, all three papers neglect the implications of new and improved technology. No contractor is likely to build a methanol-fuel complex using designs appropriate for a single-train chemical-grade methanol plant. Significant savings in capital are possible using advanced technology and the modular design now becoming available (see, for example, papers presented at the Houston World Methanol Conference, December 1989). Since these savings lower not only capital charges but also operating costs and gas costs (because of higher feedstock utilization efficiencies), the effect is cumulative.

Finally, even if it is accepted that the production of high-volume methanol fuel will be practical in the long term, the question remains of how to bring about a transition from "chemical scale" plants. The existence of a fuel infrastructure/fuel demand impasse cannot be denied. The practical problems to be overcome will be severe to potential service station owners and fuel distributors who will need to raise capital to service an uncertain market. As McGart-land points out, this is the area where government support is greatly needed. If environmental needs are to be met, the means to surmount this "activation barrier" must be made available.

Part VI

Public Acceptance and Implementation

Evolving Market-Driven Approaches to Implementing Methanol and Other Low-Emissions Technologies

CARL B. MOYER

Abstract

Although there is an emerging consensus that alternative fuels will have a role to play in some future air quality programs, the best way to implement their use is not yet clear. Several technologies—including that of methanol, ethanol, natural gas, propane, electrical energy, and reformulated gasolines— appear to have some promise. However, there is so much uncertainty in estimates of the benefits, costs, and consumer acceptance of each of these fuels that it is difficult to select a single implementation approach. An optimal implementation plan should recognize

This paper describes evaluations made by the California Advisory Board on Air Quality and Fuels. However, the interpretations and summaries of the Advisory Board given here are solely those of the author and do not necessarily reflect the views of the Advisory Board. The author wishes to thank Michael D. Jackson of Acurex, who reviewed this paper and provided many helpful comments.

the uncertainties by including enough flexibility to adjust as more is learned and as technologies improve. It appears that the production of a significant number of alternative-fuel vehicles can be encouraged by incentives, such as the current Corporate Average Fuel Economy incentives in the Alternative Motor Fuels Act of 1988. To stimulate the production of alternative-fuel vehicles further, the state of California is embarking on a program of tighter emissions standards, combined with credits for the reactivity and toxicity benefits of alternative fuels.

Federally mandated production of alternative-fuel vehicles needed to meet local air quality goals was proposed by the Bush administration in the summer of 1989. A mandate may ultimately be needed, at least in areas that lack authority to set emissions standards for vehicles. To stimulate the use of alternative fuels, California is proposing that fuel suppliers be required to sell the alternative fuels or to provide gasoline with equivalent benefits. This approach appears to offer sufficient flexibility, as well as some valuable incentives to develop better fuels. It is also politically acceptable to the various suppliers of alternative fuels. If other states or regions were to implement alternative-fuel programs as rapidly as California is doing, flexible approaches would probably be recommended. However, if alternative-fuel programs outside of California are delayed for several years, it may be possible for other states to use financial incentive or mandate approaches after California has, in effect, pioneered and demonstrated the technologies and confirmed their costs.

Introduction

There is an emerging consensus that methanol and other alternative motor fuels will have an important role to play in future air quality programs. Although automobile and truck emissions have been reduced substantially from pre-control levels, many urban areas are still struggling to meet health-based air quality goals for ozone and other photochemical pollutants. Projected growth in vehicles, trips, and vehicle-miles is a considerable obstacle for improvements in air quality and will create additional pressures for further reductions in per-mile emissions from vehicles. In addition, expected programs to limit such air toxics as benzene will create new demands for reduced emissions and new requirements for regulating the composition of hydrocarbon fuels.

In this context alternative fuels such as methanol, ethanol, natural gas, and electric power may be needed to help meet air quality goals. Such fuels, when used in properly designed vehicles, can reduce emissions of photochemically reactive components substantially, compared with emissions from gasoline-fueled vehicles using comparable emissions-control technologies. Detailed evaluations of the air chemistry in the Los Angeles basin suggest that methanol substitution would provide ozone benefits equivalent to removing about 50 percent of the gasoline-fueled vehicles from the population, as described by Armistead Russell elsewhere in this volume and in Russell et al. (1989). The ozone benefits of methanol may be less in other urban areas, as discussed by T. Y. Chang in this volume, but the combined pressures to reduce ozone and air toxics may eventually require alternative fuels to be used in several areas of the United States.

Studies of the cost-effectiveness of alternative fuels show mixed results. A recent study by the U.S. Congress Office of Technology Assessment provided a broad range of cost-effectiveness estimates for methanol, including some costs that are quite high, compared with the costs of measures already implemented (U.S. Congress, 1989). Increasingly, however, cost-effectiveness assessments indicate that, at least under some conditions, cost estimates for alternative fuels are close to those of other measures already implemented or about to be implemented. The paper by Albert McGartland in this volume provides some examples of this. The comprehensive evaluation by the California Advisory Board on Air Quality and Fuels recently concluded that methanol's cost-effectiveness is close to that of other measures being planned for the Los Angeles area, even when the only benefit attributed to methanol is ozone reduction

(California Advisory Board on Air Quality and Fuels, 1989). Moreover, methanol's cost-effectiveness improves when its costs are allocated to more than one benefit (Moyer, Unnasch, and Jackson, 1989).

Other alternative fuels have not been studied as extensively as methanol, but it seems clear that vehicles powered by natural gas, propane, and electric power can make contributions to air quality programs, at least in some applications or market niches. Ethanol tends to be more expensive than the other fuels at this time, but technological improvements in ethanol production technologies might make ethanol more competitive. Furthermore, ethanol has already obtained some market share as a blend constituent for gasoline, either used directly or as an ethanol-based ether. Finally, various "reformulated gasolines" may also be able to compete with the chemically simpler fuels by reducing the photochemical reactivity of emissions from gasoline-fueled vehicles.

It is impossible to foresee at this time the future role of each of the alternative fuels. Even with known costs and vehicle performance, it is difficult to predict market acceptance. The technology of fuel production will evolve, reducing the costs of the alternative fuels. The technology of vehicles will also evolve, changing vehicle costs, range, and performance. Therefore, any plan to implement alternative fuels should be sufficiently flexible to respond to technological changes as they occur.

General Implementation Issues for Alternative-Fuels Programs

A plan to implement alternative fuels should generate the production of vehicles that use the new fuels and must then ensure that the new fuels are available in the marketplace and are actually used, in preference to less desirable fuels. The best implementation plan should probably not try to decide on the mix of alternative fuels on the market, in view of the evolving technological development of both fuels and vehicles. The implementation plan should recognize the differing abilities of each alternative fuel to compete in the marketplace. Some of the alternatives may be restricted to certain niches. Propane and compressed natural gas, for example, may have higher initial costs but lower per-mile operating costs. These fuels, therefore, may tend to be more successful where annual mileage is high. Electric vehicles have limited range with current battery technology, but they can be cost-effective for frequent short trips.

Some Proposed Implementation Plans for Vehicles Capable of Using Alternative Fuels

The only measure designed to stimulate sales of alternative-fuel vehicles and enacted so far is the Alternative Motor Fuels Act of 1988. Under this act, new vehicles capable of using methanol, ethanol, and natural gas are considered to use reduced amounts of gasoline, in the calculation of a manufacturer's corporate average fuel economy. (Electric vehicles always have qualified for this kind of treatment in the calculation.) To meet CAFE standards, some manufacturers have had to adjust product offerings and to use marketing incentives. In some years these measures have been costly. It appears that the CAFE incentives will provide a powerful inducement for some manufacturers to offer alternative-fuel vehicles. The California Advisory Board estimated that the incentives should make about 600,000 alternative-fuel vehicles available each year, once the manufacturers have developed enough engine models (California Advisory Board on Air Quality and Fuels, 1989a). It is expected that most of these vehicles will be flexible-fuel vehicles, capable of using either methanol or gasoline or any combination of the two.

The state of California has recently announced plans to implement a program that would relax emissions standards for vehicles using the reactivity benefits of alternative fuels (CARB, 1989a). Under this program, alternative-fuel vehicles could be certified at a higher emissions level than gasoline vehicles, according to the degree of photochemical benefit offered by the alternative fuel. This approach may result eventually in a considerable number of alternative-fuel vehicles if the costs of meeting the very stringent standards for gasoline vehicles become too great.

Other methods of stimulating the sales of alternative-fuel vehicles have also been proposed. In the Los Angeles area the South Coast Air Quality Management District has proposed a measure that would require vehicle fleets to buy some alternative-fuel vehicles when adding new vehicles to the fleet. Clean Air Act revisions proposed in the summer of 1989 by the administration included requiring the sale of one million alternative-fuel vehicles per year.

Financial incentives have also been suggested. The Andrews bill proposed in 1989 (H.R. 2269) provides investment tax credits for certain equipment required by some alternative fuels. California already provides a tax credit for the costs of converting to methanol. A California bill (SB 155: Leonard), proposed in 1989, would establish emissions fees favoring alternative-fuel vehicles. Finally, electric and gas utilities have argued that, in view of the environmental benefits of electricity and natural gas, the extra costs of natural gas compressor stations, needed for vehicle refueling, and the costs of electric batteries should be included in the utility rate base for rate-making purposes, thereby spreading these costs from the vehicle operators to all consumers.

The future outcome of these various measures and proposals is uncertain. It seems likely, however, that some combination of financial incentives, emissions standards strategies, and mandates will begin to provide a number of alternative-fuel vehicles by the mid-1990s. Perhaps one million alternative-fuel vehicles per year would be added to the rather small number of natural gas, propane, and electric vehicles currently in use. Most of these vehicles would be new vehicles, supplied by manufacturers or assemblers, rather than conversions or retrofits, although the number of conversions and retrofits may be sizeable, especially in the early years.

It is likely that most of the vehicles will be dual- or flexible-fuel vehicles (in the case of electric vehicles, the dual-fuel approach is usually termed "hybrid"), especially in the early years when alternative-fuel supplies may be scattered and too thin to justify dedicated vehicles or to ensure consumer acceptance (Koyama and Moyer, 1989). If the use of alternative fuels is motivated by air quality programs, the availability of these fuels may be restricted to those regions that require them in their air quality management programs. It may be difficult to bring about a transition to vehicles dedicated to operating on alternative fuels. Therefore, implementa-tion plans for alternative fuels must consider the likelihood that the majority of alternative-fuel vehicles will be capable of using gasoline, which is universally available.

This is in contrast to the shift to unleaded gasoline that occurred in the mid-1970s in the United States. In this case the benefits of shifting to unleaded gasoline were judged to be universal, and lead pollution was therefore seen to be less of a regional problem. Furthermore, the catalytic control of vehicle emissions required the sustained use of unleaded fuels because catalysts were easily poisoned by lead. Future programs will need to address the question of fuel choice by consumers in a less constrained situation.

Approaches to Implement the Use of Alternative Fuels

Approaches for supplying alternative fuels and ensuring that they are used can employ the same kinds of measures discussed above to stimulate the supply of vehicles capable of using the fuels: financial incentives (rebates, tax incentives) and disincentives (taxes, emissions fees), fuel "quality" or property standards (analogous to emissions standards for vehicles), and mandates (sales requirements). In addition, in the case of fuels, there is the option of the government's purchasing (or even producing) the fuels and distributing them.

All of these approaches have been used by various countries. The U.S. unleaded gasoline program was essentially a mandate. Financial incentives are being used to introduce unleaded fuel in Europe. They have also been used in New Zealand, to promote the use of natural gas as a vehicle fuel, and in Canada, to promote the use of natural gas and propane (California Advisory Board on Air Quality and Fuels, 1989b). Fuel standards were used in the U.S. phasedown of lead in unleaded gasoline, and in several programs requiring oxygenates in gasoline, as in Denver and Phoenix. Government-supplied fuel has even been used to stimulate substitutes for conventional gasoline. The state of California is now supplying methanol provisionally, in conjunction with the demonstration of methanol-fueled vehicles. This approach was also used for the very large ethanol program in Brazil (Trindade and deCarvalho, 1989; Sperling, 1987).

The four general approaches differ greatly in the degree of government involvement. In the fuel standards approach the government merely determines environmental constraints on fuel properties and leaves the choice of fuel technologies, pricing and marketing, and investment in fuel production to industry. This approach would achieve maximum flexibility if it allowed "averaging" of the properties over all of the fuels sold, so as to bring forth a "mix" of fuels that met the standards on an average of all fuels sold. In a mandate program the government goes further and requires certain specific fuels. By providing financial incentives, the government influences prices of selected fuels. Finally, in a program in which the government supplies the fuel, the government makes investments in facilities for producing and distributing the selected fuels. Figure 1 provides a simple graphical depiction of the various approaches described here.

Each of the approaches has advantages and disadvantages, and each may be suitable under some particular set of circumstances. Figure 2, taken from recent work by the California Advisory Board, provides a tabular summary of some of the key advantages and disadvantages of each of the four general approaches (California Advisory Board on Air Quality and Fuels, 1989b). In general, one can say that increased government involvement can bring about a quicker and more definite change in fuel use, at the risk of implementing solutions that are perhaps not optimal over the long haul, especially if technologies and costs change. More market-oriented

Figure 1

Alternative-Fuel Implementation Paths: Definitions and Lead Responsibilities

Approach		Environmental Constraints	Fuel Choices	Marketing, Pricing	Production Investment
Approach	Fuel Standards	Government	Industry	Industry	Industry
	Mandates	Government	Government	Industry	Industry
	Financial incentives	Government	Government	Government	Industry
	Government-provided fuel	Government	Government	Government	Government

Source: California Advisory Board on Air Quality and Fuels, *Draft Final Report*, vol. 5, Mandates and Incentives, Sacramento, Calif. 1989

Figure 2
Four Approaches to Methanol and Other Alternative Fuels

	Advantages	Disadvantages
Fuel standards with fuel pool averaging	• Protects environmental objectives but allows flexible, evolving response • Brings industry expertise to purchasing, marketing, pricing • Keeps opportunity open for cleaner forms of gasoline that could be used in older cars—not entirely linked to new vehicle technologies • Good experience with Federal lead averaging program during lead phasedown in 1980s • Utilizes market forces to bring forth most cost-effective fuels mix	• Complex regulation difficult to design and administer • Not a "fail-safe" approach—will need to be supplemented by a back-up approach • May not be sufficient incentive to allow CNG, LPG, and electric power to compete • Extra costs may be regressive
Mandated sales of methanol and other specific fuels	• Produces results • Historical experience in U.S. with mandates for unleaded gasoline in 1970s and oxygenates in 1980s • Brings industry expertise to purchasing, marketing, pricing • Built-in efficient form of taxing gasoline • Government can mandate government fleets and show leadership	• Mandated fuels might obtain an undesired advantage over other fuels • Have to be sure vehicles are available capable of using the mandated fuels • Reduces incentive to develop better alternatives • May lock-in some fuels too early • Substantial market intervention by government • Extra fuels costs may be regressive

Figure 2
(Continued)

	Advantages	**Disadvantages**
Financial incentives: subsidies for favored fuels and taxes and fees for "conventional" gasoline and diesel fuel	• Experience in other countries shows that subsidies and disincentives can work, once "tuned" • Subsidy has encouraged ethanol use in the U.S.	• New taxes on gasoline might be regressive • Adds new bureaucracy to administer • Difficult to find good government expertise in fuel pricing • Possible windfall for manufacturers of favored fuels
Government-supplied fuels	• Achieves lower costs by making a long-term commitment • Places any costs on society as a whole—less regressive than fuel taxes and mandates	• May mean committing to methanol and other fuels too early, before new and potentially better alternatives are developed • Require entrepreneurial/ business skills in government, which are rare • Puts government in competition with private fuels industry • Requires new government bureaucracy

Source: California Advisory Board on Air Quality and Fuels, *Draft Final Report*, vol. 5, Mandates and Incentives, Sacramento, Calif. 1989

approaches, such as the averaging of emissions standards as well as extending credits for and the trading of emissions reductions below the standards, offer better cost-optimization in the long run, but they are hard to relate to definite milestones and objectives.

Environmental regulation in the United States has generally employed "command and control" prescriptive approaches. The regulations responded to definite deadlines and generally contemplated known control technologies with reasonably clear performance and costs. Market-oriented approaches that allow the marketplace to tailor the approach have been less common, although "bubbling" and averaging and extending credits and trading have sometimes been used. Averaging, extending credits, and trading were used, for example, in the lead phasedown program.

The California Advisory Board appears to be the only group that has conducted an extensive comparison of the four general implementation paths for the introduction of alternative motor fuels. The Advisory Board not only considered the theoretical advantages and disadvantages of the four approaches, but also took considerable notice of the current political and policy "climate" in California (California Advisory Board on Air Quality and Fuels, 1989b). The Advisory Board not only concluded that there were theoretical objections to mandates, financial incentives, and government-supplied fuel (given the probable rapid improvement in the alternative-fuel technologies), but that these measures would also face an uncertain political future. The mandate of a single alternative fuel would encounter great resistance, and not only from those alternative fuels not favored. However, the mandate of several fuels would leave unsolved the problem of how to allocate the market mix of the alternatives. Financial incentives for alternative fuels would presumably require new taxes. Although it seemed possible that some "revenue neutral" tax plan would receive voter approval, such a tax plan would have to be in addition to a major new excise tax on gasoline for the renewal and expansion of the roads in California. Finally, it seemed unlikely that the voters would be enthusiastic about a major plan to have the state supply alternative fuels, and any such proposal would be vigorously opposed by the existing fuels industry.

The Advisory Board devised its own approach based on fuel standards. The Advisory Board assumed that certain fuel properties were closely related to the photochemical reactivity of vehicle emissions and to the emissions of such toxic pollutants as benzene.

This assumption was supported by emissions research by the California Air Resources Board (CARB) for methanol, natural gas, and other alternative-fuel vehicles (CARB, 1989). It seemed feasible to establish limits or caps on these properties and to apply these caps, on an average basis, across all fuels sold. Marketable credits would be earned by sales in excess of target levels, or by fuels with better-than-average properties. These credits would help stimulate the entry of better fuels into the marketplace and would possibly stimulate the development of even better fuels.

The proposal for average fuel standards, called "fuel pool averaging" by the Advisory Board, was modeled on the federal lead averaging program. In effect, this program created a market for alcohol blenders who took advantage of the "lead credits" generated by adding alcohol to leaded gasoline to reduce the average lead content (while preserving octane value). The approach also was influenced by the oxygenate programs of Denver and Phoenix. Although these programs do not use averaging, they avoid specifying fuel composition. Instead, these programs require a specified oxygen content but allow suppliers to meet the goal with whatever combination of alcohols and ethers has the lowest costs. In addition, the fuel pool averaging proposal partially resulted from informal California proposals to present an evolving California benzene standard for gasoline in the form of a gasoline "benzene pool," with averaging, similar to the federal lead program. This approach was suggested by Leon Vann of the California Energy Commission, since it might allow a lower average benzene level in the total gasoline pool while stimulating the introduction of reformulated gasolines and gasoline blends.

Fuel pool averaging was further developed by the Advisory Board to include all motor fuels. In addition, the regulated quantities were extended to include photochemical reactivity of emissions as well as air toxics and, potentially, of any other fuel property closely related to emissions or environmental effect. (In fact, fuel pool averaging could, if desired, include properties related to global warming and energy security, although the Advisory Board did not specifically recommend including such considerations in a fuel properties regulation.)

The Advisory Board had several objectives in mind in proposing the fuel pool averaging concept. These included creating a "level playing field" on which all fuels could compete according to their environmental merits, costs, and consumer acceptance; avoiding preselecting any "winners"; creating an incentive to develop still better fuels in the future; avoiding taxes and mandates; recognizing that estimates of the costs of the various alternative-fuel technologies vary, and hence favoring a scheme in which fuel suppliers, as the best qualified experts, would be making judgments about the most effective fuels to be supplied to meet environmental goals; and creating a regulatory structure in which proposed reformulated gasolines could compete as alternative fuels (viewed in another way, this objective could perhaps be summarized as creating direct pressure for improvements in gasoline).

Generally, fuel pool averaging seemed to meet these objectives fairly well, at least when considered in a very broad and generalized form. However, when the Advisory Board began to discuss possible approaches in more detail, many difficult questions arose. Which geographic areas would be covered by the regulation, and how would geographic "hot spots" be avoided? Which fuel properties could reasonably be related to photochemical reactivity of emissions? (This is an especially difficult question for gasolines. Specific olefins? Aromatics?) Who would be regulated: producers, wholesalers, retailers, or users? What measurements would be made? How would the program be monitored and enforced? What is the risk of errors and fraud? How would the caps or limits be set? Would these vary with time (as would seem logical as the population of vehicles capable of using alternative fuels increased)? How would extending credits and trading them work? Would interpollutant trades be allowed? Could photochemical reactivity be traded for toxics? Would "scoring" be differentiated, or "pass-fail"? What if a cleaner form of gasoline was developed that had environmental benefits when used in existing on-road vehicles, and did not require the phase-in of new low-emissions vehicles? How would such gasolines be scored? What if these gasolines had less benefit per vehicle but more benefit overall through use in a greater number of gasoline vehicles, compared with the available number of alternative-fuel vehicles? Would these gasolines generate credits? How could fuel suppliers develop strategies without knowing which vehicle technologies would be selected by vehicle manufacturers?

The Advisory Board discussed these questions at some length before deciding that formulating a regulation was the proper function of the California Air Resources Board in a formal rule-making process. No serious obstacles developed in the preliminary discussions. As an additional check on the probable success of a fuel pool averaging concept, the Advisory Board asked for public comment (California Advisory Board on Air Quality and Fuels, 1989). Generally, reaction was favorable, at least with regard to the obvious goals of the proposal. Speakers from the utility industries generally felt that fuel pool averaging would not provide sufficient incentives for natural gas and electric power and that these technologies needed additional regulations or financial incentives to overcome market barriers and diseconomies of small-scale production. Informally, representatives from almost all of the alternative-fuels industries expressed doubts about whether the petroleum industry would ever consider buying any marketable credits earned by other industries. It was also pointed out that the pace with which the program was adopted would be critical in determining its fairness and cost-effectiveness. In particular, reducing caps on benzene in the fuel mix or on photo-chemical reactivity would rapidly have two negative effects: favoring methanol simply because its technology is more advanced than that of natural gas and electric technologies, and possibly adding substantial costs if rapid pace first forced major and expensive changes in gasoline that were then followed by a phase-out of gasoline in favor of alternatives as they became more practical and competitive.

In general, however, the Advisory Board concluded that a fuel pool averaging approach had far more merits than any other approach. The Advisory Board concluded its process by recommending that the California Air Resources Board (CARB) develop by September 1990 a specific rule incorporating fuel pool averaging concepts (California Advisory Board on Air Quality and Fuels, 1989c).

Approach of the California Air Resources Board (CARB) to Alternative-Fuels Regulations

The CARB staff has held three public meetings for low-emissions vehicles and fuels that might be needed for such vehicles (CARB, 1989a). The CARB staff has combined the vehicle-related rules and fuel-related rules into a single complex package. The CARB vehicle emissions rules employ the split standard approach, in which reactivity credit is given to any alternatives, or to gasoline, that have emissions with less photochemical reactivity. This approach will either stimulate the supply of alternative-fuel vehicles or drive gasoline-fueled vehicles to much lower emission levels, as first "low emissions" standards and ultimately "ultra-low emissions" standards are phased in. The current proposal requires vehicles to meet "ultra-low" levels of 0.04 grams per mile of hydrocarbons, one-tenth of the current federal standard, beginning with a rapid phase-in after the year 2000.

To the extent that the vehicles require specific fuels in order to meet emissions standards, the fuel portion of the proposed California rules requires that fuel suppliers ensure that these fuels are sold, or, alternatively, that new reformulated gasoline provides an equal benefit. Thus the proposed rule incorporates the fuel pool averaging concept recommended by the California Advisory Board.

On December 14, 1989, the California Air Resources Board approved the staff proposal incorporating split-emissions standards and fuel pool averaging, after hearing public comment from the industries affected. In general, public comment agreed with the general philosophy of the proposed rules but disagreed with the schedule and pace. Figure 3 shows the author's rough estimate of the effects of the California approach.

Prospects for Programs in Other
Areas of the United States

The revisions to the Clean Air Act proposed by the administration in the summer of 1989 mandated the sale of alternative-fuel vehicles in those areas having difficulties in meeting the ozone standard. Sales of the fuels required by these vehicles were to be ensured by the states as part of their implementation plans. Presumably, each state would need to conduct an evaluation of the approaches best suited to local conditions.

Since the introduction of the Bush proposals, and as the debates about revisions to the Clean Air Act have proceeded, alternative proposals have been made. The outcome is highly uncertain, but it is possible that, ultimately, the administration plan will prevail

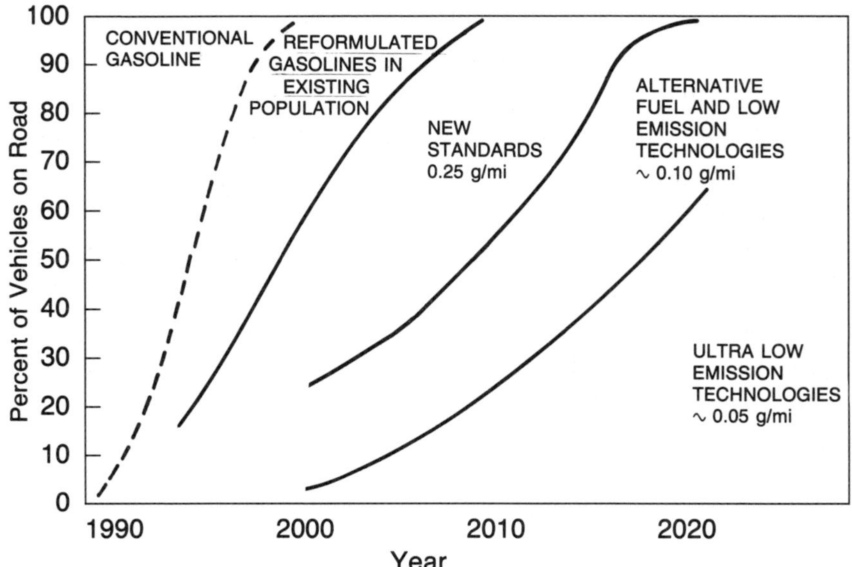

Figure 3
Best Current Estimate of Emissions-Control
Technologies During Coming Decades

Source: Acurex Corporation

with only slight modifications. Some sort of government mandates for alternative-fuel vehicles is likely, although these may be contingent upon the specific requests of the states to meet implementation plan requirements. Assuring use of the needed fuels may remain a part of the implementation plans of the states. Whether the various states and regions ultimately elect to follow the current California approach and use fuel averaging will depend on local conditions and preferences and the pace at which other regions adopt alternative-fuel strategies. If deployment is delayed, this might favor more use of mandates and financial incentives after California has, in effect, pioneered and demonstrated the technologies involved. Prompt deployment, on the other hand, might favor a California-like fuel standards program.

References

California Advisory Board on Air Quality and Fuels (1989). An Implementation Plan for Public Review. Sacramento, Calif.

______ (1989a). Draft of Final Report. Vol. 4, Economics. Sacramento, Calif.

______ (1989b). Draft of Final Report. Vol. 5, Mandates and Incentives. Sacramento, Calif.

______ (1989c). Report to the California Legislature. Vol. 1, Executive Summary, Final Report. Sacramento, Calif.

CARB (1989). Definition of a Low-Emission Motor Vehicle in Compliance with the Mandates of Health and Safety Code Section 39037.05. California Air Resources Board, El Monte, Calif.

______ (1989a). Low-Emission Vehicles/Clean Fuels and New Gasoline Specifications. California Air Resources Board, Progress Report, Sacramento, Calif.

Koyama, K., and C.B. Moyer (1989). The Transition to Alternative Fuels in California. In *Alternative Transportation Fuels: An Environmental and Energy Solution*, edited by D. Sperling. New York: Quorum Books.

Moyer, C.B., S. Unnasch, and M.D. Jackson (1989). Air Quality Programs as Driving Forces for a Transition to Methanol Use. *Transportation Research*, 23A.

Russell, A., et al. (1989). Estimate of the Air Quality Impacts of Alternative Fuel Use. California Air Resources Board. Sacramento, Calif.

Sperling, D. (1987). Brazil, Ethanol and the Process of System Change. *Energy* 12.

Trindade, S.C., and A.V. deCarvalho (1989). Transportation Fuels Policy Issues and Options: The Case of Ethanol Fuels in Brazil. In *Alternative Transportation Fuels: An Environmental and Energy Solution*, edited by D. Sperling. New York: Quorum Books.

U.S. Congress (1989). Catching Our Breath: Next Steps for Reducing Urban Ozone. Office of Technology Assessment, OTA-0-412, Washington, DC.

Consumer Demand for Methanol

DANIEL SPERLING
DAVID HUNGERFORD
KENNETH KURANI

Introduction

Consumer purchase behavior is fundamentally conservative—in that most consumers will not buy a non-petroleum fuel or vehicle unless there is a strong economic incentive to do so. This purchase behavior, however, is not fixed and unchangeable. Consumer purchasing behavior can be greatly influenced by messages transmitted by government and industry and communicated from person to person—messages that emphasize such values as environmental preservation and that influence beliefs about the future, including the relative cost and availability of fuels and the resale value of vehicles powered by non-petroleum fuels.

The authors gratefully acknowledge funding from the Policy Integration Office, U.S. Department of Energy. David Greene, Carl Moyer, and Barry McNutt provided valuable insights during the preparation of this paper.

This paper explores the conditions under which consumers are willing to purchase non-petroleum fuels and vehicles in order to enhance the understanding of consumer purchasing behavior. The role of fuel prices, fuel availability, engine power, and air pollutant emissions is examined as well as the role of government in the fuel and vehicle purchase decision. After careful analysis, the authors are skeptical about the hypothesis that new fuels must be introduced incrementally and that methanol must pass through an extended transition period of flexible-fuel vehicles.

Dedicated (single-fuel) vehicles can be optimized to take advantage of the unique attributes of a particular fuel and thus result in fewer emissions, more power, and less cost. Therefore, a quick transition to dedicated vehicles would be highly superior to prolonged reliance on inherently inferior flexible-fuel vehicles. This paper discusses why concerns about consumer acceptability of dedicated vehicles are exaggerated and how marketing and investment strategies can be designed to overcome the reluctance of consumers to purchase dedicated methanol vehicles.

This paper is neither about the economics of methanol and other alternative-fueled vehicles nor about the legitimate profit-making concerns of fuel and vehicle suppliers. The underlying premise of the paper is that low-emitting fuels and vehicles are desirable and inevitable—and that it is only the time frame of such a transition that is debatable. The paper focuses on how behavior might or could change and on the role government and industry can play in altering behavior and attitudes. This understanding of behavior and attitudes will provide the basis for minimizing the financial and temporal investment in flexible-fuel vehicles and help to create an efficient strategy for introducing cleaner burning, dedicated methanol vehicles.

Analytical Approaches

Consumer purchase decisions are the result of a complex process in which many factors are considered. Even the simple question of cost is not so simple. To calculate true life-cycle cost figures for their non-petroleum fuel and vehicle, for instance, consumers must annualize the additional capital costs and must estimate their vehicles' future resale value as well as the differences in trajectories of future fuel prices, and so on. The problem is that there is little history or past experience with alternative fuels upon which to draw in making predictions and understanding behavior.

In the absence of direct experience, this paper relies on the following analytical approaches (Greene, 1985; Train, 1986): (1) "revealed preference," whereby direct observations are made of individuals making decisions in analogous or similar situations, and (2) "stated preferences," whereby hypothetical questions are posed to individuals to determine under what conditions they would purchase a non-petroleum fuel or vehicle. The first method requires detailed surveys and insights as to what can be generalized versus what is peculiar to a society or individual. For example, would early adopters of methanol vehicles in the United States behave similarly to early adopters of natural gas vehicles in New Zealand? Would early adopters of methanol vehicles in California behave similarly to purchasers in Michigan? Would early adopters behave similarly to later adopters? Although the second method of stated references tends to overstate demand, with experience and skill it can be useful.

The findings and hypotheses presented here come from revealed preference data of ethanol car purchases in Brazil and natural gas vehicle purchases in New Zealand and Canada and from surveys conducted at the University of California, Davis, during the past few years. These surveys include an in-person survey of refueling behavior of vehicle owners in northern California in 1984, a 1985–86 mail survey of 535 diesel car owners in California, focus group interviews of natural gas vehicle owners in Vancouver, and a 1989 mail survey of 2,000 households in New York State and California to test stated preferences for new fuels and market analogies between premium gasoline and methanol purchases.

Variables in the Fuel and Vehicle Purchase Decision

Under what conditions would the owner of a flexible-fuel vehicle purchase methanol fuel instead of gasoline? The person would consider fuel price, fuel availability at retail fuel stations; power and performance; driving range per tank of fuel; perceived safety; starting ability in cold weather; perceived effect on engine life and maintenance; environmental effect; and social messages, including the level of commitment by government and industry. Most owners of flexible-fuel vehicles would probably be convinced, on the one hand, that methanol would give more power than gasoline and that it would be environmentally superior. On the other hand, they would probably believe that methanol would cost more, provide less range, and be more difficult to start in cold weather and that it might shorten engine life and increase maintenance costs. Perceptions about safety would probably be mixed, since the dangers are different and difficult to compare.

The vehicle purchase decision would be based on the same variables listed above for the fuel purchase decision, plus vehicle purchase price and expected vehicle resale value. The variables with positive and negative influences relative to gasoline in the fuel purchase decision, as discussed above, would have consistently negative and positive influences in the methanol-fueled vehicle purchase decision. The difference is that these influences are more salient in the vehicle purchase decision because of the lack of flexibility. Once a dedicated vehicle is purchased, no other fuel can be used. In contrast, the driver of a flexible-fuel vehicle can switch between gasoline and methanol.

Findings

Because the relationships between decision variables is not yet understood well enough, this paper does not attempt to represent the fuel and vehicle purchase decisions by means of a mathematical model. It does, however, offer a series of insights that provide guidance for fuels policies and provide a basis for further investigations.

The Importance of Fuel Prices

Posted fuel price is used by consumers as the principal indicator of economic attractiveness. A survey of diesel car owners indicated that consumers will determine the economic attractiveness of methanol (and other non-petroleum) vehicles based mostly on posted fuel prices at fuel stations (Kurani and Sperling, 1988). As shown in Table 1, diesel cars gained significant market share in the United States in a short period of time, with sales increasing from practically zero in 1976 to more than 6 percent of new car sales in 1981. Diesel cars became popular almost solely because of their low fuel cost. The fuel cost advantage, about 25 percent through

Table 1
U.S. Diesel Car Sales and Diesel and Gasoline Fuel Prices

	Diesel Car Sales	Sales as Percent of Total U.S. New Car Sales	Avg. Retail Diesel Fuel Price (cents/gal)	Unleaded Regular Price (cents/gal)
1970	5,000	Neg.		
1976	22,735	0.2	52.8	61.4
1977	37,498	0.4	57.9	65.6
1978	114,880	1.1	59.9	67.0
1979	271,052	2.6	84.2	90.3
1980	387,049	4.3	112.4	124.5
1981	520,788	6.1	131.0	137.8
1982	354,690	4.4	127.4	129.6
1983	197,710	2.1	125.4	124.1
1984	150,548	1.5	131.2	121.2

Neg. = < 0.05 percent

Source: Oak Ridge National Laboratory, *Transportation Energy Conservation Data Book*: 7th and 8th eds., ORNL-6050 and ORNL-6205 (Springfield, VA: NTIS, 1984 and 1985)

the early 1980s, was due to a fuel efficiency advantage of roughly 15 percent and a cost-per-gallon advantage of roughly 10 percent.

In 1978 the cost of fuel for a typical diesel car was about 2.5 cents/mi versus 3.4 cents for a comparable gasoline car (assuming 20 mpg for the gasoline car). In 1980 the cost advantage was still about 25 percent: 4.8 cents versus 6.2 cents. If the vehicles traveled 15,000 miles in 1980, the cost advantage for the year would have been $210 (more than enough to offset the somewhat higher purchase price of diesel cars).

When the difference between diesel and gasoline prices began to shrink, diesel car sales began to drop. However, even when diesel fuel prices began to exceed unleaded regular gasoline prices in 1983, diesel cars still had a fuel cost advantage because of their superior fuel efficiency. Although other factors played a role in the collapse of the U.S. diesel car market—in particular, technical problems with General Motor's diesel cars and diminished concern for energy— consumer surveys indicate that fuel prices were the most important factor both in the diesel car surge and in the subsequent collapse (COMSIS Corporation, 1986; Greene, 1986; Kurani and Sperling, 1988).

This finding suggests that great care should be taken in determining how to advertise fuel prices. The simplest procedure when dealing with alternative fuels, including natural gas and electricity, would be to advertise and post prices on a per-gasoline-equivalent gallon basis, although it would be to the advantage of methanol to post prices on an unadjusted per-gallon basis.

Fuel Availability Is Critical

Concerning fuel availability, at least 10 to 15 percent of fuel stations must supply methanol before dedicated vehicles are introduced. Surveys conducted in New Zealand (Harris, Arnoux, and Phillips, 1980; Harris et al., 1984) found that initially fuel availability was one of the most important factors that discouraged people from converting to natural gas fuels. They also found that, for natural gas vehicle owners, fuel availability concerns were about equal in importance to vehicle performance, the most important negative factor associated with ownership.

It is clear that fuel availability is a barrier, but it is not clear how large the barrier is. Studies of the diesel car phenomenon in the United States provide a quantitative indication of how many stations are needed. Until the mid-1970s the only diesel cars in the United States were those sold by Mercedes Benz and, for a short while, Peugot. They represented less than 0.1 percent of the car population. In 1976, just before Volkswagen and General Motors began mass marketing diesel cars, diesel fuel was being sold at about 9 percent of the retail fuel outlets in California (Table 2). As the market share of diesel cars in California increased from 1 percent of new car sales in 1977 to a peak of 9 percent in 1981, the proportion of retail fuel outlets serving diesel fuel increased to 22 percent.

Even with this fairly large number of fuel outlets, both diesel and gasoline car drivers expressed substantial concern about diesel fuel availability. In a 1986 random sample survey of 535 diesel car owners in California, 39 percent said that at the time they had purchased their vehicles they had been somewhat (27 percent) or very (12 percent) concerned about diesel fuel availability (Sperling and Kurani, 1987). This level of concern was found to be relatively unchanged between those who purchased their vehicles in the early years (1977–81), when fewer outlets existed, and those who had purchased their vehicles in later years, when a much larger network of outlets existed. The survey also found that those with pre-purchase concern about fuel availability in fact encountered about the amount of difficulty they had expected. The relationship between prepurchase concern and post-purchase difficulty was unaffected by the year of purchase.

These findings indicate that concern for fuel availability did not decrease as the network of fuel outlets expanded. The explanation for this apparently counterintuitive finding is that the initial group of diesel car buyers accepted the fact that few fuel outlets existed and that they would have to accommodate themselves to

Table 2
Diesel Fuel Stations and Non-Commercial Diesel Vehicle Sales in California

| | Diesel Fuel Stations | | Non-Commercial Diesel Vehicle Sales | |
	Authors' Estimate	Percent of Retail Stations Serving Diesel (Based on Authors' Estimate)[a]	No. of Vehicles	As Percent of Total Sales[b]
1976	1,200	9	–	–
1977	1,300	10	9,000	1
1978	1,440	12	20,500	2
1979	1,650	15	33,100	3
1980	2,000	19	44,400	6
1981	2,250	22	60,000	9
1982	2,500	24	48,000	7
1983	2,500	23	34,400	4
1984	2,500	25	18,000	2
1985	2,500	–	15,100	1

[a] The total number of retail fuel outlets in California decreased from 13,066 in 1976 to 10,771 in 1980, and remained at about that level through 1985. Retail fuel outlets are defined as those establishments that derive at least half of their retail sales from motor fuel sales.

[b] Estimated total passenger vehicle sales for California were in excess of 1.13 million in 1978, declined sharply in 1980, reached a low of approximately 650,000 in 1982, and gradually increased to 1.04 million in 1985.

Source: Daniel Sperling and Kenneth S. Kurani, "Refueling and the Vehicle Purchase Decision: The Diesel Car Case," SAE Paper 870644, 1987

this reality. As diesel car ownership expanded beyond a core group into the general car-buying public, however, limited fuel availability was not so readily tolerated.

Consistent with the explanation that a tolerant core group exists is a finding that the general gasoline car-owning public would be even more reluctant to accept the inconvenience of limited fuel availability. In response to survey questions asking whether they would consider purchasing a diesel car, 33 percent of 1,530 gasoline car owners said fuel availability would be a "very important" concern and another 28 percent said they would be "somewhat concerned" (Sperling and Kitamura, 1986). This total of 61 percent concerned drivers was much higher than the 39 percent response by diesel owners, and it reinforces the suggestion that as one moves beyond the self-selected early adopter group, tolerance for limited fuel availability diminishes sharply.

Because diesel fuel had been available at retail outlets for many years (to serve long-haul trucks), and because diesel cars get about 50 percent longer driving range than gasoline cars (and 100 percent or more than that of alcohol-fueled vehicles), diesel cars are not necessarily a good test case for non-petroleum vehicles.

Even so, the following conclusions can be drawn. First, there are a small number of individuals who are willing to put up with the inconvenience of a limited number of fuel outlets if the vehicles offer some positive attributes, such as lower cost or higher performance. This group of early adopters, however, is probably less than 1 percent of the car-buying population. Second, to gain significant market share (more than about a 1 percent share of new car sales) in a targeted region, at least 10 to 15 percent (and probably much more) of the fuel outlets must supply that fuel.

The cost of establishing a network of methanol fuel stations of this scale is relatively modest. The cost of retrofitting one retail fuel station for methanol use, including the installation of a new underground tank, is about $50,000 (Sperling, 1988). The total cost of establishing an initial network of 15 percent of California's stations would therefore be only about $75 million. This is a one-time cost that represents a tiny fraction of the almost $10 billion in revenue per year gained from gasoline sales in California.

To conclude, with respect to fuel availability, unless the sales of methanol vehicles are somehow mandated, a sizable network of fuel outlets must be put in place before methanol vehicles can be mass marketed, even if the vehicles have a flexible-fuel capability. The cost of doing so is not exorbitant.

Additional Power Is Important to Consumers

Moreover, they are willing to pay for it. Vehicle owners in California and New York were asked whether they would be willing to switch to a fuel that gave them 10 percent more power (roughly the amount of additional power gained from using methanol instead of gasoline in a flexible-fuel car) if the fuel was priced a specified bid amount above the gasoline they normally bought (Hungerford and Sperling, 1990). Eight different bid amounts, one to each respondent, were offered at random to the sample population. As shown in Figure 1, about 60 percent of Californians and 67 percent of New Yorkers were willing to pay 2 cents/gal (gasoline basis) for the additional power; more than 15 percent were willing to pay an additional 40 cents. Indeed, market sales of sports cars and larger engines have been increasing in recent years,

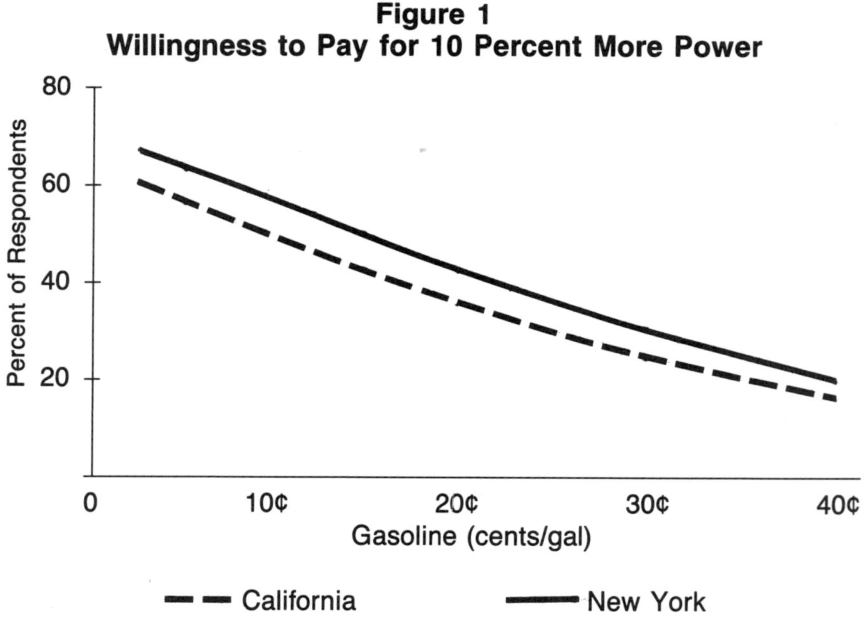

Figure 1
Willingness to Pay for 10 Percent More Power

Source: University of California, Davis

corroborating these stated measures of willingness to pay. (The authors are now attempting to quantify the relationship between the willingness-to-pay responses and vehicle sales data.)

Consumers Are Willing to Pay for a Low-Polluting Fuel

Air quality could also be important in vehicle and fuel purchase decisions. In the same survey mentioned above, in parallel questions, respondents were asked whether they would switch to a fuel that produced less air pollution if it was priced a specified amount more than gasoline. Surprisingly, the willingness to pay for low-polluting fuels was slightly greater than the willingness to pay for more power (Figure 2). Moreover, the willingness to pay for low-polluting fuels cut across the entire population; it was not correlated with income and was not much greater in California than in New York.

Figure 2
Willingness to Pay for Less Air Pollution

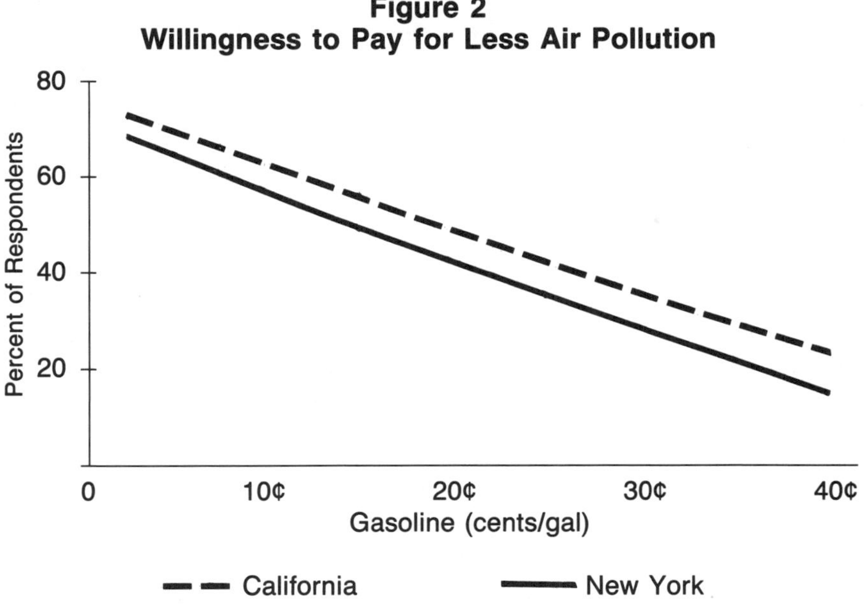

Source: University of California, Davis

The stated willingness to-pay-measure for low-polluting fuel appears valid, since the responses regarding cleaner fuels are comparable to those for engine power. Willingness to pay for power has been borne out in the marketplace. This willingness to pay for low-polluting fuels, however, does not mean that a motorist would select the more expensive low-polluting fuel when faced with a choice at the fuel pump. For example, in 1970, when environmental consciousness was at its zenith, most of the major oil companies began marketing low-lead or no-lead gasoline, primarily on the basis of the air quality benefit of eliminating lead. The unleaded gasoline sold for only 1 to 4 cents/gal more than leaded gasoline, and yet sales were less than 3 percent of gasoline sales in 1971 and did not exceed 5 percent until catalytic converters were widely introduced on vehicles in 1975 (Sperling and Dill, 1988).

The widespread willingness to pay for low-polluting fuels may therefore be interpreted as a willingness to pay if the cost burden is shared by all. It is a textbook "free rider" problem. One approach for transforming this high willingness to pay into a politically acceptable initiative might be to place a surcharge on dirtier fuels as a means of subsidizing low-polluting fuels. It would be most acceptable if that surcharge was specifically targeted to supporting low-polluting fuels and cleaner air.

Is Methanol a Premium Fuel?

The demand for methanol will depend on how the fuel is positioned in the marketplace: Will it be perceived as a premium fuel, such as premium gasoline, or an inferior fuel, such as diesel fuel for autos? Figure 3 shows the approximate relationship between fuel sales and fuel prices for diesel cars and premium gasoline. Diesel cars were clearly perceived as inferior. When diesel fuel was 10 cents/gal less expensive than gasoline, roughly 5 to 10 percent of the car-buying public was willing to buy a diesel car. But when the price increased to parity with gasoline on a physical gallon basis, diesel car sales plummeted.

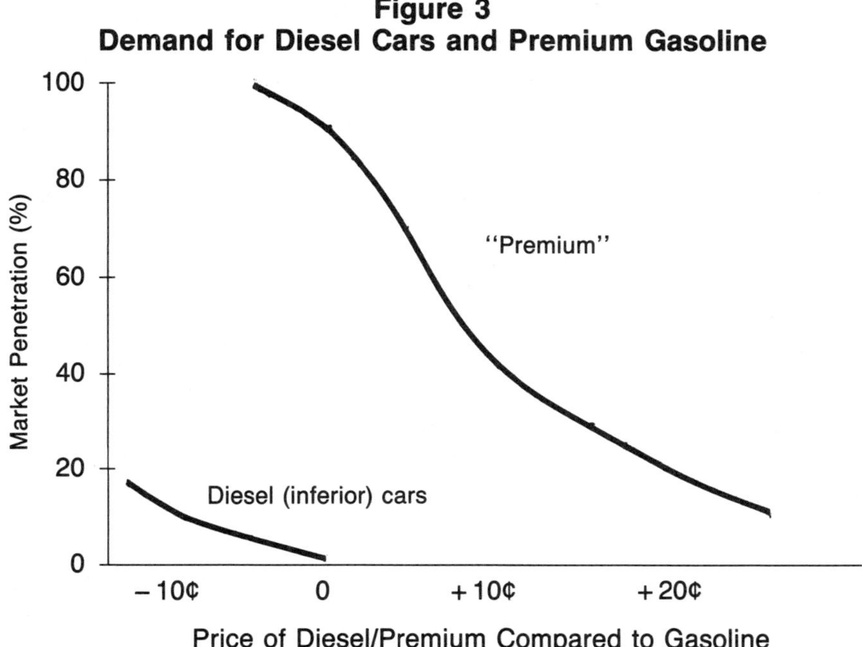

Figure 3
Demand for Diesel Cars and Premium Gasoline

Source: University of California, Davis

The high-octane gasoline story has been very different, especially during the last few years of the 1980s. High-octane gasoline accounted for about 30 percent of gasoline sales in 1989, even though it cost 10 to 15 cents/gal more, and even though it offered absolutely no benefit to half or more of those buyers (Hungerford and Sperling, 1990). Many people clearly perceive high-octane gasoline as a premium fuel, a notion not discouraged by oil companies.

The key to the success of methanol (and other alternative fuels) is to position it as a premium fuel. That can be done through the media and by government and industry pronouncements. It is a social message transmitted to consumers.

Commitment of Car Buyers to Methanol

Will owners of flexible-fuel vehicles operate their vehicles on methanol? The answer lies in part in how consumers acquire their vehicle. The New Zealand surveys cited earlier (Harris, Arnoux, and Phillips, 1980; Harris et al., 1984) and a Canadian survey (Canadian Facts, 1986) found that people who converted their vehicles to operate on both natural gas and gasoline relied very heavily on natural gas and very little on gasoline. The Canadian survey found that despite limited fuel networks, 83 percent of convertees did 70 percent or more of their driving on natural gas. The 1984 New Zealand survey found that 73 percent of the vehicles ran at least 90 percent of the time on compressed natural gas (CNG) and that one-third ran exclusively on CNG. Once they incurred the large initial conversion cost, these motorists had a large economic incentive to operate their flexible-fuel vehicle on natural gas because the fuel cost was one-half to two-thirds that of gasoline. Owners of flexible-fuel methanol-gasoline vehicles, on the other hand, will not have this incentive because methanol will cost more than gasoline.

Yet the commitment to natural gas by natural gas vehicle owners apparently was more than just a financial outlay. It represented a personal commitment as well. The Canadian survey (based on focus group interviews) found that drivers of natural gas vehicles were "carrying a minimum (often a very bare minimum) of gasoline in their tanks, purely for unforeseen (and highly undesirable) emergency situations, and are intent on using nothing but natural gas as their fuel." It was found that "every time they are forced to resort to the gasoline switch, then, they are forced to admit a small defeat, and to the extent that this happens they are unhappy with natural gas vehicles" (Canadian Facts, 1986).

There is an important lesson here for flexible-fuel cars. If they are mixed into the fleet of new cars, and consumers are not initially aware of what they are buying and/or do not have to pay extra for the flexible-fuel car, then it seems likely that they will not have any particular commitment to methanol—in other words, these owners will not be inclined to use methanol fuel in the vehicle. If, however, flexible-fuel cars were advertised as such, and cost a little more, then those who did buy the vehicle would be making a commitment to methanol and would be much more likely to purchase methanol fuel for the vehicle, all else being equal.

An Exceptionally High Level of Quality Must Be Maintained

For methanol to be a success, an exceptionally high level of quality must be maintained with all aspects of the new fuel—with the vehicles, fuel distribution system, and fuel itself. This is because consumers and the media will be scrutinizing it for any missteps or flaws. Two examples illustrate the importance of exceptionally high quality.

First, in Brazil, ethanol car sales boomed shortly after being introduced—increasing from zero percent to 73 percent of new car sales in less than a year. Then, surprisingly, sales crashed to 9 percent of new car sales within seven months. Several reasons explain this collapse (Sperling, 1987). One of the most important was that stories had started circulating in the media and person to person that ethanol cars did not run well—that they often stalled or would not start and that the fuel was bad for the vehicles. The real problem was with vehicles that were retrofitted after-market by amateurish mechanics. Although the problem was not with factory-produced ethanol vehicles, that distinction was not usually made, not even in media stories. In this case the reputation and sales of ethanol vehicles were destroyed by concerns about quality, concerns that were mostly irrelevant to manufactured ethanol vehicles.

Another story illustrating the importance of quality, and the high level of scrutiny applied to new vehicles and fuels, is that of the diesel car in the United States. The second most important reason for the demise of diesel cars, after fuel price shifts, was its poor reputation for quality (Kurani and Sperling, 1988). This reputation was due to problems with only one engine of one manufacturer—the

5.7-liter diesel engine used in a General Motors' diesel car. Other diesel engines used in GM cars—the 4.3-, 5.0-, and 6.2-liter engines—had no significant problems, nor did any other diesel engines. Yet car buyers perceived that all GM diesel cars were of inferior quality, and many people also believed that all diesel cars were inferior. The problem with one engine tarnished all diesel cars.

Role of Government Is Crucial in Reducing Uncertainty

A corollary to this finding is that consumer perceptions of safety, reliability, quality, and so on are not stable or strongly held. These perceptions can be changed fairly easily, especially in the negative direction.

Uncertainty about the future is possibly the single most important consideration in the consumer's vehicle purchase decision. It can work either for or against a new fuel. In addition, demand may vary greatly from one region to another and from time to time.

In the Brazilian experience ethanol car sales zoomed to 73 percent of sales in one year. Why? Because people were convinced ethanol would be less expensive and in greater supply than gasoline. The subsequent collapse of the market occurred because of renewed uncertainty about ethanol's future. At the height of the booming sales the government became concerned that ethanol fuel supplies would fail to keep pace with vehicle sales and that the ethanol subsidies were getting out of hand (Sperling, 1987). As a result, the government started to discourage ethanol car sales by raising ethanol fuel prices and reducing various incentives. Some officials also started to question publicly the wisdom of ethanol fuels.

The message the consumer received was that maybe ethanol was not the fuel of the future after all, maybe ethanol cars would be a misbegotten adventure soon abandoned and forgotten—and they, the consumers, would be left with a valueless car for which they could not find any fuel. The point is that consumers were very responsive to signals—first that ethanol was the fuel of the present as well as the future, and later, that ethanol was being de-emphasized and perhaps phased out.

Obviously, government will have to play a major role if new fuels are to be introduced in a timely manner. It will have to reduce the cost of buying and operating a new-fuel vehicle to that of gasoline vehicles, and it will have to reduce uncertainty and risk for the buyer. The Brazilian experience demonstrates that governmental action can be extremely effective.

Most impressive, given other experiences with natural gas vehicles, is that the Brazilian government was able to reduce consumer uncertainty to such an extent that gasoline-fueled cars were perceived to be riskier investments than ethanol-fueled cars! Brazilians were convinced in 1980 that gasoline would be a scarcer and more expensive commodity than ethanol. Substantial subsidies and incentives were needed to make ethanol attractive and to convince people of the government's determination to phase out gasoline, but another major factor was public persuasion. In summary, continuing public pronouncements by officials, backed by a steady stream of incentives and subsidies, were effective in virtually eliminating people's concern about the future price and availability of ethanol.

Conclusion

These analyses of the behavior and attitudes of the car-owning public should be generalized with care, since behavior and attitudes vary with time, space, and individuals. Variations occur because of differences in climate (cold start problems), topography, learning and information dissemination effects, the role of governments in changing perceptions of vehicle owners, refueling attitudes and behavior of drivers, relative price differences, and perceptions of safety and reliability. More research is needed to learn how consumers value different attributes, so that region-specific policies and initiatives can be designed that would most effectively attract new buyers.

Clearly, though, consumer behavior toward new fuels and vehicles tends to be conservative because of perceived risk and uncertainty. While it may be true that people have fundamental wants and desires with respect to fuels and vehicles, those wants and desires cover a broad range of acceptability. Some fuels and vehicles, such as electric vehicles, may fall outside the range of acceptability for many car owners, but that is not the case with methanol (and natural gas vehicles). The authors believe that virtually all consumers would be willing to accept methanol vehicles—if they were strongly encouraged to do so in the form of price signals and social messages.

Thus, conservativeness of consumers can be reduced, perhaps dramatically, by concerted and consistent government and industry campaigns aimed at reducing perceived risk and uncertainty. With appropriate and credible signals, new fuels and vehicles could be seen as superior goods worth a higher price, even if they have inconveniences, such as shorter driving ranges.

References

Canadian Facts (1986). Management Summary: Natural Gas for Vehicles—Conversion Motivation Study. Burnaby, British Columbia: British Columbia Hydro Gas Operations.

COMSIS Corp. (1988). A Case Study of the Commercialization of the Diesel Engine in the Domestic Light Duty Vehicle Market, prepared for the U.S. Department of Energy, Office of Vehicle and Engine R&D, Washington, DC (draft).

Greene, D.L. (1985). Estimating Daily Vehicle Usage Distributions and the Implications for Limited-Range Vehicles. *Transportation Research* 19B:4.

Greene, D.L. (1986). The Market Share of Diesel Cars in the USA, 1979–1983. *Energy Economics*.

Harris, G., L. Arnoux, and P. Phillips (1980). *CNG in Auckland: Survey and Analysis*. Auckland, New Zealand: New Zealand Energy Research and Development Committee, University of Auckland, Pub. P44.

Harris, G., et al. (1984). *CNG Market Development Study*. Auckland, New Zealand: New Zealand Energy Research and Development Committee, University of Auckland, Pub. P86.

Hungerford, D., and D. Sperling (1990). Demand for Premium Gasoline. University of California, Davis (manuscript in preparation).

Kurani, K.S., and D. Sperling (1988). The Rise and Fall of Diesel Cars: A Consumer Choice Analysis. *Transportation Research Record* 1175.

Sperling, D. (1987). Brazil, Ethanol, and the Process of System Change. *Energy* 12.

Sperling, D. (1988). *New Transportation Fuels: A Strategic Approach to Technological Change*. Berkeley, CA: University of California Press.

Sperling, D., and J. Dill (1988). Unleaded Gasoline in the US: A Successful Model of System Innovation. *Transportation Research Record* 1175.

Sperling, D., and R. Kitamura (1986). Refueling and New Fuels: An Exploratory Analysis. *Transportation Research* 20A.

Sperling, D., and K. S. Kurani (1987). Refueling and the Vehicle Purchase Decision: The Diesel Car Case. SAE Paper 870644.

Train, K. (1986). *Qualitative Choice Analysis: Theory, Econometrics, and an Application to Automobile Demand*. Cambridge, MA: MIT Press.

Commentary

CHARLES E. RISCH

Alternative fuels are not being market driven. That is to say, people are not coming into Ford showrooms asking for alternative-fuel vehicles. The factors encouraging alternative fuels are air quality benefits, energy security, and, more recently, global warming. As Carl Moyer mentioned, the transition to a new fuel requires that many forces be overcome. Certainly, fuel availability and cost are part of the problem, as well as the availability and cost of vehicles to use the fuel. Another item of great interest is the cost of the fuel supply infrastructure. From an automotive perspective, stringent emission standards that exceed current capabilities can certainly be an inhibiting force for any fuel, particularly if insufficient lead time is provided. Examples of this include the particulate standards for diesel fuel use (some of which may be withdrawn from production as early as 1991), the total hydrocarbon standard for vehicles that run on gaseous fuels, and the formaldehyde standard for methanol vehicles. (The California Air Resources Board is already considering tighter standards for methanol vehicles even before the first production vehicle has rolled off the assembly line.) Another uncertainty that both the auto and oil industries face is continuity of political support. Great effort must be exercised to minimize false starts or mixed signals in order to limit investment losses.

If environmental concerns dictate a transition to alternative fuels in the transportation sector, incentives will be necessary for the auto industry, the fuel industry, and, most importantly, for the consumer who will need a reason for buying the vehicle and using the fuel. The fuel must eventually pay for itself—there should not be a need for long-term financial support. Several fuels may coexist; some in widespread markets and others in niches. The concept of fuel pool averaging, as presented by Carl Moyer, is a good way to permit flexibility.

The oil and auto industries need a lead time to avoid major disruptions, as well as to avoid the premature introduction of a product that might damage the reputation of either the fuel or the company involved. Dan Sperling mentioned several examples of the complications that have resulted from introducing an unacceptable product. Both the vehicles and the fuels should be allowed to evolve in a mutually supportive manner.

Concern for a fuel's reputation is the primary reason for preferring market-based incentives to mandates. Incentives allow companies to make decisions based on a product's readiness and the market's interest in a new fuel. While the approach presented by Carl Moyer resembles a mandate, it does not designate a specific fuel and includes reformulated gasoline. The transfer of credits within the fuel pool averaging program is absolutely necessary to allow the market to pick the best approach.

The questions raised by Carl Moyer are valid and must be resolved before a fuel pool averaging program can be implemented. Resolving issues such as pollutant weighing, control categories, and compliance determination will not be easy and may take longer than anticipated.

Dan Sperling very aptly stated, "If no one buys the fuel or vehicles, then all else is for naught." Consumers must be assured that they are getting good value in order to get them to buy alternative-fuel vehicles and to use alternative fuels. Sperling suggests that the initial vehicles introduced should be dedicated to the alternative fuel. Experience, however, shows that such an approach will meet great public resistance. Ford's dedicated methanol Escorts in California encountered a resistance due to the reluctance of consumers to use the vehicles even when they had more than enough driving range. Fear of running out of fuel discouraged many drivers from using the dedicated fuel cars. It was this experience that led Ford to consider the flexible-fuel vehicle which uses either methanol or gasoline. These vehicles can be introduced before the fuel infrastructure is in place. As the fuel infrastructure develops, the vehicles can be calibrated for the new fuel and eventually dedicated vehicles can be introduced.

Consumers' interest in flexible-fuel vehicles is growing along with a growing awareness of environmental issues. In 1970, only 27 percent of the people in a Gallup poll thought environmental issues were of concern and this percentage has now increased to 73 percent. Ford found encouraging signs for potential sale of flexible-fuel vehicles in a recent survey. On a rating scale of excellent, very good, good, fair, and poor, the survey found that 77 percent of those responding thought a flexible-fuel vehicle was a good idea. Among college-educated participants, 83 percent thought the flexible-fuel vehicle was a good idea. Eighty-three percent of people under 35 also supported it.

Consumers are willing to pay extra for a premium fuel if there is a real or perceived value. Overall, 25 percent of drivers purchase premium gasoline today. Among new car owners, the rate is 43 percent. Better performance when operating on methanol is an advantage for the flexible-fuel vehicle, but it is not likely to be a strong selling point because the difference will not be big enough for the average driver to notice.

There are also factors that will discourage the purchase of flexible-fuel and dedicated vehicles. As Dan Sperling mentioned, the higher cost is apt to decrease interest in these vehicles. Ford found that 64 percent of those surveyed were interested in purchasing a flexible-fuel vehicle when the price was the same as that of a gasoline vehicle. With an increase in price of $500, the interest level dropped to 37 percent. When a 20 percent operating cost penalty was added, the interest level fell to 12 percent.

Flexible-fuel and dedicated methanol vehicles are projected to have similar costs. These vehicles, when using M85 or an equivalent methanol formulation, are expected to have costs that are modestly above those of gasoline vehicles. However, vehicles using 100 percent methanol (M100) are expected to be more expensive than either flexible-fuel or dedicated vehicles using M85, due to the cold-start system that would have to be added. Furthermore, a practical cold-start system has not been found for M100 vehicles.

Our surveys have also found concerns about durability and reliability when any new technology is introduced. Thus, market acceptance will be difficult for the first few model years. Another deterrent for customer interest in methanol vehicles will be the reduced driving range from the lower energy density of methanol. For the retail customer, the issue will be one of inconvenience. From a fleet standpoint, however, the concern will focus on time and cost.

In conclusion, the widespread introduction of an alternative fuel will require that flexible-fuel vehicles be used in the transition period. Above all, such a vehicle must provide value to the consumer. Consumers will have to want to buy the alternative-fuel vehicle and use the alternative fuel.

Editor's Note: After the conference Charles Risch made available the following letter from David L. Kulp, manager of fuel economy, planning, and compliance, Ford Motor Company, to Eugene Durman of the Environmental Protection Agency's policy office, which discusses cost estimates for flexible-fuel vehicles based on Ford's research and development efforts to date.

Ford Motor Company
Environmental and Safety
Engineering Staff

Dearborn, MI
September 26, 1989

Dear Mr. Durman:

During our September 6, 1989, discussion on alternative-fuel vehicles, you requested clarification of Ford's statements regarding the estimated complexity and customer cost of both flexible and dedicated methanol vehicles in comparison to a gasoline vehicle. This information is provided in response to that request.

In conjunction with Ford's 1985 efforts toward preparation of 30 research and demonstration Crown Victoria FFVs designed to operate on M85, E85 or gasoline, our engineers estimated that the incremental cost to the customer of the FFV might be approximately $150 to $300 at full production volumes (100,00 to 200,000 units). This assumed that all technical issues would be resolved inexpensively, existing emission control technology could be employed, engine optimization would equally favor each of the fuels (methanol, ethanol, and gasoline), and all open issues had been identified.

New projections of customer costs for the FFV feature will not be available for several years, pending the completion of new development work and feedback from field demonstration programs. At a minimum, some of the more recent initiatives for stringent formaldehyde control, better understanding of changes needed for resolution of the technical problems, can be expected to increase the cost of an FFV.

"Fully optimized, dedicated" methanol vehicles have been considered by many researchers to be the logical goal of an alternative fuel program. In part, this goal has been sought because the engine and vehicle can be modified to take greater advantage of the fuel's unique properties (high octane and lean burn) to achieve fuel economy near to that of gasoline vehicles. It has been our belief that the potential customer cost of M85 dedicated vehicles will be about the same as that of an FFV. In comparison to an FFV, all of the changes in materials, injectors, cold-start system, unique emission controls, spark plugs, etc., required for an FFV are needed for the dedicated vehicle, with the major exception of the optical fuel sensor. Offsetting the deletion of the fuel sensor will be several cost additions to produce the "fully optimized" M85 vehicle:

- Higher compression head/pistons
- Strengthened crank and/or bearings
- Possible unique catalyst or an EGO sensor

The use of M85 fuel, or a methanol fuel providing similar characteristics, is crucial to the cost-effectiveness of the above estimates. This fuel provides sufficient volatility and hydrocarbons to facilitate cold start at temperatures normally encountered in the U.S. In addition, the 15 percent gasoline provides flame visibility for firefighting and recognition. M85 also has color, taste, and odor that deter accidental ingestion and prolonged skin contact. This fuel also has lubrication qualities. An additive, which may replace the gasoline component of M85, is likely to evolve concurrently with efforts to improve commercial acceptability, reduce reactivity, and lower costs of the final product. We support such efforts, provided our key concerns are addressed—those concerns being the preservation of the flame visibility, fuel identification, and satisfactory enhancement of cold start.

To achieve appropriate cold-start characteristics with M100 would require major engine and/or vehicle modifications with significant cost penalties. Typical considerations are secondary cold start, fuels with separate storage, supply and control requirements, manifold/fuel heaters with dual battery (24-volt starting system) or other changes to the intake manifold, and fuel pre-treatment systems. Previous Ford development work was unable to identify any practical system that would package and function in light-duty applications. Therefore, we believe that a dedicated, M100 vehicle, even if practical for light-duty applications, will exceed the cost of either M85 option by a substantial margin. Even if neat methanol is viable, it will require additives, at a minimum, for odor, color, and taste to avoid misuse—these additives may impact vehicle and emission characteristics.

Should you have any questions on this information, please contact me.

Sincerely,

David L. Kulp
Mgr, Fuel Economy
Planning & Compliance

Commentary

JERROLD L. LEVINE

Regardless of the outcome of the alternative-fuel debate, the oil industry will supply whatever motor fuels are required—gasoline, reformulated gasoline, methanol, compressed natural gas, or whatever is in demand. However, before the oil industry can provide what the market requires, many very difficult and practical marketing issues must be resolved. An old marketing executive for whom this commentator once worked was fond of saying: "Young man, the major problem is how to get the dogs to eat the dog food!" This statement might be paraphrased: "You can bring a horse to water, but how do you get him to drink the methanol?"

The administration's proposal sets a production quota of one million alternative-fuel vehicles per year. However, these vehicles may not appeal to consumers, and no one can be required to buy them. The Soviet Union establishes production quotas and produces goods whether or not people want to buy what the state produces; there is no concept of a market-based economy. The United States is not about to go down that road.

The oil industry has the same problem with alternative fuels that auto manufacturers have with alternative-fuel vehicles. The problem is especially difficult if the vehicles are to be dual-fueled and the consumer has the choice of whether to purchase gasoline or a higher-costing alternative fuel. The automobile industry, of course, receives CAFE (corporate average fuel economy) credits for producing alternative-fuel vehicles, which may amount to several thousand dollars per vehicle. However, the administration's proposal stipulates that 30 percent of an oil company's total sales, in a non-attainment area, must be methanol—which means that it might have to be sold at a loss. This situation will certainly occur if the legislation allows a 50,000-gallon-per-month exemption for moderate-sized service stations. Those companies that are not required to sell methanol could price their gasoline competitively. Companies that had to sell methanol at a loss might try to recover by raising their gasoline prices, but they would be unable to do so because of competition from the smaller stations.

One option for an oil company would be to build a methanol plant. However, this would be risky since methanol might not always be the fuel of choice. Each methanol plant would require about $1 billion of capital investment, and about 25 plants would eventually be required under the administration's program. These plants are not likely to be built in Saudi Arabia, since the Saudis do not pay taxes. To minimize the capital costs of producing methanol some creative accounting, such as not paying taxes, would be required to make methanol competitive with gasoline.

Oil companies, such as Amoco, could have one or more billion-dollar "white elephants" 10 years from now. Oil companies already have shale oil investments that are white elephants. There are also white-elephant coal-gasification plants around the country. In addition, there are diesel passenger cars that are white elephants. But none of these white elephants cost individual companies as much as $1 billion apiece. In addition, supplying diesel fuel involved only minimal capital investments in service stations and distribution systems. A large supply of methanol fuels would require many billions of dollars of infrastructure. If after five or ten years methanol faded out, as did shale oil, coal gasification, and diesel cars, the industry would be left holding a multibillion-dollar bag. No responsible corporate management can take on that kind of risk since it is contrary to a corporation's obligation to its shareholders.

Dr. Moyer compared fuel pool averaging with the lead phasedown. However, the lead phasedown was relatively easy, since there was only one fuel property to measure. At the same time, the U.S. Environmental Protection Agency had to monitor only four lead suppliers (to check against refinery reports), and these four dwindled down to only one. In spite of this, the EPA had hundreds of violations.

Under fuel pool averaging several properties of gasoline would be regulated, including benzene, aromatics, olefins, oxygenates, Reid vapor pressure (RVP), and lead. There would be no way of checking what refiners claimed was the quality of their products. This could change almost daily within each refinery, depending on the crude oil being processed and the operating conditions within a unit. In addition, a large group of gasoline blenders buy domestic and imported gasoline wholesale and blend in various components, such as butane, lead, or alcohol. These gasoline blenders could all gain theoretical credits and there would be no way of telling whether their claims were legitimate.

In the crude oil entitlements program of the 1970s there were so-called daisy chains through which crude oil was traded back and forth between several mythical companies. The net result was that one million barrels a day of older, low-priced oil disappeared, while one million barrels a day of new, high-priced oil magically materialized. This commentator fears that the trading of credits in a fuel pool averaging program could easily foster a similar degree of large-scale cheating.

Dan Sperling stated that 50 percent of consumers *say* they would pay five to seven cents or more per gallon for an environmentally superior fuel. Amoco's market research obtained a similar result. However, people can say that they support the environment, but when individuals make their own purchase decisions, they will ask themselves: "What is the effect of this one purchase? How much improvement in the environment will occur from purchasing this tankful of environmentally superior fuel? Is it worth an extra dollar that I pay for this tankful?" This is similar to voting in a national election. People always ask themselves how much a single vote counts. The answer appears in national voter turnout statistics, which show that a very large percentage of the population does not think a single vote counts very much—and there is no price to be paid for voting. Do we really believe that consumers as a whole will spend an extra dollar or more per tankful when a single purchase does not really make much difference?

Whoever markets an alternative fuel will try to promote it as a superior product. The additional power alternative fuels offer may be a selling point, but oil companies will definitely emphasize the environmental benefits. Recent examples are Arco's EC-1, Sun Oil's and Diamond Shamrock's plans for new fuels, and Amoco's plans to sell low-volatility gasoline in Chicago. However, you can't fool all of the people all the time, and this is even more true in business than it is in politics. To promise environmental benefits and then deliver a product that offers only a small improvement at 10 to 25 cents a gallon additional cost is not a winning market strategy. For example, Chicago was out of compliance with air quality standards for only two hours during all of 1989. If the industry sells fuel for 10 to 25 cents a gallon more, to gain two hours of better air quality, many consumers will see through that fairly quickly. They will resent the price rise and criticize it as another consumer rip-off by the big oil companies.

The burden of improving air quality seems to have been placed on the oil companies, since it is the oil companies that will actually tax the consumer. The cost of methanol to each motorist might be as much as $250 per year. It would be preferable if motorists were required to send a $250 check to the Internal Revenue Service as their clean fuel tax. The IRS could then reimburse oil companies for the cost of providing alternative fuels. This arrangement would be more palatable to the oil industry, but how well would the consumer accept it? Again, "For two hours a year of benefit and $250 annual additional cost, would the dogs eat the dog food?"

This author would opt for something with much less risk, much less up-front cost, and a balance between environmental benefit and cost to the consumer. Compressed natural gas for fleets would cost less and would rely on domestic supplies, thereby enhancing energy security with a much smaller capital investment. Much of the infrastructure is already in place. There would be less effect on consumers, and the environmental benefits would be as great as or greater than those of methanol. In this regard, the Hall-Fields amendment seems to be a good approach.

Reformulated gasoline in new low-emissions vehicles can give the same low rates of emissions as EPA's special methanol vehicles. A mistake that EPA makes is to compare the tampering and other problems of existing gasoline-fueled vehicles with prototype methanol-fueled vehicles that really do not exist. EPA also assumes that these special methanol-fueled vehicles will never be tampered with, be poorly maintained, or have emissions-control systems that will decline with age.

The government should give the oil-auto industry program to assess the benefits of reformulated gasoline, along with new automotive technology, a chance. The joint program is studying the various properties of gasoline, as well as the properties of methanol and of compressed natural gas, and is considering new and improved vehicles. In the end the oil and auto industries will sell methanol or compressed natural gas or whatever mix of fuels and vehicles seems to be best. A free market can be counted on to sort that out in an efficient manner. A Soviet-style command economy always mismanages this kind of decision.

While oil companies will sell the fuels that are in demand, the industry would like the costs and benefits of the entire program to be well understood. The industry hopes to minimize some of the huge risks associated with the introduction of an alternative fuel.

Part VII

Summary and Conclusions

CHAPTER 16

The Methanol Debate: Clearing the Air

JEFFREY ALSON

Introduction

The year 1990 is the twentieth anniversary of both Earth Day, the public celebration that crystallized the environmental ethic for millions of Americans, and the Clean Air Act, which institutionalized the United States' goal of clean and healthy air for every citizen. Considerable progress toward this goal has occurred, but much remains to be done. The Clean Air Act amendments, almost certain to be adopted in 1990, will be the blueprint for achieving clean air for the foreseeable future. An intriguing feature of the Clean Air Act debate concerns various proposals to promote the introduction of new and less polluting transportation fuels.

The views expressed in this paper are those of the author and do not necessarily represent the views or policies of the U.S. Environmental Protection Agency.

This paper describes and evaluates the current public debate on methanol, a leading candidate clean fuel.

Urban Ozone and the Administration's Clean Fuels Program

As discussed in other papers, ground-level ozone is the United States' most intractable urban air quality problem and ozone reduction is one of the top priorities in the new Clean Air Act. In July 1989, President Bush submitted to Congress an innovative plan to introduce less polluting fuels into nine of our most polluted cities as part of his comprehensive Clean Air Act proposal. The underlying principles and detailed description of the proposal can be found elsewhere (Rosenberg, 1989), but the concept embodied in the proposal is to require that a portion of new vehicles in our dirtiest metropolitan areas be significantly cleaner than other new vehicles in order to help these areas move toward ozone attainment. The proposal does not mandate any particular fuel or technology. Rather, it would require the U.S. Environmental Protection Agency (EPA) to establish emissions performance standards for these cleaner vehicles. EPA has testified before Congress (Rosenberg, 1989) that such standards would provide hydrocarbon emissions reductions of 30 to 40 percent in 1995 and 80 to 90 percent in the post-2000 time frame. (Air toxics standards would require similar reductions.) Automobile manufacturers would be able to provide any combination of fuel and engine technology that could meet these emissions standards. EPA believes that methanol, ethanol, natural gas, propane, and electricity could meet these emissions standards, and it is possible that reformulated gasoline could as well.

Some have accused the proposal of not being fuel neutral, of being a "methanol mandate." This charge is completely without basis. The proposed legislation clearly establishes a level playing field that will permit all of the less polluting fuels now under consideration a fair and equal opportunity to compete. Of course, this does not necessarily mean that every candidate low pollution-emitting fuel will capture an equal portion of the clean fuels market. The fuels will compete in the marketplace, and the most cost-effective fuels will win out. Ironically, many of those who charge that the administration's proposal is a "methanol mandate" themselves represent organizations that have vested interests in one of the candidate less polluting fuels.

Methanol as a Passenger Car Fuel—What We Know and Don't Know

Vehicle Emissions and Urban Air Quality

This conference volume fortunately includes papers presented by individuals who represent the opposing sides of the methanol emissions and air quality debate. Phil Lorang of EPA presented the case for methanol's benefits, while Tom Austin of Sierra Research presented the case against methanol.

Three fuels are of interest here: gasoline, M85 (85 percent methanol, 15 percent gasoline), and M100 (pure methanol). Most of the early development work that domestic automakers have done with methanol in "gasoline clone" vehicles has used M85. By adding 15 percent high-volatility gasoline, cold starting is facilitated and flame color is improved. By making the methanol fuel more like gasoline, however, the methanol-fueled vehicle emissions become more like gasoline-fueled vehicle emissions as well. Accordingly, other solutions to the cold start and flame color problems are being sought. Unfortunately, domestic companies have yet to devote many resources to M100 development. Once they do, solutions should be readily available. EPA and the administration have argued that, solely on an environmental basis, M100 is the preferred fuel. Thus, the real issue here, for long-term urban air quality, is gasoline versus M100.

Lorang's paper summarizes the argument for M100's in-use emissions benefits. Critical to his case is the argument that when emissions-control systems on gasoline-fueled cars fail they can add significantly to the overall in-use emissions inventory. This is because of certain properties of gasoline, notably its volatility (propensity to evaporate) and reactivity (propensity to form ozone). The great advantage with methanol is that even when a methanol-fueled vehicle emissions-control system fails, the failure will be much less detrimental to the environment. This is due primarily to methanol's much lower volatility (4.5 psi versus 9 to 11 psi for gasoline) and lower reactivity. (A series of photochemical models suggests that the methanol molecule is about one-fifth as reactive, per unit mass, as a typical gasoline hydrocarbon.)

Its much lower volatility means that M100 will produce much lower levels of uncontrolled evaporative emissions, and the lower reactivity of M100 will significantly reduce the ozone-forming potential of both the evaporative and exhaust emissions. Lorang states that M100 vehicles would reduce ozone-forming potential by between 67 and 80 percent, compared with the future gasoline-fueled vehicles that will be required by the new Clean Air Act amendments. The benefit could well be larger if assumptions about in-use gasoline emissions reductions prove too optimistic.

Austin argues that methanol's benefits are overstated or uncertain. In his oral presentation he focused on M85, claiming that M100 vehicles are somehow infeasible. Austin gave data from one methanol flexible-fuel vehicle on M85, compared it with the best gasoline-fueled vehicle emissions, and concluded that M85 offered no benefit. In this author's opinion, that is an inappropriate way to compare the emissions potential of the two fuels. First, Austin focuses on only one M85 vehicle, ignoring a much larger group of more favorable M85 data. Second, he uses the very best gasoline emissions data, ignoring the super and high gasoline vehicle emitters that Lorang pointed out are the problem. Third, gasoline engines have been optimized for emissions for 20 years, with hundreds of millions of dollars spent on research and development. Little has been done to optimize M85 or M100, beyond making minor modifications to gasoline vehicle clones. The question is whether methanol-fueled vehicles can use emissions-control hardware similar to that of gasoline-fueled vehicles. If they can (and there is no logical reason why they cannot), then the lower reactivity of M85 emissions clearly promises some benefits, and the much lower reactivity and volatility of M100 offers very large benefits. Apparently, Austin does not believe that methanol-fueled vehicles can use emissions controls to the same effectiveness as gasoline vehicles. Why he thinks this is not clear.

Conference participants, in general, agreed that methanol's hydrocarbon emissions reductions would translate into lower urban ozone levels. Ted Russell of Carnegie Mellon University (in a study that was performed for the California Air Resources Board and that is the most comprehensive and defensible study of methanol multi-day ozone episodes) projected a reduction of as much as 16 percent in Los Angeles with M100 and of about half that with M85, based on complete methanol displacement of gasoline. The M100 benefit was equivalent to taking about 70 percent of the gasoline-fueled vehicles off the road—an impressive result, indeed. Tai Chang of Ford Motor Company shared his results of a 20-city study involving less sophisticated modeling. His results were lower than Russell's but directionally consistent. It was agreed that methanol use would be most beneficial in areas with low hydrocarbon-to-nitrogen oxides ratios and least effective in areas with high ratios. (In the latter areas all hydrocarbon reductions are less effective in reducing ozone.)

A final emissions issue is air toxics. Toxic air pollution can cause cancer, respiratory disease, and birth defects, and vehicles account for slightly more than half of all exposure. The most important air toxics from gasoline-fueled vehicles are benzene, butadiene, polycyclic organic matter, refueling vapors, and formaldehyde. Many methanol critics have addressed this issue in an incomplete way by focusing exclusively on tailpipe formaldehyde emissions. These emissions, however, are just a small portion of overall vehicle-related formaldehyde exposure, and formaldehyde is just one of many vehicle-related air toxics. Compared with gasoline-fueled vehicles, methanol-fueled vehicles do emit more "direct" formaldehyde emissions, but they yield far less of the "indirect" formaldehyde that is formed in the atmosphere via the transformation of reactive hydrocarbon emissions. EPA analyses suggest that overall vehicle-related formaldehyde levels would remain about the same if M100 displaced gasoline (a conclusion confirmed by Russell's modeling as well). M100 vehicles, however, would emit essentially no benzene, butadiene, polycyclic organic matter, or gasoline vapors, toxic compounds that are believed to be responsible for as many as several hundred cancer incidences per year, as shown in the paper by Paul Machiele of EPA.

Fuel Economics

The relative future economics of gasoline and methanol are by their very nature uncertain. The last two decades have demonstrated that world oil prices are impossible to predict. If the future price of an established commodity cannot be predicted with certainty, it is even more difficult to predict the price of a new commodity just entering the market and competing with the established commodity. Even those who project lower methanol-fuel prices must admit that methanol could not possibly compete with oil from the Middle East if that oil's price was based solely on the costs of extraction and shipping plus a "reasonable" profit. Ironically, the unstable price of world oil is both a major argument in favor of developing a new-fuels infrastructure and the major inhibitor to private sector investments that would develop just such an infrastructure.

Methanol-fuel cost studies can be divided into two groups: (1) those projecting that methanol could be delivered to the United States for 30 to 40 cents/gal and, taking into account distribution, markup, and sales taxes, could compete with gasoline from oil in the $15/b to $20/b range, and (2) those projecting that methanol could cost in the 50 to 60 cents/gal range, requiring the price of oil to rise to $30/b or more before methanol would be competitive. The former case was made in the paper by Mike Lawrence of Jack

Faucett Associates, while the later case is exemplified by the paper from James Sweeney of Stanford University.

There is a misperception that these two groups disagree at all steps of the analysis. On several basic inputs, such as plant cost, non-gas operating cost, ocean transport, local distribution, and taxes, nearly all of the analyses are in agreement. The major areas of disagreement are natural gas price, capital recovery rate (which includes return on investment and taxes), and service station markup. Table 1 summarizes the underlying assumptions for the different estimates.

As shown in many of the papers, so many countries are in need of capital and have so much natural gas for which there is no other use, that market pressures to maintain low gas prices will probably be strong. Practical evidence of this is the widespread investment in new methanol production capacity, despite methanol prices averaging less than 40 cents/gal during the last few years. Appropriate assumptions regarding capital recovery rates and dealer markup are very sensitive to the nature of the low pollution-emitting fuels market. If methanol entered the transportation fuels market as a low-volume specialty fuel with an uncertain long-term future, the "high price" assumptions might well be valid. However, if methanol was the fuel of choice and automakers were mass-producing methanol vehicles (as in the context of the administration's proposal), there would be a large and stable demand and market forces would tend to validate the "low price" assumptions. Of course, the fuel neutrality of the administration's program provides insurance in this regard: if the methanol skeptics are correct and methanol fuel prices are not competitive, other fuel alternatives (such as direct use of domestic natural gas) would surely capture the less-polluting fuels market.

Methanol Supply and Energy Security

In general, participants agreed that methanol fuel feedstocks are plentiful and that methanol would contribute to improved U.S. energy security. The administration's program would require

Table 1
Areas of Departure in Methanol-Fuel Cost Studies

Parameter	"Low Methanol" Prices		"High Methanol" Prices	
	Value	Basis	Value	Basis
Natural Gas Price	$0.50–$1.50 per mmBtu	Huge reserves in large number of cash-hungry countries with no other market	$1.00–$3.00 per mmBtu	Increased demand that allows countries to charge what they want
Capital Recovery Rate	15–20%	10% real return on investment, relatively low tax rates, and investment incentives	20–35%	10–15% real return on investment, relatively high tax rates
Service Station Markup	5–7 cents per gal	Dealer would maintain equal profit per vehicle mile, as with gasoline, plus recover station modification costs	11–13 cents per gal	Dealer would make same or more profit per gallon and almost 3 times the profit per vehicle mile, as compared with gasoline

Source: EPA Office of Mobile Sources, 1989

approximately one trillion cubic feet of natural gas per year at equilibrium. According to Carmen Difiglio of DOE, nearly 4,000 times this much gas exists in proved worldwide reserves. Difiglio also offered several reasons why methanol would improve U.S. energy security: (1) methanol feedstocks are more geographically diverse than are crude oil reserves, (2) the methanol fuel market would be more competitive than the oil market, (3) a methanol cartel would be less likely to try to drive up prices by limiting or with-holding fuel production, given the sunk investment in methanol production plants, and (4) the existence of a liquid fuel competitor in the largest oil-importing country in the world could potentially restrain world oil prices. This latter point is not easily quantified, but it is potentially a huge benefit of any alternative-fuels program. Every one cent/gal reduction in gasoline price translates to a $1 billion annual savings for U.S. consumers.

Global Warming

Global warming continues to be an issue on which there is much agreement. Mark DeLuchi of the University of California, Davis, calculates that the use of methanol from new remote natural gas fields, in vehicles with a 15 percent improved energy efficiency relative to gasoline, would yield the same overall global warming effect as gasoline from crude oil. Methanol from current coal-conversion technology would increase greenhouse gases by about 80 percent. Yet the development of methanol coproduction with electricity could reduce this to equivalency with gasoline from crude oil (as would the addition of carbon dioxide collection and disposal techniques to conventional coal-conversion technology). Methanol has some potential to reduce greenhouse gases: for example, in the near term by using natural gas that is currently vented or flared, or in the long term by using biomass feedstocks for at least a portion of methanol production or by taking advantage of methanol's superior combustion properties to improve automotive fuel economy significantly. It is agreed, however, that if greenhouse-gas reduction was the exclusive goal, a move to non-fossil energy sources would be preferred.

Health and Safety Issues

The health and safety issues (other than air quality) associated with methanol generated surprisingly little debate at the conference. All too often, health and safety concerns are voiced without any reference to the similar concerns that exist with gasoline. Paul Machiele's paper evaluates methanol's risk, compared with gasoline. Machiele shows that, because of the differences in physical and chemical properties, M100 use would result in a much lower frequency of vehicle-related fires (primarily due to methanol's much lower volatility and much higher flammability limit) and a much lower hazard when a fire does occur (primarily due to methanol's much lower heat rate), compared with gasoline. He projects that complete displacement of gasoline with M100 could reduce vehicle fire-related fatalities by several hundred per year.

Other safety concerns with methanol appear to be far less important. Simple vehicle design modifications should mitigate concerns associated with a flammable M100 fuel tank mixture, and low-level fuel additives are expected to provide taste, odor, and a visible flame. The human body metabolizes methanol, and inhalation and skin contact should not be a problem except in extreme situations. Direct ingestion of methanol must be avoided, however, as methanol is more directly toxic than gasoline. Most gasoline ingestion exposures involve children when gasoline is stored in the home, or occur when adults are siphoning gasoline. It is accepted that methanol should not be used for home and garden appliances and that anti-siphoning devices should be installed on automobile fuel tanks. These precautions would greatly minimize potential ingestion exposures.

Consumer Acceptance and Implementation

Carl Moyer presents a matrix of alternative-fuel implementation paths, based on whether government or industry is permitted to take the lead role in four areas: environmental constraints, fuel choices, marketing and pricing, and fuel production investment. Much of the debate, indeed, comes down to the relative authority between government and the private sector.

In general, the greater the role that government is given, the more certain the result and the more inflexible the options for industry. Conversely, the more flexibility given the private sector, the greater its ability to select the most cost-effective approaches but the lower the likelihood of major structural changes by the industry. Accordingly, if the policy goal is to achieve the most cost-effective urban emissions reductions, then the government role should be minimized and the private role maximized. The likely outcome would be a reformulated gasoline that would give some emissions reductions but would not provide the same energy security benefits as alternative fuels. If the policy goal is to achieve the broader benefits associated with opening the transportation fuels market to new fuels, then the government role should be maximized.

Dan Sperling discusses the factors affecting consumer acceptance of and demand for new vehicle fuels. He concludes that concerns about consumer skepticism of dedicated methanol vehicles are exaggerated. Based on evidence from Brazil and elsewhere, he argues convincingly that a "concerted and consistent" government program, emphasizing both price signals and the environmental benefits of methanol, can shape consumer acceptance. His conclusions support EPA's view that consumer acceptability can be achieved and that the transition period for flexible-fuel vehicles can be relatively short if government signals are strong and stable.

Making Sense Out of the Current Debate

What if Gasoline Were the New Fuel?

Much has been written about the "chicken-and-egg" problems that any new transportation fuel faces. A fundamental parameter of a truly free market is easy entry into that market, and the United States' huge and complex transportation sector clearly presents formidable market-entry barriers. For an interesting perspective on the difficulties that any new fuel would face, suppose that abundant petroleum supplies had not been discovered until recently and that the U.S. transportation infrastructure had long been optimized for alcohol fuels made from renewable feedstocks (Sperling, 1989).

Billions of dollars would have been invested in thousands of local plants for producing fuel. National networks of oil refineries or product pipelines would not exist. Vehicle engines would be designed for high-octane, high-performance, and low-emissions alcohol fuels.

Would petroleum be able to break into this established market? Gasoline would offer some benefits. Its higher energy density would yield longer vehicle driving ranges and minimize the capacity of the fuel distribution system. The petroleum industry could also claim that gasoline would be cheaper for the consumer. Yet the obstacles would be overwhelming. Health and safety advocates and regulators would vehemently oppose gasoline because of its high volatility, flammability, and explosivity; the release of "stored carbon" to the atmosphere; the high toxicity and smog-forming potential of gasoline hydrocarbon emissions; and the general uncertainty about what gasoline actually is due to the variable nature of crude oil composition and refinery operations. Automakers and consumers would oppose gasoline because cars would have to be redesigned to accommodate the lower-octane, lower-power fuel. No petroleum company would be able to obtain financing to build refineries or pipelines because of market uncertainty. And, to be sure, the American Alcohol Institute would mount a political and media blitz to crush the fledgling petroleum industry. The relevant point here is that tremendous institutional barriers would confront any new transportation fuel, regardless of the fuel's merits.

The Protagonists in the Methanol Debate[1]

The primary supporters of methanol use as an automotive fuel have been a diverse group of government agencies, such as EPA, DOE, CARB (the California Air Resources Board), and CEC (the California Energy Commission). These agencies' missions are to provide clean air and secure energy supplies. It should go without saying that government agencies do not have an economic interest in which fuels are used. A unique aspect of the debate is that there is really no private sector lobby support for methanol. There is a small methanol chemical industry, of course, and several companies have recently formed the American Methanol Institute, but chemical producers have been far more supportive of methanol as a gasoline blending agent than as a gasoline replacement. In fact, it is unclear what role chemical producers would play in a large M100 fuel market. It is quite possible that large methanol-fuel production plants could threaten the economic viability of today's smaller chemical plants.

The primary opponents of methanol are the domestic oil companies and their leading trade association, the American Petroleum Institute (API). The oil industry is generally considered to be one the most powerful lobbies in the country. This author believes that many domestic oil companies are so satisfied with the United States' exclusive reliance on petroleum for its transportation fuels, that they perceive methanol as a threat.

The government agencies have, in this author's opinion, attempted to consider not only methanol's benefits but also its drawbacks. For example, EPA and CARB have extensively analyzed the formaldehyde issue; DOE, CEC, and EPA have all addressed methanol as it pertains to global warming; and EPA has been adamant about the need for engine research, so that the minor barriers related to the use of M100 can be overcome.

This author believes, however, that the oil industry's role has not been so constructive. While formal public statements of industry spokespersons have been less adversarial, some in the industry have attempted to tilt the media debate by focusing exclusively on methanol's "negatives" or "uncertainties" while ignoring all positive attributes. A few examples follow:

One, Tom Austin of Sierra Research, an outspoken critic of methanol's environmental impacts, has been hired by API to speak about methanol (*Los Angeles Times,* 1989). Two, representatives of oil companies repeatedly raise the issue of formaldehyde, focusing exclusively on the fact that vehicle tailpipe emission levels from methanol will be higher than those from gasoline, while ignoring the projections that overall formaldehyde exposure would remain constant and overall exposure to all air toxics would significantly decrease with methanol substitution for gasoline. Three, oil company statements often list numerous "safety concerns" associated with methanol. Yet, M100 is clearly a safer vehicle fuel than gasoline, and would likely reduce annual vehicle fire deaths by several hundred. In fact, it is used in Indianapolis race cars partly for this reason. Four, some in the industry charge that EPA is unrealistic in its methanol fuel economics projections and have implied that all other studies have yielded much higher values. In fact, a number of other independent studies of the issue (DOE, CEC, and a forthcoming OTA report on methanol) have produced projections very similar to those of EPA. It is this author's opinion that some in the oil industry are not being fair in their analysis of methanol as an alternative fuel.

The Automotive and Environmental Communities

While the positions of environmental and energy agencies and the oil companies are clear, and in fact have defined the very nature of the current methanol debate, the automobile industry and the environmental community have not yet taken a firm position. The positions that these groups ultimately take may well determine the outcome of the low polluting fuels debate in Congress.

The public posture of the automotive industry toward less pollution emitting fuels has been to waver between weak opposition and ambivalence. This author believes, however, that the introduction of one or more new transportation fuels would greatly benefit the industry in the long run. It is often forgotten that the oil price shocks of the 1970s were very damaging to domestic automakers—the instability and uncertainty of future gasoline prices led to major perturbations in consumer behavior, and foreign manufacturers captured market share, which they have largely retained. One or more new fuels would provide competition to gasoline and would broaden and diversify the United States' energy supply, thus reducing the possibility of major fuel price shocks in the future.

In addition, a longer-term perspective would have to consider the alternatives that the auto industry might otherwise face. Clearly, environmental demands on the automobile industry will continue to increase, and the industry may have a far better chance to satisfy stringent environmental goals with low polluting fuels such as methanol than to do so by continuing to try to clean up gasoline. If new transportation fuels are inevitable, for whatever reason, it would be far better to begin the transition in an orderly manner that gives the industry a choice in the selection of technology, than to delay and face rigid mandates that could involve fuels or power sources that the industry might consider unacceptable. Methanol not only can provide desired environmental and energy security benefits, but also is a high-quality, high-performance automotive fuel that would allow automakers to produce attractive vehicles. Finally, domestic firms have the opportunity to take a global technology lead in the emerging worldwide alternative-fuels market. However, it appears that the perceived short-term drawbacks associated with a commitment toward lower pollution-emitting fuels (new investment requirements and a reluctance to make a commitment before a possible compromise involving less polluting fuels and lower gasoline-fueled vehicle standards) are taking precedence over these long-term benefits.

If the automobile industry's position on less polluting fuels is somewhat surprising, the environmental community's position is even more puzzling since such groups should be a natural constituency for less polluting automotive fuels. A few environmental groups have been enthusiastic supporters, a few have been opponents, and the majority have avoided any position whatsoever. It may be that the heightened interest in what many perceive to be critical long-term air pollution issues has diluted the focus of some environmentalists on current urban air pollution problems. In this view, lower pollution-emitting fuels, as now being envisioned, do not resolve these long-term issues and, given the limited resources of the environmental community, may not be "worth fighting for."

Conclusions

This author, speaking for himself and not for EPA or the administration, also believes that U.S. society may ultimately have to consider more structural solutions to two types of air pollution problems. First, a few major metropolitan areas may never be able to achieve clean and healthy air without switching to technologies that are near-zero urban polluters, such as hydrogen or electric vehicles or high-density mass transit. Los Angeles is certainly the best example, but there may be others. People may have to accept lifestyle restrictions if they want to live in certain cities and enjoy the amenities such cities offer. Second, if the scientific basis for concerns about global warming continues to strengthen, then the United States, as the world's largest energy user, will clearly have to consider alternatives to its reliance on fossil fuels for transportation, such as solar or biomass-based fuels.

It is important to recognize that although switching to methanol fuel will not magically resolve all urban air pollution problems or make much difference in the near term on global warming, it certainly is consistent with such long-term objectives. As discussed, M100 is an inherently cleaner fuel than gasoline, and its use will greatly reduce the vehicle contribution to urban ozone and toxic pollutant levels. Methanol is a good intermediate step to hydrogen fuel, because the type of engines that would use hydrogen would likely resemble optimized M100 engines. In fact, one option for M100 is to break it down thermally, on board the vehicle, to hydrogen and carbon monoxide, which yields an even less polluting and more efficient vehicle fuel. Methanol also ranks high as a possible energy source for fuel cells, a promising electric vehicle technology. With respect to global warming, in the short term methanol offers some benefits by providing a use for remote natural gas now vented or flared. In the long term methanol provides the opportunity for greatly improved automotive fuel economy (Gray and Alson, 1989) as well as the development of biomass-based methanol or ethanol (engines for the fuels would be nearly identical).

In short, methanol not only offers significant urban air quality benefits in the near term but establishes the foundation for potential solutions to long-term air pollution problems as well. Allowing the status quo to continue while holding out for the "perfect" long-term solution may simply result in losing the only opportunity to make progress toward the long-term solutions environmentalists seek.

Still, several unknowns remain. It is, and will always be, impossible to know exactly how large the ozone benefits of a switch to methanol would be in any given metropolitan area. It is also impossible to say whether methanol would be more or less expensive than gasoline, both because there is some debate about future methanol prices and because of the great volatility in world oil prices. Finally, it is impossible to say at this time what fuel or power sources will be used in the U.S. transportation sector in the second half of the twenty-first century. Nevertheless, a switch to methanol is generally consistent with most long-term paths supported by environmentalists.

The public policy issue, of course, is what should be done, given what is and is not known about methanol. The United States has a historic opportunity to improve its urban air quality and national energy security with a minimum of technological and economic risk. Yet the opposition of some major oil companies could make it very difficult for methanol or any other alternative automotive fuel to compete in the marketplace without federal leadership. Polls have consistently shown that the U.S. public is willing to pay for environmental controls, and recent polls have shown that most citizens are willing to pay extra for alternative transportation fuels in particular (USA Today, 1989). The issue is, for the first time, on the national agenda. It is now up to the country's elected leaders.

Note

1. Editor's Note: The following two sections reflect the author's personal views and are not based on material presented at the conference.

References

Gray, C., and J. Alson (1989). The Case for Methanol. *Scientific American*. November 1989.

Los Angeles Times. (1989). Oil Industry Enlists Former EPA Official in Methanol Fight. August 31.

Rosenberg, G. (1989). Testimony of the Assistant Administrator for Air and Radiation, U.S. Environmental Protection Agency, before the Senate Committee on Energy and Natural Resources. October 17, 1989.

Sperling, D. (1989). *New Transportation Fuels: A Strategic Approach to Technological Change*. Los Angeles: University of California Press.

USA Today. (1989). Drivers Want Fuel Choices. November 7.

CHAPTER 17

Methanol: A Skeptical View

JAMES J. MacKENZIE

Introduction

Many issues related to the use of methanol have been covered in this volume including the costs and consumer acceptance, as well as the potential effects on air quality, national security, and global warming. On some topics (such as using coal as a feedstock) there has been a fair amount of agreement; on others, a good deal less. The principal differences arise in areas in which there are still major technological uncertainties and a lack of solid information.

In such circumstances analysts feel free to make assumptions favorable to their viewpoint and then—sometimes using quite sophisticated tools—to draw the supporting conclusions. One must always be aware, however, of the shaky nature of such analysis: precise conclusions rigorously elicited from uncertain assumptions are themselves uncertain.

Emissions and Effects on Air Quality

Nowhere are the differences of opinion greater than in the discussion of the effect of emissions on air quality. The alternative views seemed to stem primarily from a lack of firm data on the likely emissions of both methanol-fueled and gasoline-fueled vehicles. Indeed, in the methanol debate it is not always certain which kind of vehicles are being discussed—M85 or M100 vehicles. Phil Lorang of the U.S. EPA indicated that, in its consideration of methanol-fueled vehicles, EPA clearly prefers vehicles that burn pure (M100) methanol. But how could M100 vehicles be introduced into the market? Automakers are not enthusiastic—to say the least—about manufacturing M100 vehicles. Apparently, none are planning to make M100 cars. Such vehicles would be hard to start in cold weather, and until there is a methanol infrastructure and methanol is generally available, they will be hard to sell.

The discussions we heard at the conference about clean air were characterized by pessimistic or optimistic assumptions. When analysts do not have the hard facts, they make assumptions. This is certainly not new. Indeed, Mark DeLuchi clarified the rules of the game most candidly when he demonstrated how significantly the calculated greenhouse effects of methanol-fueled cars could be changed just by slightly "tweaking" various parameters—all within legitimate ranges, however.

The levels of pollutants emitted are uncertain, not just from M85 or M100 vehicles but from the various mixtures that would result as the vehicles were driven first on one fuel and then on another. The level of emissions from new gasoline-fueled vehicles is also uncertain. According to Tom Austin of Sierra Research, "EPA's projections for methanol-fueled vehicle exhaust emissions are reasonable," but EPA "is projecting that gasoline-fueled vehicles will exceed the standard they are certified to meet by more than 100 percent." He goes on to say that "all of the. . . emissions benefits for methanol projected by EPA result from EPA's more pessimistic assumption regarding the performance of future gasoline vehicles in actual use."

In any case, it is clear that methanol is not an inherently less polluting fuel. Much new technology will be required to reduce the formaldehyde emissions from methanol-fueled vehicles to levels low enough to make these cars competitive in the clean air contest with gasoline-fueled vehicles. Methanol-fueled vehicles still need significant improvement before formaldehyde can be reduced to acceptable levels.

Even if methanol-fueled vehicles could be made less polluting with respect to hydrocarbon emissions, they would not "work" everywhere. Their effects in a given region would depend on the local ratio of hydrocarbons to nitrogen oxides. In some regions, reducing emissions of reactive hydrocarbons (the principal expected benefit of methanol-fueled vehicles) might be an ineffective tool. However future cars are powered, as the paper by Sandy Sillman and his collaborator points out, nitrogen oxide reductions would be needed to cut regional, rural ozone levels. Such reductions would be required, for example, to reduce the concentrations of background ozone that affect cities in the eastern United States.

The effects of deterioration in vehicle emissions performance are also uncertain. It has been indicated that dedicated methanol vehicles that emit little formaldehyde when new have dramatically increased emissions at higher mileage. The question also arises: Can NO_x emissions be sufficiently controlled? According to a paper by researchers at the California Air Resources Board, which was presented at the Air and Waste Management Association meeting in June 1989, work is still needed to reduce NO_x and formaldehyde emissions from methanol-fueled vehicles (Dunlap, Cross, and Drachand, 1989). What effects will reductions of nitrogen oxides have on vehicle efficiency and the economics of methanol? How much potential is there to reduce further emissions from gasoline-fueled vehicles? Are we succumbing to an unwarranted enthusiasm that new technology will magically solve our problems, even though experience indicates that things usually do not work out as well as had been hoped? (Recall "cold fusion"?)

One thing is certain: there is no way to know the answers to these questions without more research. The lack of data explains why major differences were expressed at the conference and in the papers in this volume. What is needed is more reliable information; fewer assumptions and more facts. But even though there are essentially no data on the performance of M100 vehicles, the clean air effects of introducing millions of these vehicles costing billions of dollars are being calculated as though their performance were really known.

Methanol technology will improve, as has the technology of gasoline-fueled vehicles. Such improvement, however, is not the issue. The performance of gasoline-fueled vehicles will also improve. The important questions are: How will the gap between these different kinds of vehicles change? What will be the effects on air quality? What will the costs be? According to Albert McGartland of ABT Associates, the economic uncertainties of methanol are so great that no meaningful estimate of its cost-effectiveness as a means of reducing ozone can be made. In my opinion, the most sensible approach to reducing ozone concentrations is not being discussed. The government should pursue the more mundane policy of placing a bounty on older, more polluting cars and find ways to get them off the road and replaced with newer models with considerably lower emissions.

The regional air quality benefits of switching to methanol are necessarily uncertain, less because of a lack of modeling capability than because of uncertainties about vehicle performance and emissions from other sources. One modeling exercise, conducted by Ted Russell of Carnegie Mellon, described a base case for Los Angeles in which all stationary sources were assumed to meet stringent limitations that would substantially reduce overall ozone concentrations from present levels by 2010. As a result, reductions from introducing methanol-fueled vehicles would represent a fairly significant fraction of the remaining pollution. This base case included no methanol-fueled vehicles, and gasoline-fueled vehicles that were much less polluting than they are now. In Russell's methanol scenario, fully one-half of the light-duty vehicles would be capable of running on M85 by 2010. (What fuel they would actually be using is, however, an unanswered question.) Ozone concentrations were calculated to be reduced by 0.02 ppm from assumed peaks (present ozone concentrations in the Los Angeles basin reach 0.36 ppm), about 9 percent compared to the base case. Total ozone exposures—measured as the sum of hours of exposure that exceeds the ozone standard—were calculated to be reduced by 19 percent. Whether these four million methanol-fueled vehicles and the required fueling infrastructure could be built, and 250,000 barrels of methanol provided to fuel them, are separate issues altogether. Tai Chang's study of methanol's potential clean air effects employed so-called trajectory models and found results not nearly as encouraging as those in Ted Russell's presentation, partly as a result of the assumption that future vehicles would contribute less to the ozone problem than current vehicles. Chang estimates that full introduction of *optimized M100 vehicles* would reduce peak ozone concentrations in nine polluted urban areas between 1 and 5 percent by 2015.

National Security

The administration's original proposal called for the introduction of about 10 million alternatively-fueled vehicles over 10 years. These vehicles would displace about 330,000 b/d of gasoline, which amounts to less than 2 percent of present U.S. oil use and much less than 1 percent of present world oil demand. Small programs will necessarily have small effects. It seems doubtful to this author that a program of this modest size would have any significant effects on the world oil market and especially on OPEC's strong influence on world oil prices.

There is a general consensus among the authors that plenty of natural gas is available for use as a feedstock to make methanol. Another consensus reached is that coal would not be a significant source of methanol for a long time. According to Mark DeLuchi, vehicles powered by methanol made from coal would result in carbon dioxide emissions about 80 percent greater than those of gasoline-fueled vehicles—a "showstopper" for coal as a feedstock if one considers global warming an important problem. Where would the methanol come from? About 87 percent of the world's exportable surplus natural gas reserves are in OPEC countries or in the Soviet Union. It is certainly debatable whether the security risks of importing methanol from these countries would be acceptable to the United States. The risks would presumably depend on the size of the program. Methanol supporters have several responses to the security issue. One is that host countries where methanol plants were located would have an economic incentive to keep them operating. Although this argument seems plausible, it may not hold in the politically volatile Persian Gulf region where very powerful non-economic forces are at work. Nor would it preclude host countries from gradually ratcheting up the price of natural gas feedstocks for strategic or political reasons.

A second argument is that there are potential sources of methanol feedstocks outside the Eastern or OPEC blocs that would provide diversity, and therefore security, of supply. Perhaps so. But there is a problem here that is generally overlooked. The nations of the world will soon be entering into prolonged, difficult negotiations on a global climate convention and protocol, with the goal of reducing emissions of greenhouse gases including carbon dioxide. Natural gas is the fossil fuel with the lowest carbon dioxide emissions per unit of energy by far and could well become a highly valued fuel in these negotiations. There is no way to predict what kinds of international agreements will be reached to reduce carbon dioxide emissions. But one should not automatically assume (as global climate change moves ever more to the forefront) that low-cost natural gas will be available from these countries, many of them relatively poor, to satisfy the United States' ever growing appetite for transportation fuels.

It seems to this author that the very expensive plants needed to make methanol from natural gas would not be built without some assurance that there would be a demand for the methanol they would produce. How could this assurance be given in the case of M85 vehicles? If EPA has its way, the United States will be building and using M100 cars. This implies widespread availability of service stations carrying M100. These vehicles would not be able to switch back and forth between methanol and gasoline fuels. As a result, much of the fuel-switching benefit for national security would not materialize. If instead M85 were chosen for security and other benefits, then it is predicted that the potential air quality benefits would be cut in half.

The concept of national security was discussed but never actually defined. In any case, the methanol program is unlikely to lead to any significant reduction in the United States' energy imports, or, therefore, to its economic security. Potential security benefits seem elusive to this author. If M100 vehicles were built and methanol supply was roughly in equilibrium with demand, there would be no response available in the event of an embargo. If M85 vehicles were built, there could be some security benefit if world oil capacity was sufficiently large and an embargo did not include producers of both methanol and oil.

Methanol's Costs

Sharp differences are expressed by the authors on the expected costs of producing methanol. McGartland suggests that the most important uncertainties are the feedstock costs and the rate of return required to build the methanol-conversion facilities. Would these facilities be constructed in regions perceived by investors as being risky? If so, would the required rate of return be substantially higher than if the facilities were located, for example, in the United States? These questions are important but may be unanswerable at this time. Similarly, although the cost of natural gas as a feedstock may be estimated, the actual price required by natural gas suppliers is another matter. Estimating it is "an almost impossible task," according to Michael Lawrence of Jack Faucett Associates. James Sweeney of Stanford University concludes that methanol would not be a competitive motor fuel until crude oil prices more than double from their present level.

Michael Lawrence observes that methanol vehicles' appeal to consumers will depend on methanol (equivalent) fuel prices being within pennies of competing gasoline fuels. This certainly appears to be true for dual-fuel vehicles. It seems unlikely that consumers owning such vehicles would buy methanol blends priced substantially higher than gasoline unless they are required to. If M100 cars are eventually introduced, drivers will have to buy M100, whatever the price.

Consumer Acceptance

Would consumers willingly buy methanol-fueled vehicles? The answer is not clear but it seems unlikely, without some major and obvious benefits to consumers. As a result the introduction of these cars will almost certainly require government intervention. Carl Moyer contends that manufacturers will have a strong inducement to build alternatively-fueled vehicles because of federal legislation giving them artificially large CAFE credits. The mere availability of these vehicles, however, does not mean that consumers will buy them. Moyer outlines several possible approaches that have been considered in California to encourage the use of alternatively-fueled vehicles. Several of these—such as the government-mandated use of specific fuels, financial subsidies, and direct government involvement in fuel supply—have been ruled out due to political and economic obstacles. California has opted instead for a system of fuel standards designed to reduce emissions of compounds that lead to ozone formation.

According to Moyer, California took this approach partially to ensure a "level playing field" on which all fuels could compete according to their environmental merits, costs, and consumer acceptance. I would argue that this approach may well be suited for one set of environmental impacts, the reduction of hydrocarbon emissions, but it fails to account for an even more serious environmental threat, greenhouse warming. In their emissions of carbon dioxide, alternative fuels differ decidedly, as DeLuchi has shown, and these differences should also be taken into account in selecting future fuels.

Dan Sperling points out that there are many factors affecting consumer decision making, but fuel pump prices seem especially important. The wide availability of fuels and clear signals of support from federal and state governments would also be important. Any especially bad experiences with early models could have a very negative effect on the introduction of flexible-fuel vehicles.

Public Health and Safety

According to EPA's Paul Machiele, methanol's principal safety and health benefit appears to be reduced fuel-related fires from motor vehicle accidents. This reduction is due to methanol's lower volatility and its lower level of toxic emissions from fuel burning, compared with gasoline. Projections indicate that, relative to gasoline, a 70 percent reduction in fires could result with M100 fuel, and a 20 percent reduction with M85 fuel. Areas of basic medical research still needing study were described in detail.

Machiele believes that engineering solutions will be found to alleviate the dangers posed by methanol's reduced flame luminosity and increased toxicity. (With respect to anti-ingestion measures, one could of course make the same statement about gasoline-fueled vehicles. One is tempted to ask: If the risks of methanol poisoning can be reduced by anti-siphoning technology, why can't the risks of gasoline ingestion also be reduced? And if they can, why aren't they?) Hugh Spitzer of the American Petroleum Institute (API) raised the issue of whether the carcinogenic hazard of gasoline may be less than previously believed. Results of a forthcoming API-sponsored study on refinery workers need to be carefully reviewed and their implications assessed to determine whether the risks from so-called toxic ingredients in gasoline are less than previously believed.

Global Warming

The greenhouse issue was reviewed in some detail. Mark DeLuchi notes that there are uncertainties about the relative effects of gasoline-fueled and methanol-fueled vehicles on greenhouse-gas emissions. The best estimate, though, is that the effects of methanol-fueled vehicles would be equivalent to gasoline-fueled vehicles, provided that the feedstock was methane and not coal. According to DeLuchi, if major progress is to be made in controlling automotive carbon dioxide emissions, methanol will be unacceptable as anything but a transitional fuel. In this author's view, that is correct, and it is also the primary reason why switching to methanol fuel will eventually lead to a dead end. According to EPA's climate office, global carbon dioxide emissions will need to be reduced by 50 to 80 percent to stabilize atmospheric concentrations. This is certainly not attainable in the short term. Given the need, however, it certainly argues forcefully against developing a big fossil fuel-based synfuels industry.

The global warming problem calls for long-term thinking. Air pollution is only one of the United States' serious environmental problems, and it cannot be solved by exacerbating—or at least by ignoring—other national priorities, such as the balance of trade deficit, national security, and global warming.

In this author's opinion these linked problems can best be dealt with in the short term through improved vehicle efficiency, more stringent limits on emissions, stronger inspection and maintenance programs, less polluting (reformulated) gasolines, and a concerted effort to get older, more polluting vehicles off the road. The longer term will require the development of new energy sources that are inherently less polluting, emitting no carbon dioxide and little or none of the conventional pollutants. Candidates include hydrogen- and electric-powered cars. At the same time it is increasingly evident that the United States will have to improve the overall efficiency of its transportation system through land-use controls and the provision of alternative means of travel, especially for commuting purposes.

Reference

Dunlap, L.S., R.H. Cross, and K.D. Drachand (1989). The Air Resources Board Methanol Vehicle Emission Test Programs. Paper presented at the Air and Waste Management Association's 82nd Annual Meeting, June 25–30, 1989.

Editor's Conclusion

WILFRID L. KOHL

The papers and commentaries in this volume enhance our understanding of methanol as a possible alternative-fuel choice, while pointing out several uncertainties that remain. In terms of tailpipe emissions, methanol is not a "clean" fuel in the purest sense of that word, but it does offer a means of reducing vehicle emissions that are precursors of urban ozone in certain settings. Methanol 100 would provide the greatest emission reductions, although it seems likely that M85 flexible-fuel vehicles would be introduced first. While there is considerable uncertainty in drawing conclusions from today's atmospheric models regarding emissions of future vehicles, available modeling indicates that the use of M100 might reduce peak ozone levels in some urban areas by more than 10 percent (and perhaps by as much as 16 percent), depending upon prevailing atmospheric conditions. The range of reductions shrinks by about half for M85. In areas with higher NO_x concentrations the reduction will be considerably less. It is unlikely that methanol use will mitigate rural regional ozone.

Methanol does not offer any long-term benefits in terms of global climate change. Assuming use of natural gas as a feedstock, it emits roughly the same amount of CO_2 as conventional gasoline. The use of coal to make methanol, on the other hand, would substantially worsen greenhouse gas emissions. Any major reduction of greenhouse gases from vehicles would require either greatly improved fuel economy or switching away from fossil fuels.

This means that methanol can at best be considered a transitional fuel from an environmental standpoint. More research is definitely needed on methanol vehicle technology—especially to deal with the cold-start problem of M100 vehicles of which only a few prototypes exist. Health and safety aspects need some further scrutiny, although the main problems in this area—invisible and odorless flame, fuel tank flammability, toxicity—appear manageable or solvable and there are a few benefits.

The energy security benefits and cost estimates of methanol are more controversial, as is the question of the best way to introduce it into the market. With respect to security, it seems clear that most methanol supplies would be made from abundant foreign sources of natural gas and then imported into the United States. Under optimistic assumptions the introduction of methanol might reduce gasoline consumption and hence oil imports by one million b/d by 2010, which would exert some downward pressure on world oil prices. While a diversity of suppliers would be likely, twice as much methanol would need to be imported because of the difference in energy content. Thus, methanol would not relieve any burden on the balance of payments. To protect against supply curtailments, flexible fuel (presumably M85) vehicles would be desirable to allow switching to gasoline, but M85 vehicles would have reduced emissions benefits—an important trade-off consideration.

A variety of cost assumptions and estimates are offered in this volume, but in the final analysis—and Alson and MacKenzie agree on this point—any reliable estimates are impossible without knowing the path of future oil prices. Only under the most optimistic assumptions would methanol be competitive with gasoline in the near term. It is therefore likely that considerable government involvement would be required to introduce methanol into the market (unless methanol prices do turn out to be competitive). Given cost uncertainties, however, and the difficulty the government will have in forecasting oil prices correctly (e.g., remember the Synthetic Fuels Corporation), it does seem prudent not to mandate the use of methanol. There should be allowance for market mechanisms and competition with other fuels, so that the market can correct if other fuels and technologies turn out to be more competitively priced.

While the reader will form his/her own conclusions, it seems to the editor that, as a transitional fuel, methanol may offer enough emission reductions and other benefits to justify its consideration as a serious alternative-fuel candidate for introduction over the next decade in those areas with the most severe problems of urban ozone. However, there are still many questions. If M85 is introduced first to ease methanol into the vehicle and fuel markets, how many years will be needed before M100 vehicles are perfected and M100 fuel made available? What is the most effective role for government and for market mechanisms in the implementation process? The evolution of oil and other energy prices will be a critical variable. More research is needed on a number of questions raised about methanol in this volume.

Under the American economic and political system, it is always difficult to plan coherently for the future. If we could do this, the best approach would be to adopt a flexible strategy toward alternative fuels with short-run and long-run components. In the short to medium term the United States should continue research and development efforts focused on reformulated gasoline and other alternative fuels as well as methanol. Over the longer run, even cleaner forms of highway transportation, such as electric or hydrogen vehicles, may be needed—especially in the nation's most polluted cities—to achieve any substantial improvement in air quality while reducing emissions of greenhouse gases and enhancing the nation's energy security.

About the Authors

Jeffrey Alson

As assistant to the director of the Emission Control Technology Division, U.S. Environmental Protection Agency (EPA), Jeffrey Alson explores the potential contribution of alternative motor vehicle fuels to reducing urban air pollution. He is also responsible for developing EPA policies on methanol and other alternative transportation fuels and has written several papers on the subject. He is coauthor of *Moving America to Methanol: A Plan to Replace Oil Imports* (1985) and "The Case for Methanol," *Scientific American* (1989).

Thomas C. Austin

Tom Austin, a senior partner at Sierra Research, Inc., has assisted the firm in designing and evaluating vehicle inspection and maintenance programs, developing emissions simulation models, analyzing proposed vehicle emission standard revisions, and developing implementation plan amendments for state and local governments. Previously, Mr. Austin served as the executive officer of the California Air Resources Board and director of its 500-person staff. Dozens of regulatory programs implemented under Mr. Austin's leadership include vehicle inspection and maintenance, gasoline vapor recovery, more stringent motor vehicle emissions standards, and a variety of controls on emissions from industrial facilities. Sierra's clients include the California Air Resources Board, South Coast Air Quality Management District, Alaska Department of Environmental Conservation, the U.S. Environmental Protection Agency, U.S. Office of Technology Assessment, Motor Vehicle Manufacturers Association, the American Petroleum Institute (on methanol), Western States Petroleum Association, and other regulatory agencies and private companies.

William Becker

William Becker is executive director of the State and Territorial Air Pollution Program Administrators (STAPPA) and the Association of Local Air Pollution Control Officials (ALAPCO). These are the national associations of state and local air pollution officials, whose members have primary responsibility for implementing air pollution control programs established under the Clean Air Act. Prior to joining STAPPA/ALAPCO in 1980, Mr. Becker worked as an analyst in environmental policy for the Congressional Research Service and as the director for environmental affairs of two trade associations.

Tai Y. Chang

Tai Chang is a staff scientist in the chemistry department of the Ford Motor Company, which he joined in 1969. Since 1973, he has concentrated on the area of photochemical-dispersion modeling of urban regional air pollutants. His prior work involved research in the field of quantum chemistry. Dr. Chang holds a Ph.D. in theoretical chemistry from the University of Wisconsin and was postdoctoral research fellow at Harvard University's chemistry department from 1967 to 1969.

Mark A. DeLuchi

Mark DeLuchi is a postgraduate researcher at the University of California, Davis, where he studies alternative transportation fuels and technologies and evaluates transit and transportation planning. His most recent research has focused on alternative transportation fuels and feedstocks and the emissions of greenhouse gases from their production and use. Mr. DeLuchi completed his graduate work at UC-Davis in environmental policy analysis as a member of the graduate group in ecology.

Carmen Difiglio

Carmen Difiglio is acting director of the Office of Policy Integration, a branch of the Office of Policy, Planning, and Analysis within the U.S. Department of Energy (DOE). He has written a number of articles for publication about transportation and energy economics, including articles on the use of alternative transportation fuels. Earlier Mr. Difiglio was executive director of the Interagency Commission on Alternative Motor Fuels; director of DOE's assessment of alternative transportation fuels; chairman, section on environmental quality and the conservation of resources, Transportation Research Board, National Academy of Sciences; U.S. delegate, International Energy Agency Experts Group (Standing Group on Long-Term Cooperation); chairman, marketing and mobility panel, Interagency Task Force for Motor Vehicle Goals Beyond 1980.

Lawrence Fishbein

Lawrence Fishbein is the principal scientist at the International Life Sciences Institute of the Risk Science Institute (RSI) in Washington, D.C. A specialist in toxicological research and environmental health, Dr. Fishbein has received recognition for his work from the Food and Drug Administration. He has also served in adjunct professorships at the University of Arkansas and at North Carolina State University. Dr. Fishbein is the author or coauthor of six books and approximately 350 publications and presentations in environmental toxicology, analysis, and risk assessment.

Carol J. Henry

Carol Henry joined the International Life Sciences Institute (of the RSI) as executive director in May 1988. Dr. Henry provides scientific direction for RSI research programs. Prior to joining RSI, Dr. Henry was vice-president at ICF Clement, with responsibilities for the direction and management of scientific and policy issues concerning toxic and hazardous materials. She has over 15 years experience in toxicology and related fields, with specialized expertise in inhalation toxicology, carcinogenesis, regulatory toxicology, and risk assessment. Dr. Henry is a coauthor of numerous scientific articles, most recently in the area of inhalation toxicology.

Jerry Horn

Jerry Horn is a senior research engineer at Chevron Research Company in Richmond, California, where he is primarily concerned with alternative fuels for light-duty vehicles. His most recent work has focused on analyzing emissions from methanol-fueled vehicles, particularly flexible-fuel vehicles. Earlier, his chief area of responsibility involved motor gasoline performance and gasoline quality issues. Mr. Horn has also participated in the planning, design, and execution of vehicle emissions programs, including vehicle refueling emissions, gasoline-oxygenate blend emissions, vehicle exhaust benzene emissions, and mobile source emissions modeling. He is a member of the Society of Automotive Engineers and the Air and Waste Management Association.

David Hungerford

David Hungerford is a graduate student in the Division of Environmental Studies, University of California at Davis. He is conducting research on consumers' demand for transportation fuels, which focuses on demand for premium grade gasoline and methanol in order to devise appropriate policies to influence consumer decisionmaking.

Michael D. Jackson

As manager of alternative fuels in the environmental systems division of Acurex Corporation, Michael Jackson is responsible for research on alternative energy uses in transportation and stationary systems. He is currently involved with projects that include using methanol in buses and trucks, evaluating particulate traps for heavy-duty diesel engines, and establishing fueling stations for alternative fuels. These projects are performed for the California Energy Commission, California Air Resource Board, and the South Coast Air Quality Management District, among others. For 13 years, Mr.

Jackson has also worked on feasibility studies and full-scale demonstrations of alternative fuels and energy systems, including advanced coal conversion processes, solar energy, alternative transportation fuels, and advanced thermodynamic cycles.

Michael Kelly

As a principal of Jensen Associates, Inc., Michael Kelly has been involved in numerous studies of international gas markets and trade. His fields of expertise include strategic planning, business development, economic evaluation, and market analysis. Through extensive overseas work he has acquired detailed knowledge of energy markets in Europe, Japan, and Australia, as well as in North America. In addition to specializing in petroleum supply, distribution, and marketing, Mr. Kelly has evaluated markets for other forms of energy, including liquefied natural gas (LNG). Clients of Jensen Associates include national oil companies, LNG importers and exporters, major and independent oil companies, electric utilities, gas distribution companies, U.S. and foreign governments. Prior to joining Jensen Associates, Inc., he worked for Broken Hill Proprietary Company in Australia, where he undertook the initial market analysis for the Northwest Shelf LNG project. Mr. Kelly's work has been widely published or quoted in *Oil and Gas Journal*, *Hydrocarbon Processing*, and *Petroleum Intelligence Weekly*, among other journals.

Wilfrid L. Kohl

Wilfrid Kohl is research professor of international relations at the Nitze School of Advanced International Studies (SAIS) and director of the International Energy Program at The Johns Hopkins Foreign Policy Institute. In 1986 Dr. Kohl served as a consultant to the Office of Technology Assessment. He was a member of the Atlantic Council's 1989 study group on future U.S. energy policy. In previous positions Dr. Kohl has served on the staffs of the National Security Council and the Ford Foundation, directed the Johns Hopkins University Energy Research Institute, and headed SAIS's European Center in Italy. Dr. Kohl is the author of several books on international relations and energy policy, including *After the Second Oil Crisis: Energy Policies in Europe, America and Japan* and *International Institutions for Energy Management*. He is editor and coauthor of *After the Oil Price Collapse: OPEC, the United States and the World Oil Market*, to be published in 1990 by Johns Hopkins University Press. He has contributed articles to major journals dealing with energy policy, foreign policy, and European-American relations.

Kenneth Kurani

Ken Kurani is a graduate student in the Department of Civil Engineering, University of California at Davis. He has spent much of the last five years researching consumers' attitudes and experiences regarding transportation fuels, and alternatives to petroleum-derived fuels. He is completing a research study of New Zealand's experience with transportation fuels derived from natural gas.

Thomas J. Lareau

Tom Lareau, a senior economist in the policy analysis department of the American Petroleum Institute (API), is primarily responsible for economic studies of environmental and energy policy. Dr. Lareau has written API reports on energy demand, energy security, and the cost of acid rain control. He has prepared a comprehensive review of environmental policy, cost-benefit analysis of ethanol mandates, and a report on the cost-effectiveness of alternative fuels, including methanol. He is a specialist in applied microeconomic analysis, energy and environmental economics, econometrics, and public finance.

Michael F. Lawrence

Michael Lawrence, vice president and senior project director at Jack Faucett Associates, is responsible for industry economics studies, energy policy analysis, regulatory impact analysis, and analysis of indoor environmental pollution from radiation. He also directs a study of the economic and financial effects of changes in the air pollution regulation in the transportation sector for the Environmental Protection Agency. Mr. Lawrence has also undertaken economic and policy analyses of various supply options for the administration's alternative-fuels program. He is the author of several publications concerning the energy market and industry.

Jerrold L. Levine

As director of Amoco Corporation's corporate studies, Jerrold Levine is responsible for business planning and strategy and is involved with energy and environmental regulations affecting the company. Mr. Levine is former chairman of the American Petroleum Institute's Motor Gasoline Committee. He also serves as spokesman on many issues for both Amoco and the industry. A Chicago native, Mr. Levine holds chemical engineering degrees from Purdue University (B.S.) and the University of Michigan (M.S.).

Alan C. Lloyd

A native of Wales, Alan Lloyd joined the South Coast Air Quality Management District as its chief scientist in March 1988. Dr. Lloyd currently heads the Technology Advancement Office, which has helped promote programs in which industry and agencies jointly sponsor technology that promises to reduce air emissions. Previously, Dr. Lloyd was a senior air quality scientist at Environmental Research and Technology, where he was actively involved with technical studies for industry and government agencies.

Philip A. Lorang

Philip Lorang is chief of the technical support staff of the Emission Control Technology Division of the U.S. Environmental Protection Agency (EPA). Mr. Lorang supervises EPA activities concerned with improving the control of in-use vehicle emissions. Mr. Lorang has helped to develop appropriate policy for periodic inspection programs and for oxygenated fuel programs. He has recently assisted in the development of the administration's proposal for an alternative-fuel vehicle program.

Paul A. Machiele

Paul Machiele is project manager of the standards development and support branch of the Emission Control Technology Division, U.S. Environmental Protection Agency (EPA). For the past three years Mr. Machiele has been responsible for methanol fuel safety concerns for the Office of Mobile Sources. He is the author of an SAE paper and AICHE paper on health, safety, and environmental concerns with methanol. Mr. Machiele holds a B.S. and M.S. in mechanical engineering from the University of Michigan.

James J. MacKenzie

James MacKenzie is a senior associate in the climate, energy, and pollution program of the World Resources Institute (WRI). At WRI he has completed a policy report exploring the linkages among the problems of climate change, air pollution, and national energy security. Before joining WRI, Dr. MacKenzie was a senior staff scientist at the Union of Concerned Scientists, where he wrote on nuclear power safety, conservation, solar energy, and global oil resources. As a senior staff member for energy at the President's Council on Environmental Quality in 1979, Dr. MacKenzie cochaired the impacts panel of the domestic policy review.

Albert McGartland

Albert McGartland is an environmental economist at ABT Associates, Inc., serving as a consultant to the U.S. Environmental Protection Agency and the National Acid Precipitation Assessment Program. Before joining ABT Associates, Inc., Dr. McGartland reviewed economic analyses of proposed regulations at the Office of Management and Budget. Coauthor of a report to the President's Task Force on Regulatory Relief, which examined the costs and benefits of alternative fuels, he has also published in numerous professional journals. Forthcoming articles will appear in *American Economic Review*, *Contemporary Policy Issues*, and the *Northwestern Law Review*.

Jana B. Milford

Jana Milford is an assistant professor in the department of civil engineering and the Environmental Research Institute at the University of Connecticut. She is responsible for research and teaching in the area of environmental engineering, focusing on air pollution. Prior to her tenure at the University of Connecticut, Dr. Milford was a Congressional Fellow and analyst in the Office of Technology Assessment. There she analyzed the urban ozone problem for congressional committees interested in amending the Clean Air Act. Author and coauthor of several articles and reports, her publications and testimony include: "Ozone Control and Methanol Fuel Use," *Science* (1989), *Quantitative Estimate of the Air Quality Impacts of Methanol Fuel Use* (1989), and *Catching Our Breath: Next Steps for Reducing Urban Ozone*, OTA-0-412.

Carl B. Moyer

As the senior consulting scientist for the environmental systems division of Acurex Corporation, Carl Moyer participates in a wide range of technical projects and proposals. He manages a design study of a low-emissions gas turbine combustor and directs a study that evaluates methanol substitution for gasoline and diesel fuel as an air quality control strategy. He has supervised and conducted studies for the California Energy Commission, California Air Resources Board, South Coast Air Quality Management District, the Gas Research Institute, and major oil companies. Earlier, Dr. Moyer worked on projects in heat transfer, thermodynamics, combustion, power generation, air pollution, fluid mechanics, and material thermal response. He has authored 87 publications and presentations based on his field research. Dr. Moyer belongs to various honors societies, including Phi Beta Kappa, Tau Beta Pi, Sigma Xi, and the International Cogeneration Society.

Michael J. Murphy

Michael Murphy is a research engineer at the Batelle Memorial Institute. He has done research on energy and environmental matters, including studies on the fire, emissions, safety, and health properties of alternative-fuels. Dr. Murphy has performed health and safety analyses of alternative-fuel use, especially methanol, for the Urban Mass Transit Administration, the U.S. Environmental Protection Agency, and private industry. He holds a Ph.D. in mechanical engineering from the University of Wisconsin.

Charles E. Risch

Charles Risch is a principal staff engineer for Ford Motor Company. Working in the fuel economy planning and compliance department of the automotive emissions and fuel economy office, Mr. Risch is the contact person between Ford and the government on alternative-fuels issues. He also represents Ford on the Alternative Fuels Panel of the Motor Vehicle Manufacturers Association. Mr. Risch was chairman of the alternative-fuels session for the 1989 Society of Automotive Engineers' Congress and Exposition.

Sara J. Rudy

Sara Rudy is a research scientist in the chemistry department of Ford Motor Company, where she is involved in the area of photochemical-dispersion modeling to determine the impact of mobile source emissions on urban and regional air quality. Prior to joining Ford Motor Company in 1988, Ms. Rudy was involved in computer modeling for a civil-environmental engineering firm as well as atmospheric modeling in space physics at the University of Michigan.

Armistead (Ted) Russell

Ted Russell is an assistant professor of mechanical engineering at Carnegie Mellon University (CMU), where he is involved in studies of environmental and energy problems and of their interrelationship. One such project considered the effect of using methanol fuel in mobile and stationary applications on air quality, and is now focusing on the effects of other alternative fuels. Before coming to CMU, Dr. Russell was a senior research engineer at the California

Institute of Technology's Environmental Quality Laboratory. Dr. Russell has undertaken research projects sponsored by the U.S. Environmental Protection Agency, California Air Resources Board, and the Coordinating Research Council of the oil and automobile industries. For his research and teaching efforts Dr. Russell received the Ralph Teetor Award from the Society of Automotive Engineers and the top award in the national paper contest by the Air Pollution Control Association.

Perry J. Samson
Perry Samson is associate professor of atmospheric sciences and cochair of the Michigan Atmospheric Deposition Laboratory at the University of Michigan. He began his professional career as a research scientist for the New York State Department of Environmental Conservation in 1973. In 1979 he completed a Ph.D. in meteorology from the University of Wisconsin, Madison. His research has focused on development of methods to diagnose the transport, dispersion, and deposition of pollutants over scales ranging from meters to thousands of kilometers. Dr. Samson is a certified consulting meteorologist of the American Meteorological Society, and is a member of the Air and Waste Management Association and other professional groups.

Harry Schwochert
Harry Schwochert is manager of heavy-duty and service activities in the automotive emission control department of General Motors' environmental activities staff. He has been involved with projects related to the control of automotive emissions since joining General Motors 27 years ago. Mr. Schwochert currently manages the coordination of heavy-duty emissions and alternative-fuel and emission-related vehicle regulatory activities. He is responsible for regulatory analysis, represents the corporation in external contracts, and provides internal guidance and information.

Glyn D. Short

Glyn Short received a bachelor's degree in chemistry at the University of Bristol in 1960, before obtaining his doctorate in physical chemistry at the University of Exeter. He then did photochemical research for two years at MIT before working for the British Admiralty as a senior research fellow in photochemistry. Dr. Short has been with Imperial Chemical Industries for more than 15 years as a manager of research and development and new business ventures. He has done catalysis research and methanol synthesis. For the last three years, he has been concerned with establishing the credibility of methanol as a replacement for diesel fuel in the United States. As a director of the American Methanol Institute, Dr. Short has given many presentations, lectures, and talks on the subject of fuel methanol and methanol chemistry.

Sanford Sillman

Sanford Sillman is currently a research investigator in the Department of Atmospheric, Oceanic and Space Sciences at the University of Michigan. He received his B.A. in physics and mathematics from Brown University and an M.S. in the technology and policy program at MIT. After working for several years at the Solar Energy Research Institute, he resumed graduate study, completing a Ph.D. in applied physics at Harvard University. He has since worked as a postdoctoral research associate at Harvard and at Michigan. His research has focused on the combined impact of transport and photochemistry in large-scale ozone episodes.

Daniel Sperling

Daniel Sperling is an associate professor of environmental studies and civil engineering at the University of California, Davis, and director of the University's Transportation Research Group. Dr. Sperling is also chairman of the Alternative Transportation Fuels Committee of the Transportation Research Board (National Research Council), a member of the National Academy of Sciences Committee on Liquid Fuel Options, and a member of the editorial board of *Transportation Research Journal*. He is a frequent speaker on alternative energy, the greenhouse effect, and transportation strategies for developing countries. Among Dr. Sperling's books and numerous technical papers on transportation and energy topics are *Alternative Transportation Fuels: An Energy and Environmental Solution* (1989) and *New Transportation Fuels: A Strategic Approach to Technological Change* (1988).

Hugh L. Spitzer

Manager of health sciences at the American Petroleum Institute (API), Hugh Spitzer oversees toxicological, epidemiological, and exposure research and consults with industry representatives to identify research needs. In addition, Mr. Spitzer is a member of several committees at the Risk Science Institute, the University of Texas, and the M. D. Anderson Cancer Center. Before joining API, Mr. Spitzer worked at the Office of Technology Transfer and Regulatory Support of the U.S. Environmental Protection Agency. He is the author of numerous papers and has contributed to various books and symposia.

James L. Sweeney

James Sweeney has specialized in applying economics and mathematical modeling to natural resource issues, energy economics, and policy analysis. He is a professor in the Department of Engineering-Economic Systems at Stanford University, where he has served as director of the Energy Modeling Forum, chairman of the Institute for Energy Studies, and director of the Center for Economic Policy Research. The author of numerous articles, Dr. Sweeney is also coeditor of *Resources and Energy*. Former director of the Office of Energy Systems Modeling and Forecasting of the U.S. Federal Energy Administration, he now serves as a consultant to governments (U.S. and foreign), industry, and research organizations.

Margaret A. Walls

Margaret Walls is a fellow in the energy and natural resources division at Resources for the Future. She received her Ph.D. in economics from the University of California, Santa Barbara, in 1988. In addition to her current research on the economics and environmental effects of alternative motor fuels, Dr. Walls has written on oil and gas supply modeling and energy policy analysis.